This book is the property of:

____

Additional copies of this book and many other maintenance and trades books are always available from:

MAINTENANCE TROUBLESHOOTING BOOKS

"The Biggest Name in Little Books"

2917 Cheshire Road
Wilmington, DE 19810

mtrouble@mtroubleshooting.com
www.mtroubleshooting.com

(800)755-7672
(302)738-0532

Audel™ Mechanical Trades Pocket Manual

All New Fourth Edition

Thomas Bieber Davis

Carl A. Nelson, Sr.

Wiley Publishing, Inc.

Vice President and Executive Publisher: Bob Ipsen
Publisher: Joe Wikert
Executive Editor: Carol Long
Development Editor: Emilie Herman
Editorial Manager: Kathryn A. Malm
Senior Production Editor: Angela Smith
Text Design & Composition: Wiley Composition Services

Copyright © 2004 by Wiley Publishing, Inc. All rights reserved.

Copyright © 1986, 1990 by Macmillan Publishing Company, a division of Macmillan, Inc.

Copyright © 1983 by The Bobbs-Merrill Co., Inc.

Copyright © 1974 by Howard W. Sams & Co., Inc.

Published simultaneously in Canada

No part of this publication may be reproduced, stored in a retrieval system, or transmitted in any form or by any means, electronic, mechanical, photocopying, recording, scanning, or otherwise, except as permitted under Section 107 or 108 of the 1976 United States Copyright Act, without either the prior written permission of the Publisher, or authorization through payment of the appropriate per-copy fee to the Copyright Clearance Center, Inc., 222 Rosewood Drive, Danvers, MA 01923, (978) 750-8400, fax (978) 646-8700. Requests to the Publisher for permission should be addressed to the Legal Department, Wiley Publishing, Inc., 10475 Crosspoint Blvd., Indianapolis, IN 46256, (317) 572-3447, fax (317) 572-4447, E-mail: permcoordinator@wiley.com.

Limit of Liability/Disclaimer of Warranty: While the publisher and author have used their best efforts in preparing this book, they make no representations or warranties with respect to the accuracy or completeness of the contents of this book and specifically disclaim any implied warranties of merchantability or fitness for a particular purpose. No warranty may be created or extended by sales representatives or written sales materials. The advice and strategies contained herein may not be suitable for your situation. You should consult with a professional where appropriate. Neither the publisher nor author shall be liable for any loss of profit or any other commercial damages, including but not limited to special, incidental, consequential, or other damages.

For general information on our other products and services please contact our Customer Care Department within the United States at (800) 762-2974, outside the United States at (317) 572-3993 or fax (317) 572-4002.

Trademarks: Wiley, the Wiley Publishing logo, Audel, and related trade dress are trademarks or registered trademarks of John Wiley & Sons, Inc. and/or its affiliates, in the United States and other countries, and may not be used without written permission. All other trademarks are the property of their respective owners. Wiley Publishing, Inc., is not associated with any product or vendor mentioned in this book.

Wiley also publishes its books in a variety of electronic formats. Some content that appears in print may not be available in electronic books.

Library of Congress Control Number: 2003105639

ISBN: 0-7645-4170-6

Printed in the United States of America

SKY10064820_011224

Contents

Introduction

Mechanical Trades Pocket Manual is a concise reference for maintenance and mechanical craftspersons. It provides information for preventive maintenance and mechanical repair, and includes tricks of the trade that will help mechanics expected to perform the impossible to "get production running." While the book is primarily concerned with installation, maintenance, and repair of machinery and equipment, it also covers other fields of activity. Current methods, procedures, equipment, tools, and techniques are presented in plain, easy-to-understand language to aid the mechanic in the performance of day-to-day activities. New sections on lubrication, vibration measurement, and PM inspection show troubleshooting techniques and help the mechanic fix the real problem—not just treat the symptoms of troublesome equipment.

Drawings, photographs, and tables are used throughout this book to illustrate the text and help you perform these activities. Discussions of principles and theory are limited to the key information necessary to understand the subject matter.

While this book covers topics from shafting to bearings, carpentry to welding, measurement to machine assembly; it is impossible to give the depth of detail for each and every subject in a small pocket manual. Readers who wish to learn more about these topics should consider the *Audel Millwrights and Mechanics Guide: All New Fifth Edition*, published by Wiley Publishing, Inc. This larger format book allows each subject to be explored to a greater depth.

—Thomas Bieber Davis
Maintenance Troubleshooting

1. SAFETY

It's hard to walk past the plant gate or the security office without seeing a sign indicating how many days since the last lost-time accident at the facility. Most employers take great pride in their safety record and preach safety as a way of life. This section includes some tips and information for keeping your operations safe.

OSHA

The *Occupational Safety and Health Act* (OSHA) in the United States covers workplace conditions for employees. If you are engaged in maintenance, repair, or installation activities, your employer probably has details on the particular parts of the OSHA standards that apply to you.

As a mechanical tradesperson, you might find yourself involved in many activities that require you to think out particular safety issues for your own special type of work. For example, you might be servicing a pump and have all the parts and the manufacturer's manuals, but it would make a great deal of difference in your approach to the job if the pump were handling acid instead of water. You have to consider the safety details for each job, using your trade skills combined with your knowledge of the process, location, or conditions.

Lock Out and Tag

When you shut down a machine, process, pipeline, or electrical apparatus to inspect or perform a repair, you need to *lock out and tag* a piece of equipment so that it cannot be accidentally started or energized. Usually, the start/stop of the switchgear controlling the piece of equipment is physically disabled with a lock (Fig. 1-1).

The employee working on the equipment usually holds the key to the lock. The lock itself has a tag identifying whose lock is being used. In the case of a pipeline, the valve controlling the flow into the line is closed and a lock placed

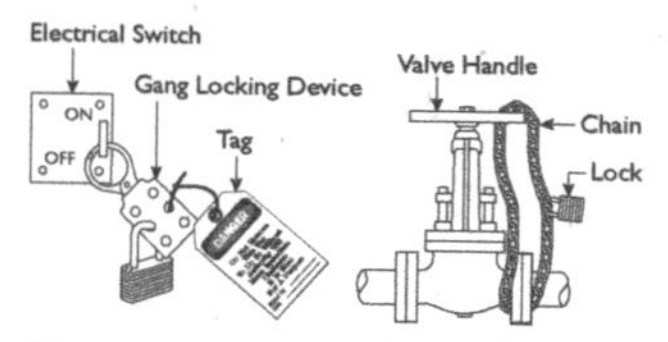

Fig. 1-1 Lock out and tag.

on the handle or bonnet with an appropriate tag attached. If two or more people are working at the same machine, then each additional person also places a lock on the equipment. A gang-locking device can be used to hold many locks on the job. In addition, it's a smart idea to try the local start/stop switch at the machine site or open the valve downstream from the locked main valve as a double check to make sure that the machine or process is "safed-out."

MSDA

Another safety issue for maintenance or installation personnel is the use of chemicals or hazardous materials. In the United States, *Material Safety Data Sheets* (MSDA) are used to identify the details of each particular chemical and list safe handling techniques. These sheets, usually available in the workplace, indicate if rubber boots, face shield, respirator, dust mask, or goggles might be required to work safely around a substance. Keep in mind that chemicals can be oil or grease, too. Removed parts of a machine often need to be cleaned after disassembly and require chemical degreasers or solvents.

Use of Protective Equipment

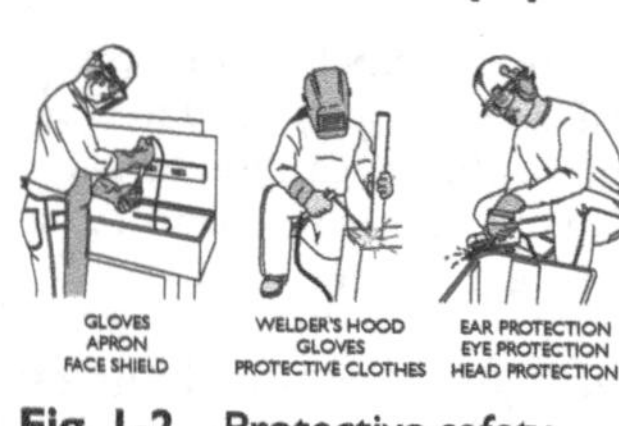

Fig. 1-2 Protective safety gear.

Often the maintenance worker needs gloves, a hard hat, steel-toe shoes, earplugs, or other articles of apparel to guard against injury to the eyes, feet, head, or ears. Jobs such as grinding, drilling, nailing, painting, or welding mandate protective gear (Fig. 1-2).

2. THE BASIC TOOLBOX

The proper use of hand tools is the key to accomplishing many successful jobs. While each mechanical trade might require specialized tools, a group of basic tools will be the heart and core of any craftsperson's toolbox. This set will allow the mechanic to perform most basic repair and installation jobs.

Wrenches—Open End, Box, Combination

Wrenches are one of the most widely used hand tools. They are used for holding and turning bolts, cap screws, nuts, and various other threaded components of a machine. It is important to make sure the wrench holds the nut or bolt with an exact fit. Whenever possible, it is important to *pull* on a wrench handle and keep a good footing to prevent a fall if parts let go in a hurry. Typical types of wrenches needed in the basic toolbox include box wrenches, open-end wrenches, and combination box/open-end wrenches from ¼ to 1¼ in. in increments of ¹⁄₁₆ in. to accomplish most standard jobs (Fig. 2-1). The same types of wrenches in metric openings would span from 7 to 32 mm. Metric wrenches are stepped in 1-mm increments, but the basic toolbox would usually omit the 20-, 29-, and 31-mm openings because they are used so infrequently.

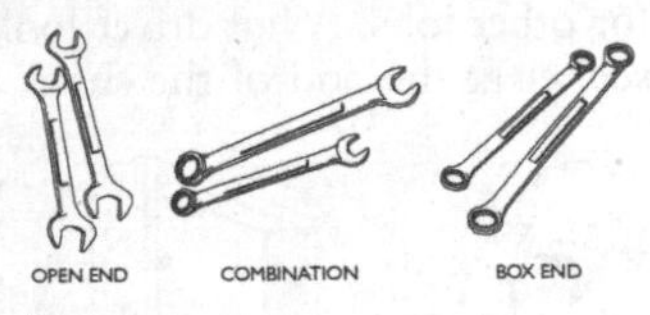

Fig. 2-1

Socket Wrenches

Socket wrenches greatly speed up many jobs. The basic toolbox should contain sockets ranging from ⁵⁄₁₆ to 1¼ in. in ¹⁄₁₆-in. increments. Usually socket wrenches are sold in sets where the square drive of the ratchet is used to identify the set. Ratchets are typically made with ¼-, ⅜-, ½-, ¾-, and 1-in. square drivers. The larger the drive, the greater the capacity of the wrench set. For most work, the basic toolbox needs to have a

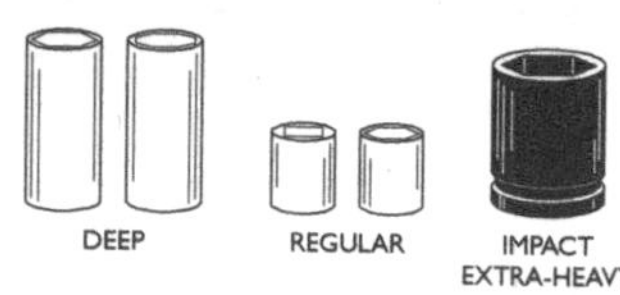

Fig. 2-2

1/4-in. drive set, a 3/8-in. drive set, and a 1/2-in. drive set. Larger drives, like the 3/4 and 1 in., can be stocked in a central tool crib or kept with a supervisor for use by any craftsperson when the need for these larger units is merited. Sockets are made in regular depth and extra deep (Fig. 2-2). Openings may be 12-, 8-, 6-point, or square for the type of work that the wrench set must handle. In addition, jobs that involve the use of an electric or pneumatic impact wrench need extra-heavy wall sockets that will hold up under heavy pounding from the tool.

Nut Drivers

Nut drivers are a must for electricians, but they are also handy for other jobs. A nut driver looks like a screwdriver but has a socket at the end of the shank (Fig. 2-3). The basic toolbox should have a nut driver set, which includes the 3/16-, 1/4-, 5/16-, 11/32-, 3/8-, 7/16-, 1/2-, 17/32-, 9/16-, 5/8-, 11/16-, and 3/4-in. sizes. Often overlooked, these tools speed up the repair or adjustment of equipment and surpass the ability of a ratchet set for loosening or tightening light- to medium-duty assemblies.

Fig. 2-3

Adjustable Wrenches

Adjustable wrenches are sometimes called "fits all" wrenches because they cover such a wide range of jobs. These tools are open-end wrenches with an expandable jaw that allows the wrench to be used on many sizes of bolts or nuts. They are available in lengths from 4 to 24 in. Some adjustable wrenches allow the jaw to be locked. These are available in lengths from 6 through 12 in. When using an adjustable wrench, always make sure you apply the force to the fixed jaw (Fig. 2-4).

Practically all manufacturers supply parts and repair kits to refurbish adjustable wrenches. The kits are relatively inexpensive, compared to the purchase of a brand new wrench. The basic toolbox should include adjustable wrenches with lengths of 4, 6, 8, and 12 in. These four sizes can cover a wide range of jobs.

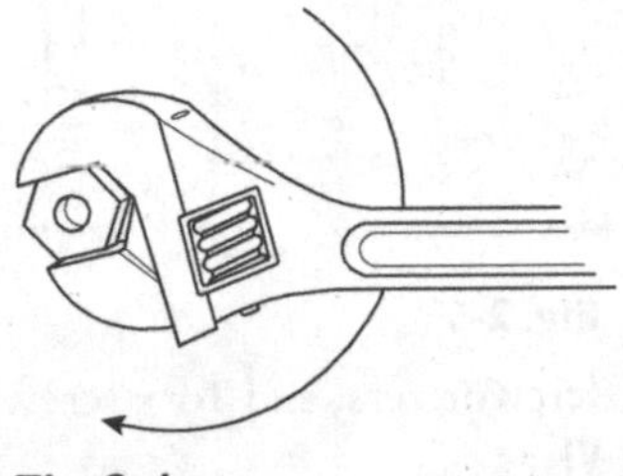

Fig. 2-4

Screwdrivers

Screwdrivers are the most used tools in a toolbox. They are intended for one simple use: driving and withdrawing threaded fasteners such as wood screws, machine screws, and self-tapping screws. Correct use of a screwdriver involves matching the size of the screwdriver to the job and matching the type of screwdriver to the head of the screw. The first screws that were developed had slotted heads. The proper screwdriver for these types of screws is called a conventional screwdriver, and it can be classified by tip width and blade length. Generally, the longer the screwdriver, the wider the tip—but not always. Cabinet screwdrivers have long shanks but narrow tips because they are used to drive screws into recessed and counterbored openings that are found in furniture and cabinets. Stubby screwdrivers are available that have very wide tips for use in tight spaces. Most conventional screwdriver tips are tapered. The tip thickness determines the size of the screw that the screwdriver will drive without damaging the screw slot. The taper permits the screwdriver to drive more than one size of screw.

The world of screwdrivers has radically changed since the slotted head screw. Probably the first "new" screw that was developed was the Philips head screw—a recessed slot design. The recessed slot screw allows a more positive non-slipping drive-up while attaching the fastener. The most common

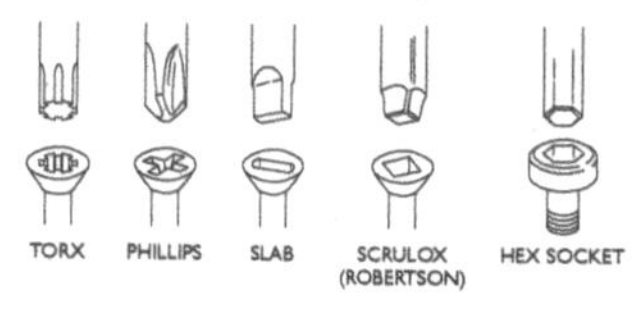

Fig. 2-5

style of recessed head is the Philips screw. Other recessed screws and screwdrivers are shown in Fig. 2-5.

Essential screwdrivers for the basic toolbox are a set of slotted screwdrivers, Philips screwdrivers, and Torx screwdrivers.

Vises

While a vise would not necessarily be part of the basic toolbox, it certainly would be essential on the tool bench. There are four basic categories of vises: machinist's vise, woodworker's vise, pipe vise, and drill press vise. Each has its purpose and benefits.

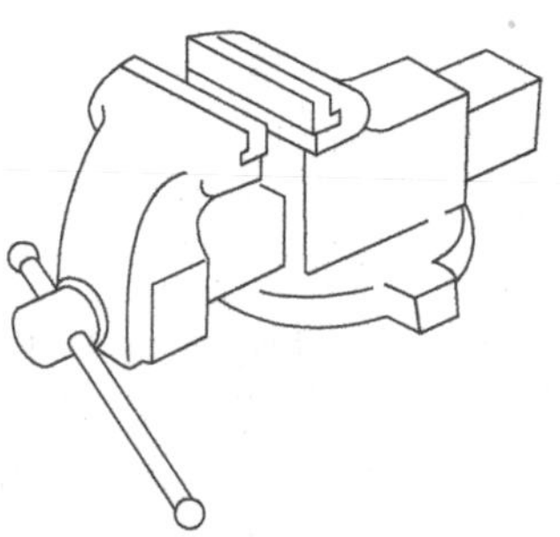

Fig. 2-6 Machinist's vise.

The *machinist's vise* is the strongest vise made (Fig. 2-6). It is designed to withstand the great strains in industrial work. Models are made with stationary bases and swivel bases and can be equipped with pipe jaws as well as interchangeable jaws. Usually jaw widths start at 3 in and go to 8 in for large jobs. Copper (or brass) jaw caps are available to prevent marring of the work.

The *woodworker's vise* is a quick-acting vise that bolts to the underside of the workbench (Fig. 2-7). These vises are equipped with a rapid-action nut that allows the movable jaw to be moved in and out quickly with the final tightening by turning the handle a half-turn or so. The jaws on the vise are large, often 7 to 8 in. The jaws are usually metal, but they are intended to be lined with wood (which allows replacement) to protect the work.

A *pipe vise* is designed to hold pipe or other round material. Most can hold pipe up to 8 in. in diameter. They are usually available with a tripod mount to allow the vise to be portable, but they can be bolted to a workbench as well. The most popular type uses a chain for clamping and can also be used to hold irregular work (Fig. 2-8).

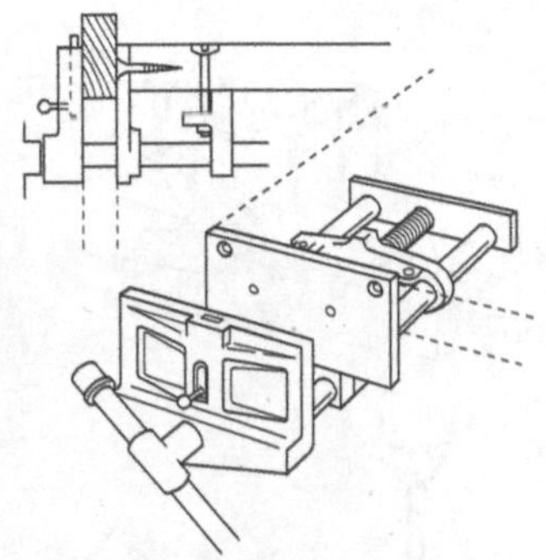

Fig. 2-7 Woodworker's vise.

The *drill press vise* is made to accept round, square, or oddly shaped work and hold it firmly in place for drilling (Fig. 2-9). The better vises often have a quick release movable jaw that allows them to be moved up to the work or away from the work without turning the handle. The handle is used for the final half-turn or so to loosen or tighten the jaws.

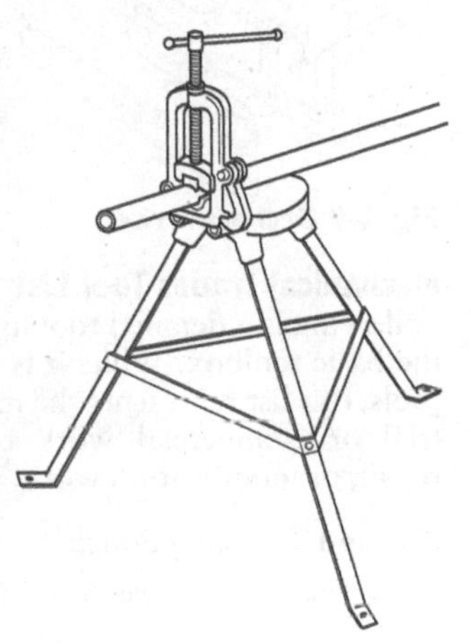

Fig. 2-8 Portable pipe vise.

Clamps

Clamps serve as temporary devices for holding work securely in place. They are first cousins to vises and are often used in a field location where vises are not available. Clamps are vital for such jobs as locking two pieces of metal together for welding (Fig. 2-10), securing two pieces that need to be held while gluing, or creating a third hand to allow ease of sawing, drilling, or other mechanical processes.

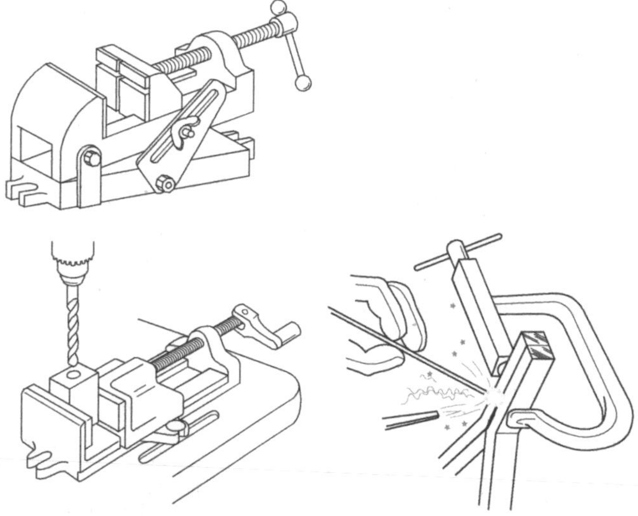

Fig. 2-9 Drill press vise. **Fig. 2-10**

Mechanical Trades Tool List

Following is a detailed tool list that can be used for stocking the basic toolbox. While it is certainly possible to add more tools, this list represents the minimum needed in most industrial or commercial work to perform basic installation, repair, or modification work.

Box and Traveling Pouch

1 Toolbox — 10-drawer top cabinet and 8-drawer bottom roll-away
1 Leather Tool Pouch with belt

Wrenches

1 set Open-End Wrenches (¼ in. × 5⁄16 in. – 1 in. × 1¼ in.)
1 set Box Wrenches (¼ in. × 5⁄16 in. – 1 in. × 1¼ in.)

1 set Combination Wrenches (1/4 – 1 5/16 in.)
1 set Open-End Metric Wrenches (7 – 25 mm)
1 set Box Metric Wrenches (7 – 25 mm)
1 set Combination Metric Wrenches (6 – 22 mm, 24 mm)
1 set Flare Nut Wrenches (3/8 in. × 7/16 in. – 3/4 in. × 7/8 in.)
1 set Flare Nut Metric Wrenches (9 × 11 mm – 19 × 21 mm)
1 Folding Allen Wrench Set (small)
1 Folding Allen Wrench Set (large)
5 Adjustable Wrenches (4, 8, 10, 12, 18 in.)

Sockets

1 set 1/4-in Drive Socket Set

- With shallow sockets — 5/32 – 3/4 in.
- With deep sockets — 3/16 – 3/4 in.
- With metric shallow sockets — 4, 5, 5.5, 6–15 mm
- With metric deep sockets — 4–15 mm

1 set 3/8-in Drive Socket Set

- With shallow sockets — 5/16 – 1 in.
- With deep sockets — 5/16 – 1 in.
- With metric shallow sockets — 9–19 mm, 21 mm
- With metric deep sockets — 4–15 mm
- With Allen Head sockets — metric and English
- With 8-point sockets (for square heads) — 1/4 – 1/2 in.

1 set 1/2-in Drive Socket Set

- With shallow sockets — 3/8 – 1 1/4 in.
- With deep sockets — 3/8 – 1 1/4 in.
- With metric shallow sockets — 9–28 + 30 mm + 32 mm
- With metric deep sockets — 13–22 mm + 24 mm

Pliers

3 pair Slip Joint Pliers (6 3/4, 8, 10 in.)
2 pair Arc Joint (7, 9 1/2 in.)
2 pair Locking Pliers (8-in. Straight Jaw)
2 pair Locking Pliers (6-in. Curved Jaw)
1 set Snap Ring Pliers

Screwdrivers and Nutdrivers

1 set Straight Screwdrivers (1/8 in. × 2 in., 3/16 in. × 6 in., 1/4 in. × 8 in., 1/4-in. stubby)
1 set Phillips Head Screwdrivers (#1, #2, #3)
1 set Torx Screwdrivers (T-10, T-15, T-20, T-25, T-27, T-30)
1 set Nut Drivers

Alignment and Prying

1 Double-Faced Engineer Hammer (48 oz)
1 Rolling Wedge Bar (16 in.)
1 set Screwdriver Type Pry Bars

Scraping, Filing, Extracting, Punching

1 Complete Set—Thread Files, metric and English
1 set Screw Extractors
1 set Punches
1 set Cold Chisels
1 set Files (for filing metal)
1 Center punch
1 Gasket Scraper (1½-in. face)

Hammers

4 Ball Peen Hammers (8, 12, 16, 30 oz)
1 Claw Hammer—Curved Claw (16 oz)
2 Soft Faced Mallets (24 and 12 oz)

Leveling and Measuring

1 Line Level
1 Torpedo Level (9 in.)
1 Level (24 in.)
1 Machinist's Scale (6 in.)
1 English/Metric Dial Caliper (0–6 in., 0–150 mm)
1 Micrometer (0–1 in.)
1 Plumb Bob (4½ in. with line)
1 Chalk Line Reel
1 Combination Square

1 Carpenter's Square
1—25 ft Tape (3/4-in. wide)
1 set Feeler Gauges (combination inch and metric)

Cutting and Clamping

1 Utility Knife
1 pair Pipe and Duct Snips (compound action)
1 Hack Saw with blades
2 "C" Clamps (3 in.)
2 "C" Clamps (4 in.)
2 "C" Clamps (6 in.)

Pipe and Tubing

2—10-in. Pipe Wrenches (Aluminum Handle)
2—14-in. Pipe Wrenches (Aluminum Handle)
2—18-in. Pipe Wrenches (Aluminum Handle)
2 Strap Wrenches (6 and 12 in.)
1 Tubing Cutter (1/4 in. × 1 in.)
1 Set Pipe Extractors

Electrical

1 pair Side Cutters (9 1/4 in.)
1 pair Wire Strippers
1 pair Diagonal Cutting Pliers (8 in.)
1 pair Long-Nosed Pliers (8 in.)
1 pair Needle-Nosed Pliers (6 in.)
1 pair Bent Needle-Nosed Pliers (4 1/2 in.)
1 Electrical Multi-Tester (Volts, Ohms, Continuity)
1 Flashlight

Troubleshooting

1 Mechanics Stethoscope
1 Infra-Red Thermometer

3. POWER TOOLS — PORTABLE

The field of portable power tools covers everything from small electric hand tools to heavy-duty drilling, grinding, and driving tools. Tool manufacturers have made the largest strides in the field of battery-powered tools.

Battery-Powered Tools

Cordless tools were a novelty when they first arrived on the scene. Some of the earliest entrants to the battery-powered tool market were lightly made and ineffective.

Fig. 3-1 Cordless drill kit.
Courtesy Milwaukee Electric Tools.

Perhaps the first successful battery-operated portable power tool used in industry was the drill motor, first introduced with two batteries and a charger (Fig. 3-1). One battery charged while the other was in use with the tool. Mechanics, recognizing the usefulness of cordless tools, demanded longer battery life between recharging and more power from the tool, but they also wanted lighter weight to promote ease of use. They also sought tools that provided more functions than just drilling.

The tool industry responded with drastic and rapid improvements. Currently, battery-powered tools are popular because of their portability, increased power, longer battery life, and overall convenience. Cordless tools are particularly effective where work must be done overhead or in hard-to-reach locations. Their weight and size have been reduced as battery technology has advanced.

Fig. 3-2 Multi-purpose kit.
Courtesy Milwaukee Electric Tools.

The list of effective cordless tools includes screwdriver/drill, saber saw, jigsaw, grinder, sander, soldering iron, and impact wrench. Many tools are now available in kits where the same battery provides power to a range of different tools (Fig. 3-2).

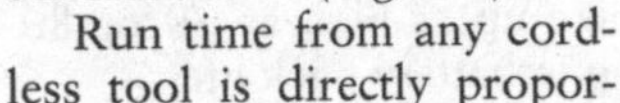

Run time from any cordless tool is directly proportional to the amp-hour rating of the battery. The amp-hour rating is like the size of a gas tank in a truck—the larger the tank (amp-hours), the farther the truck can go.

Cordless tools are usually powered by nickel-cadmium batteries (called NI-CADs) or nickel metal-hydride batteries (abbreviated Ni-Mh). A battery pack has individual cells, each putting out 1.2 volts, soldered together in series.

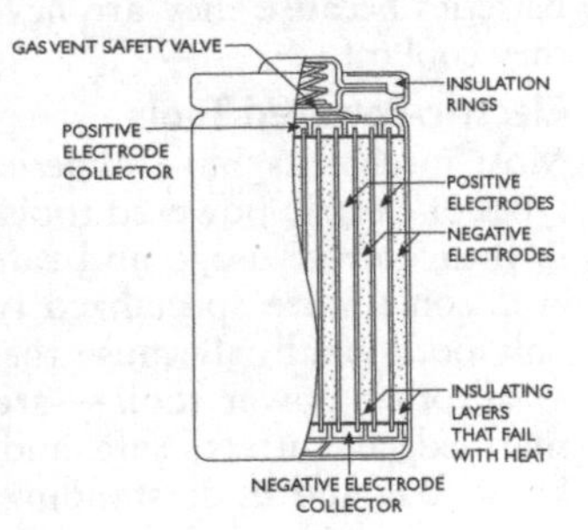

Fig. 3-3 Cutaway of a rechargeable battery.

The working life of a cordless tool battery depends, in large part, on how that battery is charged, and especially on how the cells in a battery pack rise in temperature while charging (Fig. 3-3). Heat is the enemy of both Ni-Mh and NI-CAD. Too much internal heat, and battery lifespan drops to less than half the potential 1,200 to 1,500 charge/discharged cycles. Battery chargers for heavy-duty industrial tools are now surprisingly sophisticated, monitoring dozens of parameters during each charge cycle. There are two reasons for this. The first is to minimize damaging heat build-up in cells so the battery cycle life will meet its potential. The other reason is to provide a troubleshooting diagnostic service for that time when the inevitable battery problem does arise.

To extend battery life, avoid anything that boosts tool load and current draw beyond what's necessary, thus reducing excessive cell temperatures. Keeping bits and blades sharp helps achieve this, but it's especially important to avoid prolonged stall conditions with any cordless tool; this is when the motor is loaded but the bit or blade it's driving has become stuck. This battery-frying situation can cause momentary current draw to spike up as high as 70 or 80 amps, with a corresponding drop in battery pack life.

Many "old timers" in the mechanical trades business suggest that tools should be purchased with a total of three rechargeable batteries. One battery can be in use with the tool, the recently discharged battery can sit and cool off, and the third battery (already cooled off) can be in the charger. The cost incurred by the purchase of the third battery is more than offset by the increase in the lifespan of all three batteries because they are never placed in the charger until they cool off.

Electric-Powered Tools

Most mechanics have experience with the more common types of electric-powered tools, and they are knowledgeable in their correct usage and safe operation. There are, however, some more specialized types that mechanics may use only occasionally. Because these specialized tools—as well as all other power tools—are relatively high speed, using sharp-edged cutters, safe and efficient operation requires knowledge and understanding of both the power unit and the auxiliary parts, tools, and so on.

Electromagnetic Drill Press

The *electromagnetic drill press* is the basic equipment used for magnetic drilling. It can be described as a portable drilling machine incorporating an electromagnet, with a capability of fastening the machine to ferrous metal work surfaces

(Fig. 3-4). The magnetic drill allows you to bring drilling equipment to the work, rather than bring the work to the drilling machine. A major advantage over the common portable drill motor is that it is secured in positive position electromagnetically, rather than depending on the strength and steadiness of the mechanic. This feature allows drilling holes with a much greater degree of precision in respect to size, location, and direction, with little operator fatigue.

Magnetic drilling is limited to flat metal surfaces large enough to accommodate the magnetic base in the area where the hole or holes are to be drilled. The work area should be cleaned of chips and dirt to ensure good mating of the magnetic base to the work surface. The unit is placed in the appropriate position, and the drill point is aligned with the center point location. When proper alignment has been established, the magnet is energized to secure the unit. A pilot hold is recommended for drilling holes larger than ½-in. diameter. The operation proceeds in a manner similar to a conventional drill press. Enough force should be applied to produce a curled chip. Too little force will result in broken chips and increased drilling time; too great a force will cause overheating and shorten the drill life.

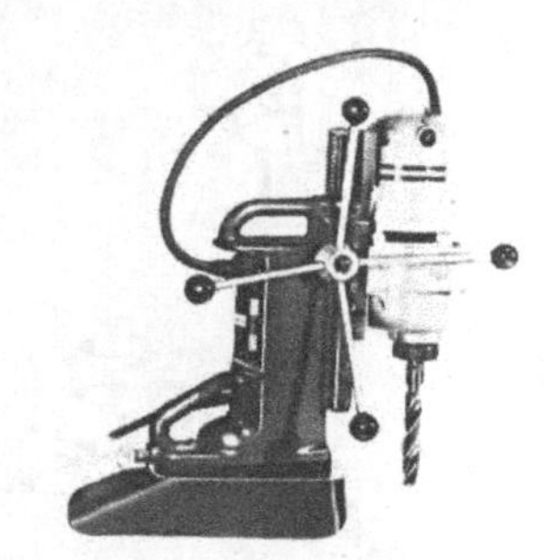

Fig. 3-4 Magnetic drill press.
Courtesy Milwaukee Electric Tools.

You can expand the capacity of magnetic drilling equipment by using carbide-tipped cutters. These tools are related to "hole saws" used for woodworking, but they possess much greater capacity and strength. These are tubular-shaped devices with carbide-tipped multiple cutting edges, which are highly efficient. The alternating inside and outside cutting edges are ground to cut holes rapidly with great precision. Hard carbide cutting tips help them outlast regular high-speed twist drills. They are superior tools for cutting large diameter holes because their minimal cutting action is fast and the power required is less than when removing all the hole material. Approximately a ⅛-in. wide kerf of material is removed and a cylindrical plug of material is ejected on completion of the cut. When used in conjunction with the magnetic drill, they enable the mechanic to cut large-diameter holes with little effort and great accuracy. Fig. 3-5 shows a magnetic drill with a hole-cutting attachment.

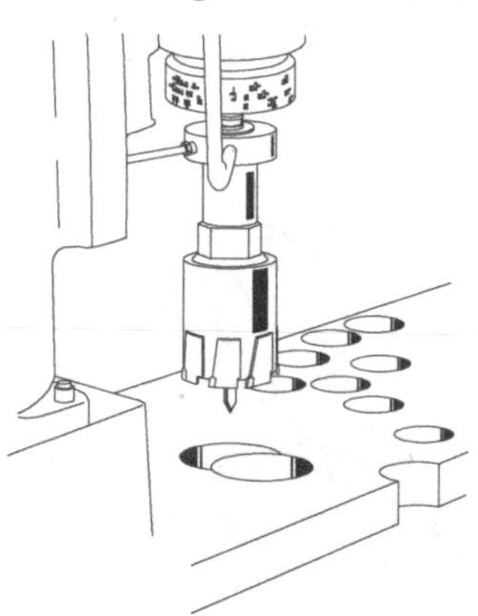

Fig. 3-5 Magnetic drill with hole-cutting attachment.

The hole cutter makes possible operations in the field that could otherwise be performed only on larger fixed machinery in the shop. The arbor center pin allows accurate alignment for pre-marked holes. As with most machining operations, a cooling and lubricating fluid should be used when cutting holes with this type of cutter. Ideally, it would be applied with some pressure to furnish a cutting action as well as cooling and lubricating. This may be accomplished by applying a fluid directly to the cutter and groove or by using an arbor lubricating mechanism. With a special arbor, it is possible to introduce the fluid under pressure. The fluid can be force-fed by a hand pump, which is part of a hand-held fluid container. Introducing the fluid on the inside surfaces of the cutter causes a flushing action across the cutting edges, which tends to carry the chips away from the cutting area and up the outside surface of the cutter.

Diamond Concrete Core Drilling

Diamond concrete core drilling is used to make holes in concrete structures. The tool that is used is the *diamond core bit.* Prior to the development of this tool, holes in concrete structures required careful planning and form preparation, or breaking away of sections of hard concrete and considerable patch-up. The tool is basically a metal tube, on one end of which is a matrix crown embedded with industrial-type diamonds distributed throughout the crown and arranged in a predetermined pattern for maximum cutability and exposure. Bits are made in two styles of construction, the closed back and the open back (Fig. 3-6).

There is a decided cost advantage in using the open-back bit, in that the adapters are reusable, offering savings in cost on each bit after the first. Also, if a core is lodged in the bit, removing the adapter makes the core removal easier. The closed-back bit offers the advantage of simplicity. Installation requires only turning the bit onto the arbor thread, without positioning or alignment issues. As the bit is a single, complete unit, there is no problem with mislaid, lost, or damaged parts.

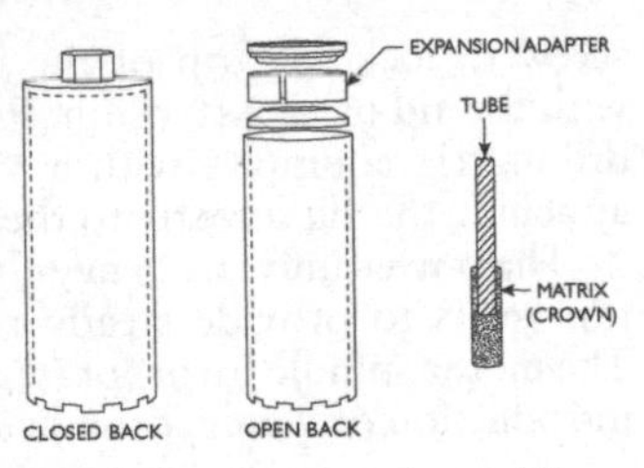

Fig. 3-6 Diamond core bits.

Successful diamond core drilling requires that several very important conditions be maintained: rigidity of the drilling unit, adequate water flow, and uniform steady pressure. Understanding the action, which takes place when a diamond core bit is in operation, will result in better appreciation of the importance of these conditions being maintained. Fig. 3-7 illustrates the action at the crown end of a diamond core bit during drilling. Arrows indicate the water flow inside the bit, down into the kerf slot and around the crown as particles are cut free and flushed up the outside surface of the bit.

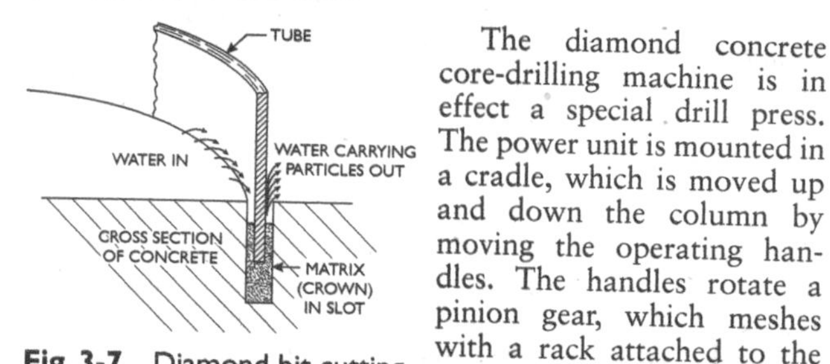

Fig. 3-7 Diamond bit cutting action.

The diamond concrete core-drilling machine is in effect a special drill press. The power unit is mounted in a cradle, which is moved up and down the column by moving the operating handles. The handles rotate a pinion gear, which meshes with a rack attached to the column. To secure the rig to the work surface, the top of the column includes a jack-screw to lock the top of the column against the overhead with the aid of an extension. Fig. 3-8 shows a concrete core-drilling rig equipped with a vacuum system, which makes attaching the rig directly to the work surface possible.

The power unit is a heavy-duty electric motor with reduction gears to provide steady rotation at the desired speed. The motor spindle incorporates a water swivel, which allows introduction of water through a hole in the bit adapter to the inside of the core bit. The power unit shown in Fig. 3-9 attaches to the cradle of the rig shown in Fig. 3-8. The diamond core bits (both open- and closed-back styles) fit the threaded end of the motor spindle.

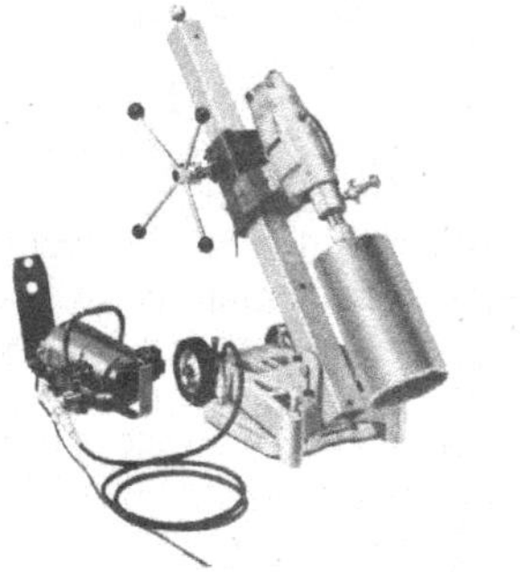

Fig. 3-8 Diamond concrete core drilling rig. *Courtesy Milwaukee Electric Tools.*

The rigidity of the drilling rig plays a critical part in successful diamond core drilling. The rig must be securely fastened to the work surface to avoid possible problems. Slight movement may cause chatter of the drill bit against the work surface, fracturing the diamonds. Greater movement will allow the bit to drift from location, resulting in crowding of the bit, binding in the hole, and possible seizure and damage to the bit.

An easy way to anchor the unit is with the jackscrew provided at the top of the column. A telescoping extension, pipe, 2 × 4, or other material cut to the appropriate length may be used. This allows the rig to be braced against the opposite wall.

The versatility of a diamond core-drilling rig may be greatly increased with a vacuum system. With this device it is possible to anchor the unit directly to the work surface, eliminating the need for extensions, braces, or other securing provisions. The rig in Fig. 3-8 is equipped with such a system. It consists of a vacuum pump unit and auxiliary hose, gauge, fittings, and more.

To use the vacuum system of attachment, first clean the area where the work is to be performed. Remove any loose dirt or material that might cause leakage of the seal of the pad to the work surface. Place the rig, with the diamond core bit on the spindle, in the desired location. Loosen the pad nuts to allow the pad to contact the work surface without restraint. Start the vacuum pump to evacuate the air from inside the pad. This produces what is commonly called the "suction," which holds the pad in place. A vacuum gauge indicates the magnitude of the vacuum produced. The graduated gauge face is marked to show the minimum value

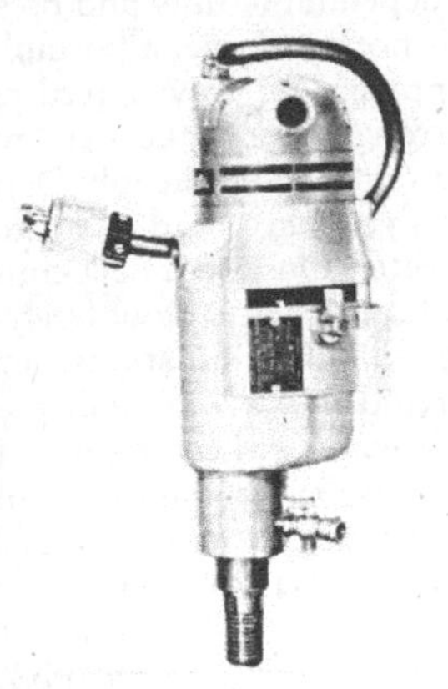

Fig. 3-9 Diamond drill power unit. *Courtesy Milwaukee Electric Tools.*

required for satisfactory operation. A clean, relatively smooth surface should allow building the vacuum value to the maximum. Should the gauge register a value below the minimum required, do not attempt to drill. Check for dirt, porous material, cracks in the surface, or any other condition that might allow air to leak past the pad seal. When the gauge reading indicates that the pad is secure, tighten the pad nuts to fasten the rig base firmly to the vacuum pad. Standing on the base of the rig is not a substitute for good pad fastening. While additional weight will add a little to the downward force, it will not prevent the rig from floating or shifting out of position.

Water—which is vital to the success of diamond core drilling—is introduced through the water swivel, a component of the lower motor housing. The swivel has internal seals, which prevent leakage as the water is directed into the spindle and into the inside of the bit. The preferred water source is a standard water hose (garden hose), which provides dependable flow and pressure. When a standard water hose is not available, a portable pressure tank (used for garden sprayers or gravity feed tanks) can be used. Whatever the arrangement, take care to ensure adequate flow and pressure to handle the job. Depending on the operating conditions, you may need to make provisions to dispose of the used water. On open, new construction, it may be permissible to let the water flow freely with little or no concern for runoff. In other situations, it will be necessary to contain runoff and find a way to dispose of the used water. You can make a water collector ring and pump for this purpose, or you can use the common wet-dry shop vacuum and build a dam with rags or other material.

When the rig is secured and water supply and removal provisions are made, drilling may commence. Starting the hole may present a problem because the bit crown has a tendency to wander, particularly when starting in hard materials and on irregular and inclined surfaces. At the start, the crown may contact only one spot, and thrust tends to cause the bit to walk. Often it is sufficient to guide the bit lightly with a board notched at the end. To start, apply light pressure to the bit. After the crown has penetrated the material, pressure may be fully applied. The feed force must be uniformly applied in a steady manner, not jerky or intermittent. Too little pressure will cause the diamonds to polish; too much can cause undue wear. To aid the mechanic in maintaining a steady pressure of the proper amount, most diamond drilling rigs incorporate an ammeter to indicate motor loading. In addition to the regular calibrations, the meter dial has a green area that shows the working range and a red area to indicate when too much pressure is being applied. Enough force should be exerted on the operating handles to keep the ammeter needle in the green area, indicating that the proper bit pressure and drilling speed are being maintained. This prevents overload and ensures optimum bit life.

The importance of constant flowing water during drilling cannot be overstressed. The water pressure and flow must be sufficient to wash cuttings from under the bit crown and up the outside of the bits, as illustrated in Fig. 3-7. The water also acts as a coolant, carrying away the heat that might otherwise cause the diamonds to polish or, in an extreme case, cause the bit to burn (to turn blue). All water connections must be tight and the water flow steady (one to two gallons per minute). Water flow and bit rotation must *not* stop while the bit is in the hole. The bit should be raised out of the hole *while turning* and then the water and power shut off. Stopping and starting the bit while in the hole could cause binding and damage to the bit. When the bit is cutting freely, the operator can feel movement into the concrete. The off-flowing water will have a slightly sludgy appearance as it carries away the concrete particles.

Diamond core bits are capable of drilling all masonry materials: concrete, stone, brick, and tile, as well as steel embedded in concrete, such as reinforcing rod and structural steel. They are not, however, capable of continuous steel cutting and so experience considerable wear and/or deterioration of the diamonds if required to do extensive steel cutting. When the bit encounters embedded steel, the mechanic will notice an increased resistance to bit feed. The feed rate should be decreased to accommodate the slower cutting action that occurs when cutting steel. When cutting steel at reduced feed rate, the water becomes nearly clear and small gray metal cuttings will be visible in the off-flowing water.

Diamond core bits will also cut through electrical conduit buried in concrete. This is a major consideration when cutting holes for changes or revisions in operating areas. Not only is there a possibility of shutting down operations if power lines are cut, but there is a possible electric shock hazard for the mechanic. Common sense dictates that you consider the possibility of buried conduit carefully when doing revision work. If the location of buried lines cannot be determined, drilling should not be attempted unless all power to lines that might be severed is disconnected.

When the concrete being drilled is not too thick, the hole may be completed without withdrawal of the bit. With greater thickness, drilling should proceed to a depth equal to at least two times the diameter of the hole. The bit should then be withdrawn and the core broken out. This may be done with a large screwdriver or pry bar inserted into the edge of the hole and rapped firmly with a hammer. The first section of the core can usually be removed with two screwdrivers, one on each side of the hole, by prying and lifting. If you find the core is embedded, the bit should be withdrawn and the core and any loose pieces of steel removed. When deeper cores must be removed, a little more thought and effort are required. Larger holes may permit reaching into them and, if necessary, drilling a hole to insert an anchor to aid in removal. If the core cannot be snapped off and removed, it may be necessary to break it out with a demolition hammer.

You should also consider that when drilling through a concrete floor, escaping water and concrete core falling from the hole as the bit emerges from the underside of the floor are possibilities. Make provisions to contain the water and catch the falling core to prevent damage and avoid injury.

Portable Band Saw

The *portable band saw*, while not in common use by mechanics, performs many cutting operations efficiently and with little effort (Fig. 3-10). Many field operations that require cutting in place consume excessive time and effort if done by hand. If acetylene torch cutting were used, the resulting rough surfaces might not be acceptable and/or the flame and flying molten metal might not be permissible. Often the object being worked on cannot be taken to another location to perform a sawing operation. In such instances, a portable band saw can do the work in the field. The portable band saw is in effect a lightweight, self-contained version of the standard shop band saw. It uses two rotating rubber-tired wheels to drive a continuous saw blade. Power is supplied by a heavy-duty electric motor, which transmits its power to the wheels through a gear train and worm wheel reduction. This reduces speed and develops powerful torque to drive the saw blade at proper speed. Blade guides, similar in design to those of the standard shop machine, are built into the unit, as well as a bearing behind the blade to handle thrust loads.

Fig. 3-10 Portable band saw.

A new generation of portable band saws with variable speed motors allows the mechanic to adapt the cutting speed to the material. This maximizes cutting performance and blade life. For example, slower speed is better for cutting stainless steel and very hard alloys.

Blades for portable band saws are available in a tooth pitch range from 6 to 24 teeth per inch. The rule of thumb for blade selection is to always have 3 teeth in the material at all times. Using too coarse a blade will cause thin metals to hang up in the gullet between two teeth and to tear out a section of teeth. Too fine a blade will prolong the cutting job, as each tooth will remove only small amounts of metal. Do not use cutting oil. Oil and chips transfer and stick to the rubber tires, causing the blade to slip under load. Chips build up on tires, causing misalignment of the blade.

Blades can be purchased with straight-pitch or variable-pitch tooth arrangements (Fig. 3-11). In addition, choices in blade materials range from inexpensive carbon steel through electron beam welded bi-metals used for tough applications. While the more exotic blades are expensive, they last a longer time and are less prone to breakage. Keep in mind that if a very smooth cut is required, variable-pitch blades produce much less vibration.

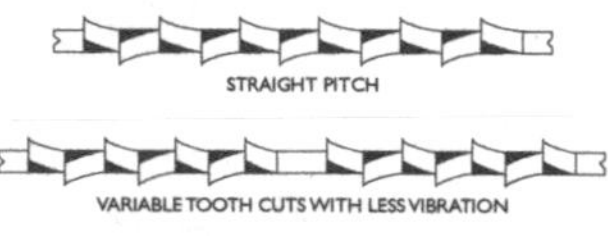

Fig. 3-11 Saw teeth.

Hammer Drill

The *hammer drill* combines conventional or rotary drilling with hammer percussion drilling (Fig. 3-12). Its primary use is the drilling of holes in concrete or masonry. It incorporates two-way action in that it may be set for hammer drilling with rotation or for rotation only. When drilling in concrete or masonry, special carbide-tipped bits are required. These are made with alloy steel shanks for durability, to which carbide tips are brazed to provide the hard cutting edges necessary to resist dulling. The concrete or masonry is reduced to granules and dust by this combined hammering and rotating action, and shallow spiral flutes remove the material from the hole.

For the most part, the hammer drill is used to drill "blind" holes in concrete that are used as a base for anchors to attach a machine base or structural member to the concrete. Holes are drilled to a prescribed depth, and the anchor is inserted and locked in place by a wedging action produced by striking the top or by a screwed arrangement.

A trick for blowing the dust out of the hammer drill hole is to use a turkey baster. Be careful not to get the dust in your eyes, though. The anchor is dropped into the hole and checked to ensure that the hole is not too shallow. Most manufacturers require that the top of the anchor should be just level with the concrete before it is set or wedged.

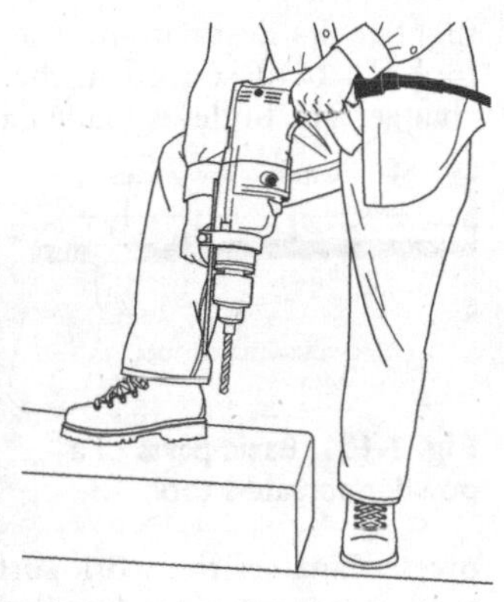

Fig. 3-12 Hammer drill.

Explosive-Powered Tools

The principal use of *explosive-powered* or *powder-actuated tools* is to fire a fastener into material to anchor or make it secure to another material. Some applications are wood to concrete, steel to concrete, wood to steel, steel to steel, and numerous applications of fastening fixtures and special articles to concrete or steel. Because the tools vary in design details and safe handling techniques, general information and descriptions of the basic tools and accessories are given, rather than specific operation instructions. Because the principle of operation is similar to that of a firearm, safe handling and use must be given the highest priority. An explosively actuated tool in simple terms is a "pistol." A pistol fires a round composed of a cartridge with firing cap and powder, which is attached to a bullet.

When you pull the trigger on a pistol, the firing pin detonates the firing cap and powder and sends a bullet in free flight to its destination. The direct-acting powder-actuated tool can be described in the same manner, with one small change. The bullet is loaded as two separate parts. Fig. 3-13 shows the essential parts of an explosive-actuated tool system—the tool and the two-part bullet (the combination of fastener and power load).

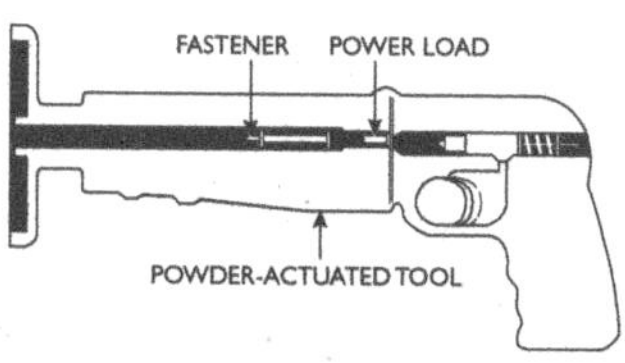

Fig. 3-13 Basic parts of a powder-actuated tool.

The fastener is inserted into the tool first. Then the mechanic inserts the power load. The tool is closed and pressed against the work surface, the trigger is pulled, and the fastener travels in free flight to its destination.

Obviously, a larger fastener will require more power to drive, and so would a fastener that must penetrate steel instead of wood. The mechanic uses a chart to match the loads with the fasteners for various penetrations of materials, fastener size, or depth. Often the end of the explosive charge is color coded to match the similar color code at the top surface of the fastener.

Safe use of these tools cannot be stressed enough. Some basic rules apply:

- Use the tool at right angles to the work surface.
- Check the chamber to see that the barrel is clean and free from any obstruction before using the tool.
- Do not use the tool where flammable or explosive vapors, dust, or similar substances are present.
- Do not place your hand over the front (muzzle) end of a loaded tool.
- Wear ear protection and goggles.

Manufacturers' tools may vary in appearance, but they are similar in principle and basic design. They all have a chamber to hold the power load, a firing pin mechanism, a safety feature, a barrel to confine and direct the fastener, and a shield to confine flying particles. There are two types of tools:

- Direct-acting, in which the expanding gas acts directly on the fastener to be driven into the work (Fig. 3-14)
- Indirect-acting, in which the expanding gas of a powder charge acts on a captive piston, which in turn drives the fastener into the work (Fig. 3-15)

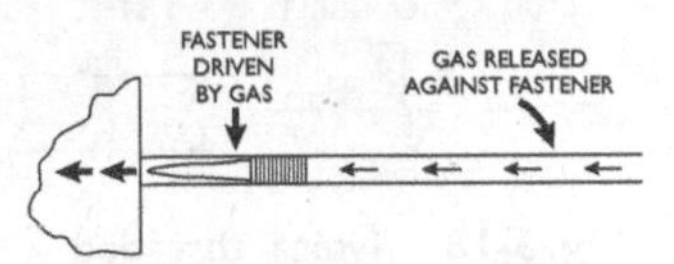

Fig. 3-14 Direct-acting operating principle.

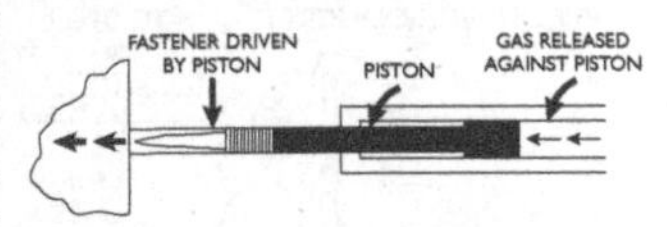

Fig. 3-15 Indirect-acting operating principle.

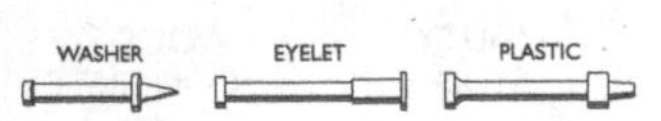

Fig. 3-16 Typical alignment tips.

The fasteners used in powder-actuated tools are manufactured from special steel and heat treated to produce a very hard yet ductile fastener. These properties are necessary to permit the fastener to penetrate concrete or steel without breaking. The fastener is equipped with some type of tip, washer, eyelet, or other guide member. This guide aligns the fastener in the tool, guiding it as it is being driven (Fig. 3-16).

Two types of fasteners are in common use: the *drive pin* and the *threaded stud*. The drive pin is a special nail-like fastener designed to attach one material to another, such as wood to concrete or steel (Fig. 3-17). Head diameters are generally 1/4, 5/16, or 3/8 in. For additional head bearings in conjunction with soft materials, washers of various diameters are either fastened through or made a part of the drive pin assembly.

The threaded stud is a fastener composed of a shank portion, which is driven into the base material, and a threaded

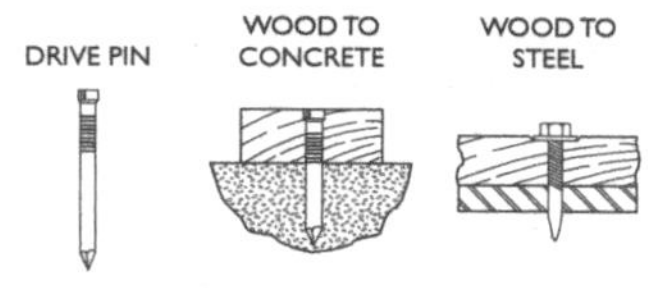

Fig. 3-17 Typical drive pins.

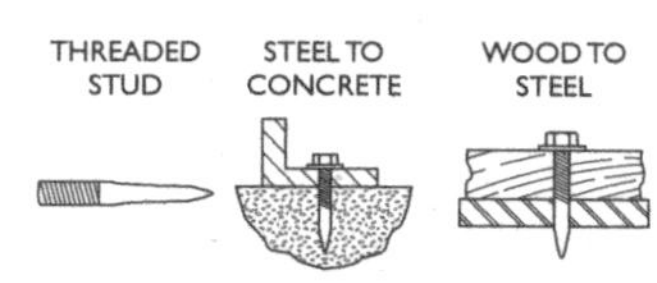

Fig. 3-18 Typical threaded studs.

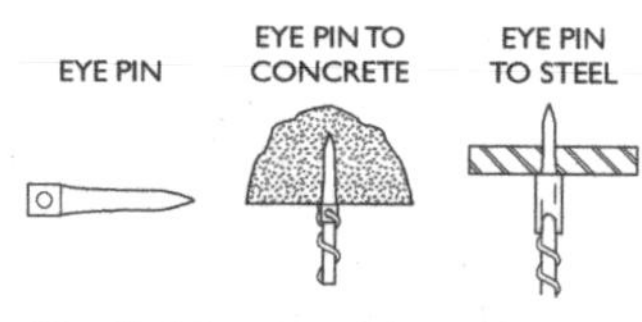

Fig. 3-19 Special eye pins.

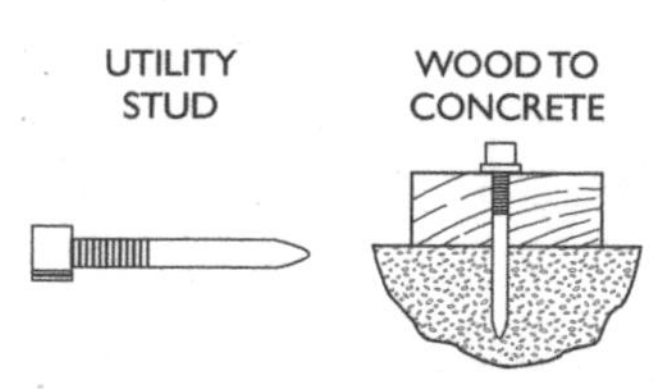

Fig. 3-20 Special utility studs.

portion to which an object can be attached with a nut (Fig. 3-18). Usual thread sizes are #8-32, #10-24, ¼ in-20, 5⁄16 in-18, and ⅜ in-16.

A special type of drive pin with a hole through which wires, chains, and so on can be passed for hanging objects from a ceiling is shown in Fig. 3-19.

Another special type of fastener, in this case a variation of the threaded stud, is the *utility stud*. This is a threaded stud with a threaded collar, which can be tightened or removed after the fastener has been driven into the wood surface. Fig. 3-20 shows the stud and its use to fasten wood to concrete.

The *power load* is a unique, portable, self-contained energy source used in powder-actuated tools (Fig. 3-21). These power loads are supplied in two common forms: cased or caseless. As the name implies, the propellant in a cased power load is contained in a metallic case. The caseless power load does not have a case, and the propellant is in solid form.

Whatever the type, caliber, size, or shape, a standard number-and-color code is used to identify the power level or strength of all power loads. The power loads are numbered 1 through 12, with the lightest being #1 and the heaviest being #12. In addition, because there are not 12 readily distinguishable permanent colors, power loads #1 through #6 are in brass-colored cases and #7 through #12 are in nickel-colored cases. A combination of the case color and the load color defines the load level or strength. The number and color identification code is shown in the following listing:

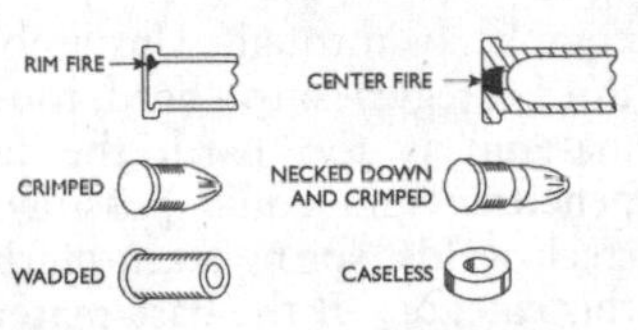

Fig. 3-21 Powder load construction.

Power Level	*Case Color*	*Load Color*
1	Brass	Gray
2	Brass	Brown
3	Brass	Green
4	Brass	Yellow
5	Brass	Red
6	Brass	Purple
7	Nickel	Gray
8	Nickel	Brown
9	Nickel	Green
10	Nickel	Yellow
11	Nickel	Red
12	Nickel	Purple

When selecting the proper power load to use for an application, start with the lightest power level recommended for the tool. If the first test fastener does not penetrate to the desired depth, the next higher load should be tried, until the proper penetration is achieved.

The material into which the fastener shank is driven is known as the base material. In general, base materials are metal and masonry of various types and hardness. Suitable base materials, when pierced by the fastener, will expand and/or compress and have sufficient hardness and thickness to produce holding power and not allow the fastener to pass completely through. Unsuitable materials may be put into three categories: too hard, too brittle, or too soft. If the base material is too hard, the fastener will not be able to penetrate and could possibly deflect or break. Hardened steel, welds, spring steel, marble, and natural rock fall into this category. If the base material is too brittle, it will crack or shatter, and the fastener could deflect or pass completely through. Materials such as glass, glazed tile, brick, and slate fall into this category. If the base material is too soft, it will not have the characteristics to produce holding power, and the fastener will pass completely through. Materials such as wood, plaster, drywall, composition board, and plywood fall into this category.

Shields and special fixtures are important parts of the powder-actuated fastening system and are used for safety and tool adaptation to the job. The shield should be used whenever fastening directly into a base material, such as driving threaded studs or drive pins into steel or concrete. In addition to confining flying particles, the shield also helps to hold the tool perpendicular to the work. Medium- and high-velocity class tools are designed so that the tool cannot fire unless a shield or fixture is attached. One of the most important acts that the conscientious user of a powder-actuated fastener system can perform is to see that the tool being used is equipped with the proper safety shield to ensure both safety and good workmanship.

Because masonry is one of the principal base materials suitable for powder-actuated tool fastening, it is important to understand what happens when a fastener is driven into it (Fig. 3-22). The holding power of the fastener results primarily from a compression bond of the masonry to the fastener shank. The fastener, on penetration, displaces the masonry, which tries to return to its original form and exerts a squeezing action on

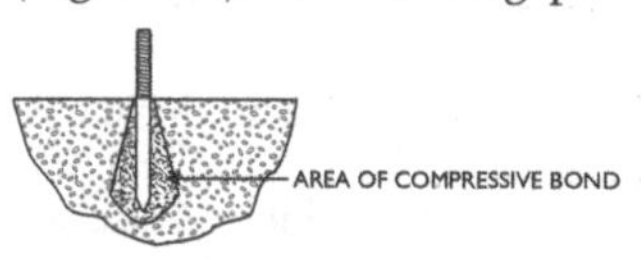

Fig. 3-22 Fastener bond in masonry.

the fastener shank. Compression of the masonry around the fastener shank takes place, with the amount of compression increasing in relation to the depth of penetration and the compressive strength of the masonry.

When an excessive direct pullout load is applied to a fastener driven into masonry material, failure will occur in either of two ways: The fastener will pull out of the masonry (Fig. 3-23a), or failure of the masonry will occur (Fig. 3-23b). This illustrates an important relationship between the depth of penetration and the strength of the bond of the fastener shank, and the strength of the masonry itself. When the depth of penetration produces a bond on the fastener shank equal to the strength of the masonry, the maximum holding power results.

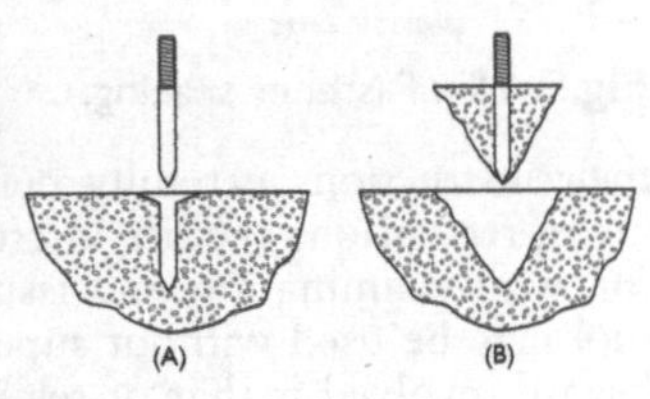

Fig. 3-23 Pullout from masonry.

Because the tensile strength of masonry is relatively low, care should be taken not to place it under high tension when driving fasteners. This may occur if fasteners are driven too close to the edge, as illustrated in Fig. 3-24.

Do not fasten closer than 3 in from the edge of the masonry. If the masonry cracks, the fastener will not hold, and there is a chance a chunk of masonry or the fastener could escape in an unsafe way. Setting fasteners too close together can also cause the masonry to crack. Spacing should be at least 3 in. for small diameter fasteners, 4 in. for medium-sized diameters, and 6 in. for larger diameters, as shown in Fig. 3-25.

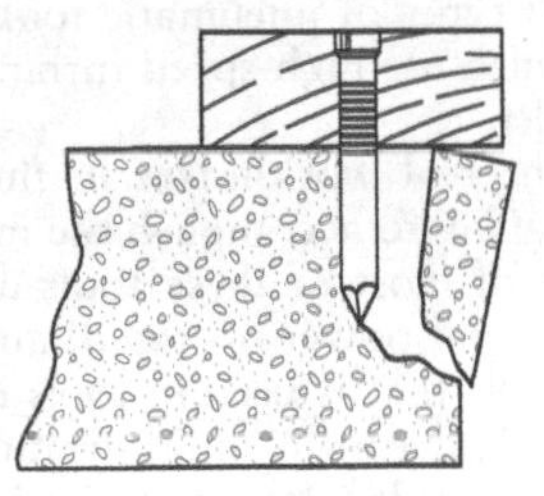
Fig. 3-24 Edge failure.

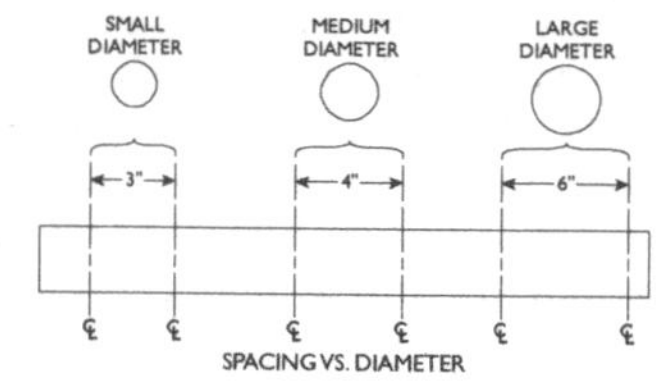

Fig. 3-25 Fastener spacing.

The potential hazards associated with the use of powder-actuated tools exceed those commonly encountered with other portable tools. Manufacturers stress safe operation in their instruction material, and it is very important that you study their instructions carefully before using the tool. In some areas, regulations require instruction by a qualified instructor, with examination and issuance of a permit before the tool may be used without supervision. Perhaps the greatest hazard involved is that of release of the fastener. You must exert careful thought and effort to ensure that the fastener is under control at all times and prevented from escaping in free flight.

Fluid Power Portable Tools

The driving force for many portable power tools is called *fluid power*. Usually portable fluid power tools are pneumatic, using air, or hydraulic, using oil.

The most popular pneumatic tools are impact types of tools: impact wrenches, jackhammers, and hammer drills. Other types of pneumatic tools deliver rotary motion, most of which use high-speed turbines such as a nut runner or air grinder.

The real workhorses in fluid power tools make use of hydraulics to accomplish the mission. The two basic components of most of these tools are the hydraulic pump, both hand- and power-operated, and the hydraulic cylinder. The principle that is used in tools involving the hydraulic pump and cylinder is the multiplication of force by hydraulic leverage (Fig. 3-26). In practice, a force is exerted on a small surface to generate hydraulic pressure. This pressure is transmitted, usually through a hydraulic hose, to a greater surface to produce a force magnified as many times as the ratio of the power cylinder surface area to the output cylinder area.

This multiplication of force is offset by the corresponding decrease in travel; that is, the small plunger in the diagram must travel 40 in. to displace enough fluid to move the large plunger 1 in.

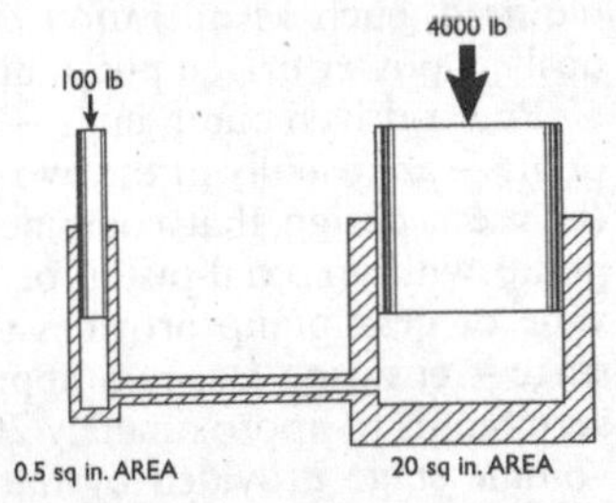

Fig. 3-26 Multiplication of force.

The primary component in a hydraulic power system is the hydraulic pump. The widely used hand-operated, single-speed hydraulic pump is a single-piston type designed to develop up to 10,000 pounds per square inch (psi) pressure. It is commonly constructed with a reservoir body, at one end of which is the power head. The head contains the power cylinder and piston, usually about ⅜ to ½ in. in diameter. Force is developed by use of a lever handle, which enables the effort applied to the handle end to be multiplied many times. Thus a force of 80 lb applied to the handle end would result in a force of 1200 lb acting on the hydraulic piston when under load (Fig. 3-27). Under this loaded condition, a pressure of 8000 psi would develop in the hydraulic system.

When time or speed is a factor, the two-speed hand pump may be used. The principal of operation is the same as the single-speed pump, except that it has the feature of a second piston of larger diameter. The two-speed operation provides high oil volume at low pressure for rapid ram approach, and when switched to the small piston it provides a high-pressure, low-volume stage for high-force operation. For applications of greater frequency or where a large volume of hydraulic fluid may be

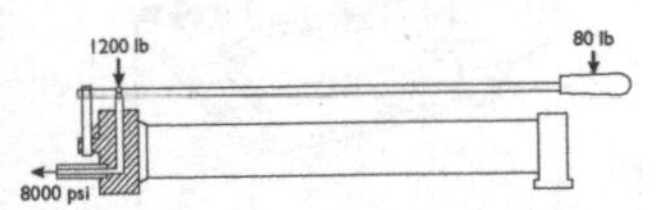

Fig. 3-27 Single-speed hand-operated hydraulic pump.

required, such as operation of several cylinders simultaneously, a power-driven pump unit may be more suitable.

Power-driven pump units—electric motor, gasoline engine, or air—are usually of the two-stage type. Many manufacturers use a design that combines a roller-vane or a gear-type pump with an axial-piston or radial-piston type. The roller-vane or gear pump provides the low-pressure, high-volume stage—ensuring fast ram approach and return. When pressure builds to approximately 200 psi, the high-pressure, low-volume stage provided by the axial piston or radial piston takes over to handle high-pressure requirements. Power pumps come equipped with various-sized reservoirs to provide hydraulic fluid for particular applications.

Fig. 3-28 shows an electric-driven, two-stage hydraulic pump for use with a single- or double-acting ram. It is equipped with a three-position, four-way manual valve for driving double-acting rams or multiple single-acting rams. Compressed-air-driven power units as well as gasoline engine-driven units in a great variety of designs are also in wide use.

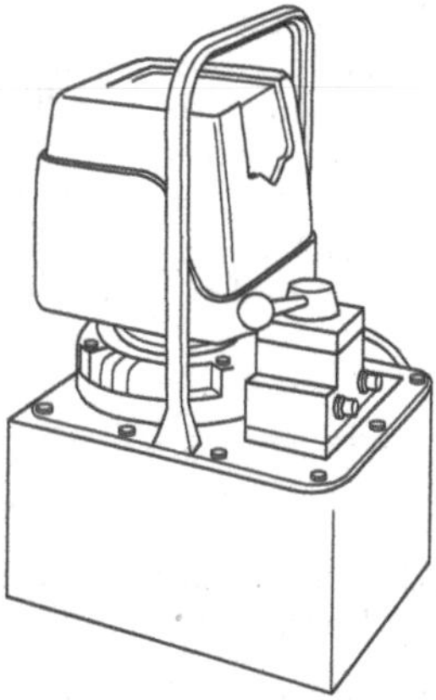

Fig. 3-28 Electric hydraulic pump.

The high-pressure hose is vital to the operation of portable hydraulic tools. Most of these hoses are designed for an operating pressure of 10,000 psi, with a bursting strength of 20,000 to 30,000 psi. The lightweight hoses are constructed with a nylon core tube and polyester fiber reinforcement, while those designed for heavy-duty, more severe service have one or more layers of braided steel webbing to help withstand the internal hydraulic pressure, as well as the external abuse that may be encountered in some applications.

The pressure gauge is a valuable, although not always necessary, component of the hydraulic power system. Its principal advantage is that it provides a visual indication of the pressure generated by the pump; thus, it can assist in preventing overloading of the hydraulic system as well as the equipment on which work is being done. Gauges, in addition to being graduated in psi, may also include a scale graduated in tonnage. This style of gauge must be matched to the cylinder diameter to obtain correct tonnage values. Correct practice among gauge manufacturers is to show a danger zone on the gauge face. This is done with a red background coloring in the area over 10,000 psi.

The second major component in a hydraulic power system is the hydraulic cylinder. Hydraulic cylinders operate on either a single- or double-acting principle, which determines the type of "return" or piston retraction. Single-acting cylinders have one port and, in their simplest form, retract due to weight or force of the load (load return). They are also made with an inner spring assembly, which enables positive retraction regardless of the load (spring return). Double-acting cylinders have two ports, and the fluid flow is shifted from one to the other for both hydraulic cylinder lifting and retraction (hydraulic return). Both single-acting and double-acting cylinders are manufactured with solid pistons or with center-hole pistons. Center-hole pistons allow insertion of pull rods for pulling applications. The single-acting, push-type, load-return style of hydraulic cylinder is shown in Fig. 3-29.

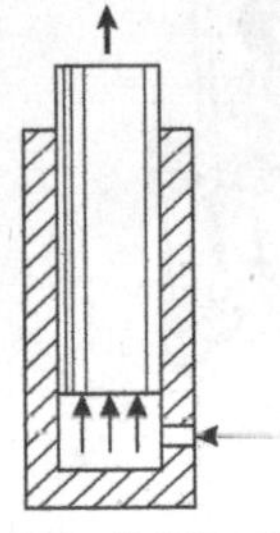

Fig. 3-29 Single-acting load-return cylinder.

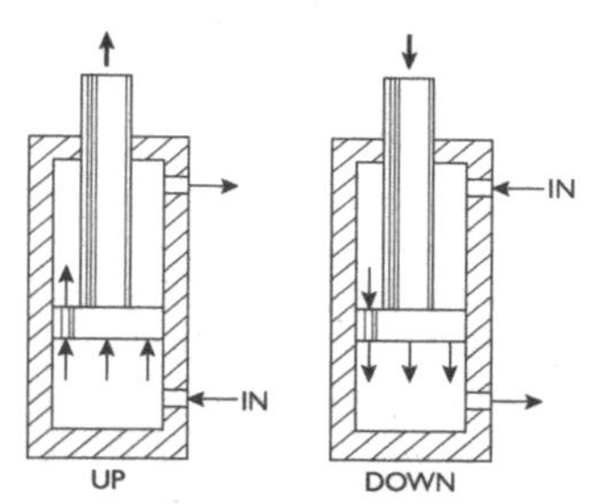

Fig. 3-30 Double-acting cylinder in lift and retract modes.

The single-acting cylinder, either load- or spring-return, is the most commonly used for hydraulic power tool operations performed by industrial mechanics. In some operations, however, power is required in both the lift and retract directions. The double-acting cylinder provides hydraulic function in both the lifting and lowering modes. Fig. 3-30 shows the principle of operation of a double-acting cylinder.

Double-acting cylinders should be equipped with a four-way valve to prevent trapping of the hydraulic fluid in the retract system. Most manufacturers build a safety valve into the retract system to prevent damaging the ram if the top hose is inadvertently left unconnected and the ram is actuated.

Fig. 3-31 Horizontal puller-press. *Courtesy Poulan Puller-Press, Inc.*

All manner of pumps, cylinders, accessories, and special attachments are available to adapt hydraulic power to a multitude of uses. Perhaps the most frequent application the mechanic is concerned with is the disassembly of mechanical components. This usually requires considerable force to remove one closely fitted part from another. One of the great advantages of hydraulic power over striking or driving to accomplish the operation is the manner in which the force is applied. Instead of heavy blows, which might cause distortion or damage components, the force is applied in a steady controlled manner. Fig. 3-31 shows a portable hydraulic power unit used with a jaw-type mechanical puller. Note that this has the advantage of mechanical positioning and adjustment plus controlled hydraulic power.

A great variety of hydraulic power accessories and puller sets—including all types of jaw pullers, bar pullers, and adapters—are available for all manner of pulling and pushing operations. Fig. 3-32 shows the use of a hydraulic unit to remove a sheave from the end of a motor shaft.

Fig. 3-32 Portable puller in field use. *Courtesy Poulan Puller-Press, Inc.*

The use of a portable hydraulic puller set greatly enhances the ability of a mechanic to tackle large jobs right at the machine site. Lifting heavy loads, moving machinery, pulling shafting or bearings, and removing pressed-on parts are all jobs where hydraulics plays an important part.

If the portable hydraulic set fails to operate properly, use Table 3-1 to help to determine the cause.

Table 3-1 Portable Hydraulic Troubleshooting Chart

Problem	*Possible Cause*
Cylinder will not advance	Pump release valve open Coupler not fully tightened Oil level in pump is low Pump malfunctioning Load too heavy for cylinder
Cylinder advances part way	Oil level in pump is low Coupler not fully tightened Cylinder plunger is binding

(continued)

Table 3-1 (continued)

Problem	*Possible Cause*
Cylinder advances in spurts	Air in hydraulic system Cylinder plunger binding
Cylinder advances more slowly than normal	Leaking connection Coupler not fully tightened Pump malfunctioning
Cylinder advances but will not hold	Cylinder seals leaking Pump malfunctioning Leaking connection Incorrect system setup
Cylinder leaks oil	Worn or damaged seals Internal cylinder damage Loose connection
Cylinder will not retract or retracts more slowly than normal	Pump release valve is closed Coupler not fully tightened Pump reservoir overfilled Narrow hose restricting flow Broken or weak retraction spring Cylinder internally damaged
Oil leaking from external relief valve	Coupler not fully tightened Restriction in return line

4. POWER TOOLS—STATIONARY

Three "must have" tools in the shop are the *power hack saw*, the floor *drill press*, and the hydraulic *shop press*.

Power Hack Saw—Reciprocating

No tool gets a workout in the shop like a power hack saw (Fig. 4-1). Cutting steel, pipe, pieces for welding fabrication, and metal conduit are all jobs for this piece of equipment.

The stock is usually held in a vice mounted on the base of the machine. An electric motor supplies the power for the machine. Usually some form of lubricant-coolant is pumped onto the workpiece to cool the cut and flush out the chips.

The strokes per minute and the feed rate can be controlled and changed depending on the type of material that is being cut. Usually the feed rate is controlled by a weight that is added or moved along a bar to increase the pressure of the saw on the stock to be cut. Power hack saw blades can be selected with various numbers of teeth per inch. Table 4-1 shows the settings for most saws to obtain quality cuts of various materials.

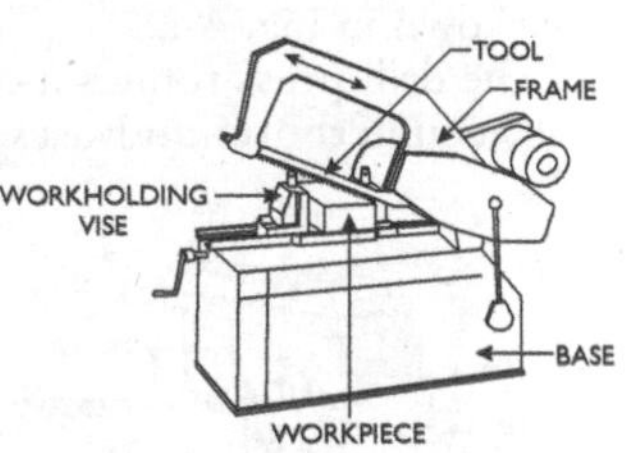

Fig. 4-1 Reciprocating power hack saw.

Table 4-1 Power Hack Saw Setup Recommendations

Material	*Teeth Per Inch*	*Strokes Per Minute*	*Feed Pressure*
Aluminum	4–6	150	Light
Brass, soft	6–10	150	Light
Brass, hard	6–10	135	Light
Cast iron	6–10	135	Medium
Copper	6–10	135	Medium
Carbon tool steel	6–10	90	Medium

(continued)

Table 4-1 (continued)

Material	*Teeth Per Inch*	*Strokes Per Minute*	*Feed Pressure*
Cold-rolled steel	4–6	135	Heavy
Drill rod	10	90	Medium
High-speed steel	6–10	90	Medium
Machinery steel	4–6	135	Heavy
Malleable iron	6–10	90	Medium
Pipe, steel	10–14	135	Medium
Structural steel	6–10	135	Medium
Tubing, brass	14	135	Light
Tubing, steel	14	135	Light

Drill Press

Round holes are drilled in metal by means of a machine tool called a *drill press*. The main components of this shop tool are shown in Fig. 4-2.

The drill press rotates a cutting tool (twist drill, countersink, counterbore) and uses pressure from the operator on the feed lever to advance the drill through the work piece. The speed of the drill and the pressure on the feed lever vary depending on the material that must be drilled. Usually the workpiece is located on the table below the drill and is locked into position. Never attempt to start a twist drill without first using a center punch to make an indentation for starting the drill point (Fig. 4-3).

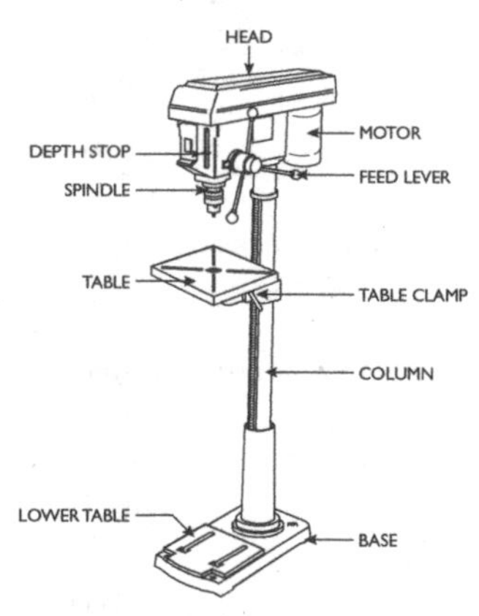

Fig. 4-2 Components of a drill press.

Often a drill press vise can be used to lock the piece in place to make sure it does not move when the hole is drilled. The use of parallels in the drill press vise (Fig. 4-4) keeps the workpiece flat and the hole square to the surface of the workpiece.

Drill speeds in revolutions per minute (RPM) for various diameter drills and typical workpieces are shown in Table 4-2. It shows figures for high-speed twist drills. For lower-quality carbon steel drills, the drill speed RPM should be halved.

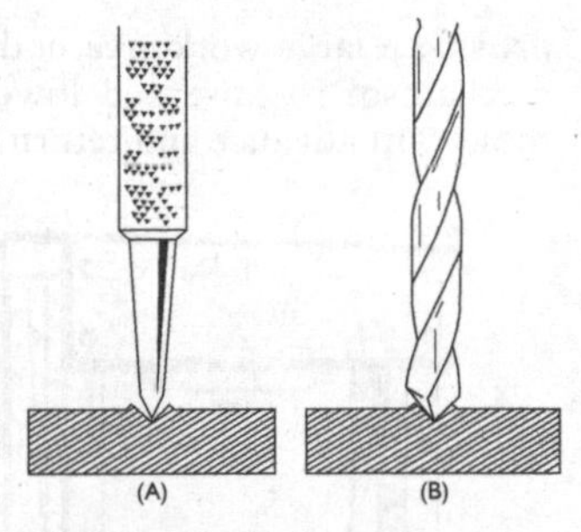

Fig. 4-3 Center punch before drilling.

Using a cutting oil or other suitable lubricant will greatly enhance the smoothness of the drilled hole and extend the life of the drill bit.

Hydraulic Press or Arbor Press

No maintenance shop can be without an arbor (mechanical) or hydraulic press. This tool removes and installs interference fitted parts (shaft and bearing, shaft and gear, shaft and sprocket, pin and hub, and others). The shop press can be used for straightening bent machine components to allow their reuse as well as to bend certain components that must be attached by clamped forces.

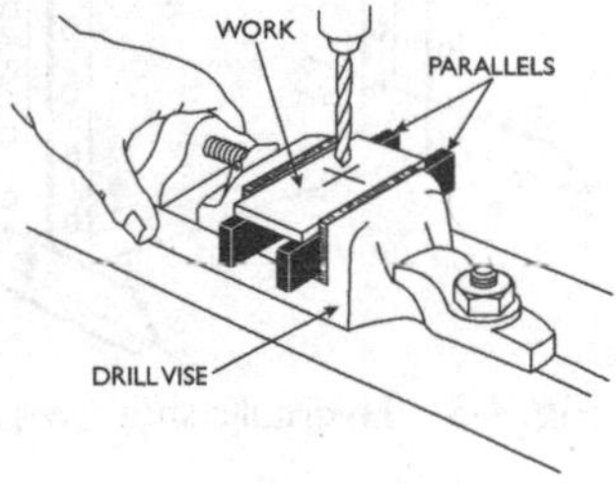

Fig. 4-4 Vise with parallels keeps work square and level.

Shop presses are both hand- and power-operated. They are almost essential for mechanical work involving controlled application of force to assemble or disassemble machinery. Most shop presses provide a large work area under the ram, a winch and cable mechanism to raise and lower the press bed quickly, and rapid ram advance and return (Fig. 4-5).

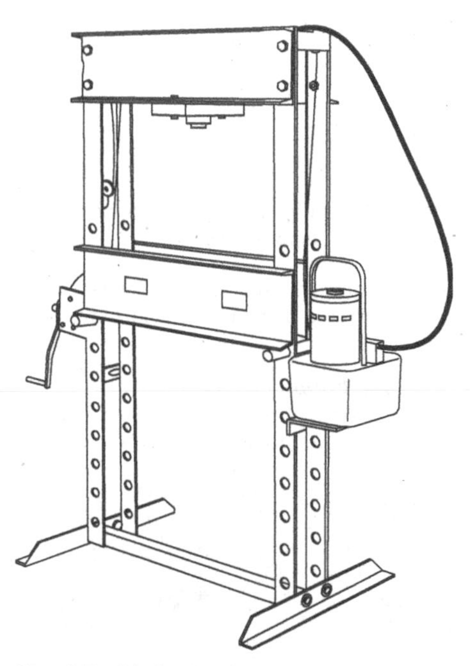

Fig. 4-5 Hydraulic shop press.

Table 4-2 Drill Speeds (in RPM)

Dia. Drill (Inches)	Plastics	Soft Metals	Annealed Cast Iron	Mild Steel	Malleable Iron	Hard Cast Iron	Tool Steel	Alloy Steel
1/16	12217	18320	8554	7328	5500	4889	3667	3056
3/32	8142	12212	5702	4884	3666	3245	2442	2038
1/8	6112	9160	4278	3667	2750	2445	1833	1528
5/32	4888	7328	3420	2934	2198	1954	1465	1221
3/16	4075	6106	2852	2445	1833	1630	1222	1019
7/32	3490	5234	2444	2094	1575	1396	1047	872
1/4	3055	4575	2139	1833	1375	1222	917	764
9/32	2712	4071	1900	1628	1222	1084	814	678
5/16	2445	3660	1711	1467	1100	978	733	611
11/32	2220	3330	1554	1332	1000	888	666	555
3/8	2037	3050	1426	1222	917	815	611	509
13/32	1878	2818	1316	1126	846	752	563	469
7/16	1746	2614	1222	1048	786	698	524	437
15/32	1628	2442	1140	976	732	652	488	407
1/2	1528	2287	1070	917	688	611	458	382
9/16	1357	2035	950	814	611	543	407	339
5/8	1222	1830	856	733	550	489	367	306
11/16	1110	1665	777	666	500	444	333	277
3/4	1018	1525	713	611	458	407	306	255

5. TWIST DRILLS

Drill Terms and Geometry

Twist drills, as we know them today, are the most common and widely used metal-cutting tools.

The parts of a twist drill are identified in Fig. 5-1. One particular component is the point angle—the angle included between the cutting lips projected on a plane parallel to the drill axis. The clearance angle, or heel of the drill point, is also shown.

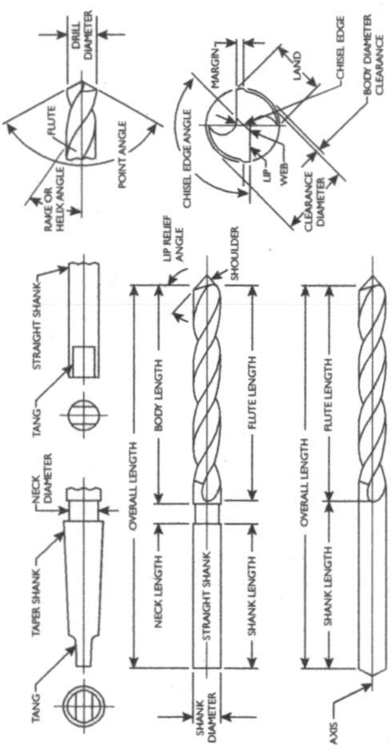

When drilling steel with a drill press, a twist drill with a point angle of 118° is recommended. Field work on hard or tough materials, especially when using a battery- or electric-powered drill motor, requires a drill point of 135°.

The heel or clearance angle for the 118° point should be about 8–12°. The 135° point should have a clearance angle of about 6–9°. The types of points, clearance, and angles—depending on the material to be drilled—are shown in Fig. 5-2.

Fig. 5-1 Parts of a twist drill.

A twist drill cuts by wedging under the material and raising a chip. The steeper the point and the greater the clearance angle, the easier it is for the drill to penetrate the work-piece. The blunter the point and the smaller the lip angle clearance, the greater the support for the cutting edges. From this logic it makes sense that a greater lip angle and a lesser lip clearance are good for hard and tough materials, and the decreased point angle and increased lip clearance are suitable for softer materials.

When drilling some softer non-ferrous metals such as brass or copper, the drill point tends to "bite in," or penetrate too rapidly. This gives tearing and a poor finish to the hole.

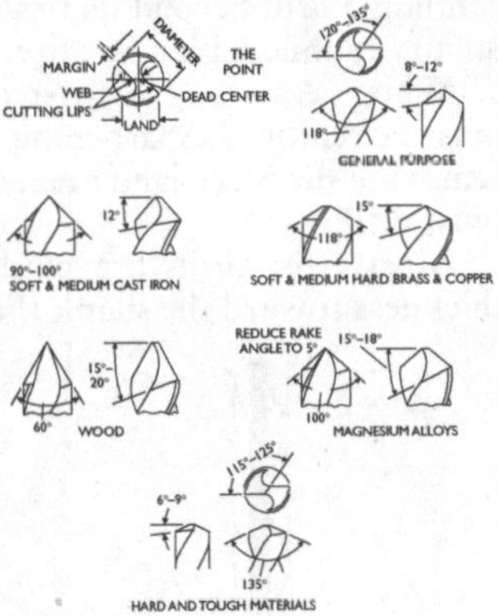

Another tricky material to drill is stainless steel. If the craftsman allows the drill to rotate without adequate pressure on the drill, the stainless will work harden and further drilling becomes almost impossible. Always put plenty of pressure on the drill when making a hole in stainless to keep this condition from spoiling the work.

Fig. 5-2 Drill point angles versus material.

Drill Sharpening

A drill begins to wear as soon as it is placed in operation. The maximum drill wear occurs at the corners of the cutting lips. The web, or chisel-point edge, begins to deform under the heat generated during drilling. The increase in wear at the corners travels back along the lands, resulting in a loss of size and tool life.

Wear occurs at an accelerated rate. When a drill becomes dull it generates more heat and wears faster. In other words, there is more wear on the twentieth hole than on the tenth, still more on the thirtieth, and this continues. As wear progresses, the torque and thrust required increase. In addition to the accelerated wear, drill breakage due to excessive torque is one of the most common drill failures. In comparison, running a drill beyond its practical cutting life is like driving an automobile with a flat tire.

When regrinding a twist drill, all of the worn sections must be removed. Sharpening the edges or lips only, without removing the worn land area, will not properly recondition a twist drill.

Most twist drills are made with webs that increase in thickness toward the shank (Fig. 5-3). After several sharpenings and shortenings of a twist drill, the web thickness at the point increases—resulting in a longer chisel edge (Fig. 5-4). When this occurs, it is necessary to reduce the web so that the chisel edge is restored to its normal length. This operation is known as *web-thinning*.

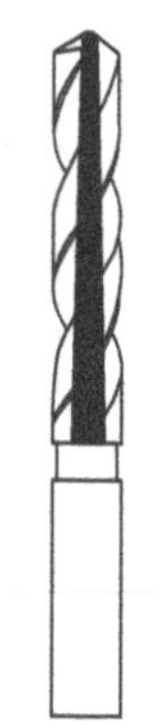

Fig. 5-3 Web increases toward shank.

There are several common types of web-thinning. The method shown in Fig. 5-5 is perhaps the most common. The length A is usually made about one-half to three-quarters the length of the cutting lip.

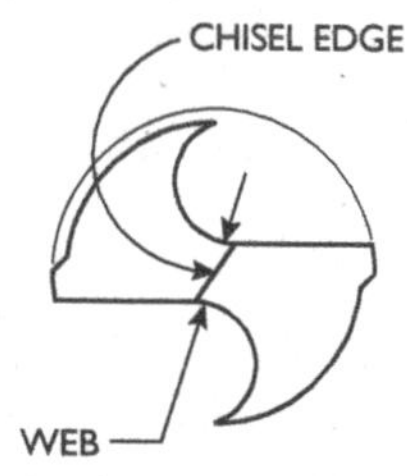

Fig. 5-4 Chisel edge.

The usual method of web-thinning involves the use of a pedestal or bench grinder. Start with a clean, sharp grinding wheel. Hold the drill at approximately 30–35° to the axial centerline of the drill (Fig. 5-6).

The corner of the wheel must be lined up with the tip of the web.

The cutting lip must be turned out about 10° to make sure the edge is not ground and the web can be ground away.

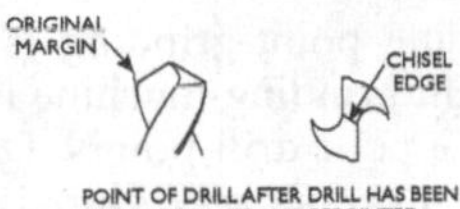

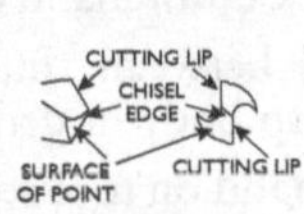

Fig. 5-5 Hold these dimensions when web-thinning.

After the worn portion of the drill has been removed and the web thinned (if necessary), the surfaces of the point must be reground. These two conical surfaces intersect with the faces of the flutes to form the cutting lips, and with each other to form the chisel edge. As in the case of any other cutting tool, the surface back of these cutting lips must not rub on the work, but must be relieved in order to permit the cutting edge to penetrate. Without such relief the twist drill cannot penetrate the work, but will only rub around and around.

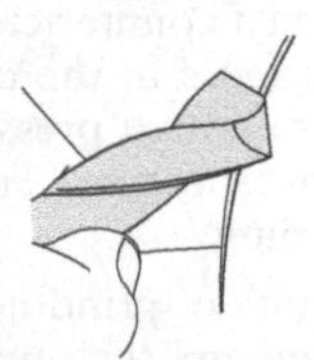

Fig. 5-6 Web thinning on a grinder.

In addition to grinding the conical surfaces to give the correct point angle and cutting clearance, both surfaces must be ground alike. Regardless of the point angle, the angles of the two cutting lips (A1 and A2) must be equal. Drill points of unequal angles or lips of unequal lengths will result in one cutting edge doing most of the cutting. This type of point will cause oversized holes, excessive wear, and short drill life.

To maintain the necessary accuracy of drill point angles, lip lengths, lip-clearance angle, and chisel-edge angle, the use of machine point-grinding is recommended. The lack of a drill-point grinding machine is not sufficient reason, though, to excuse poor drill points. Drills may be pointed accurately by hand if proper procedure is followed and care is exercised.

Grinding the Drill Point by Hand

The most commonly used drill point is the conventional 118° point. This grind will give satisfactory results for a wide variety of materials and applications.

1. Adjust the grinder tool rest to a convenient height for resting back of forehand on it while grinding drill point.
2. Hold drill between thumb and index finger of left hand. Grasp body of drill near shank with right hand.
3. Place forehand on tool rest with centerline of drill making desired angle with cutting face of grinding wheel (Fig. 5-7a) and slightly lower end of drill (Fig. 5-7b).
4. Place heel of drill lightly against grinding wheel. Gradually raise shank end of drill while twisting drill in fingers in a counter-clockwise rotation and grinding conical surfaces in the direction of the cutting edges. Exert only enough pressure to grind the drill point—without overheating. Frequently cool drill in water while grinding.
5. Check results of grinding with a gauge to determine if cutting edges are the same lengths and at desired angle and that adequate lip clearance has been provided (Fig. 5-8).

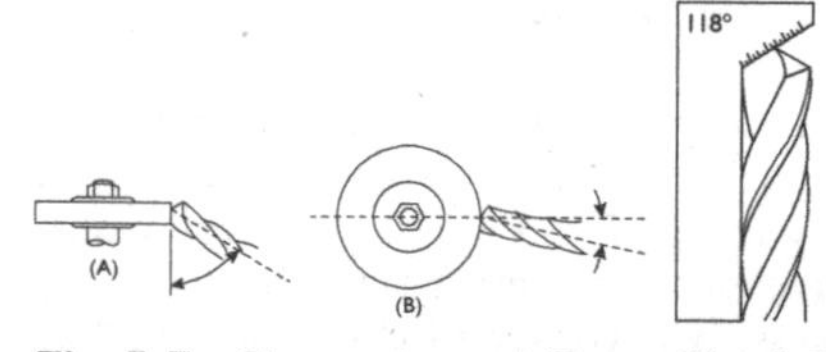

Fig. 5-7 Sharpening a drill.

Fig. 5-8 Checking the cutting edges and angle.

6. MECHANICAL DRAWING

Drawings are a graphic method of conveying information about the size, location, parts, distance, and relationship of the components of a machine, electrical system, piping run, or other devices. The axiom that *a picture is worth a thousand words* is very true for maintenance and installation work. The ability to sketch a machine or electrical schematic or, conversely, the knowledge to "read" a drawing to determine how things work is a must.

An artist uses a technique called *perspective drawing* to illustrate a scene. Perspective drawing is similar to what the eye can see. It is pictorial in nature, often influenced by artistic ability, and does not work well to describe and dimension a part of a machine. Mechanical drawing does not depend on artistic ability. It is a method of graphically representing a part or machine that anyone can learn independent of artistic skill.

Mechanical and electrical drawings follow rules or conventions. A simple example of a convention is the graphic representation of the exact location of the center point to drill a hole by the use of centerlines and dimension lines shown on a drawing (Fig. 6-1).

Obviously, after this plate is fabricated the dimension lines and centerlines will not show on the actual piece. They are represented by the special lines—called centerlines and dimension lines—on the drawing.

Lines on a drawing define the shape, size, and details of an object. Using conventions to control these lines, it is possible to describe an object graphi-cally so that it can be visualized accurately.

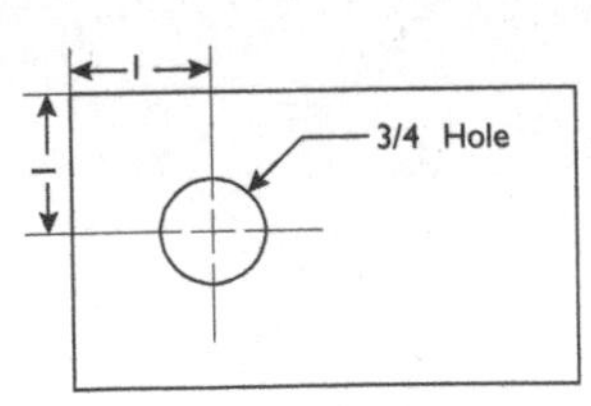

Fig. 6-1 Drawings follow simple rules.

The weight of a line in a sketch is also important. Weight can be considered the thickness and boldness of a line. Lines that describe the outline of an object are drawn with thick, bold lines, and lines describing the dimensions or centers of objects are drawn with thinner, lighter lines. The types of line conventions are shown in Table 6-1.

The most widely used system of mechanical drawing is called *orthographic projection*. This projection technique is the basis on which industrial drawings are prepared worldwide.

Graphic projection is a process whereby a drawing is made by rays of sight that are "projected" onto a viewing or picture plane. Orthographic is derived from the Greek word *orthogonious*, which means perpendicular. Orthographic views are therefore placed perpendicular to each other.

Table 6-1 Line Conventions

Line Type	*Description*
Visible lines (or object lines) Thick Visible or Object Line	*Visible lines* are used to represent all edges and visible outlines that are not hidden in a particular view. These lines should be drawn as thick, continuous lines.
Hidden lines Medium Hidden Line (Invisible Outlines)	*Hidden lines* appear as short dashes of medium thickness. They are used to show those surfaces, edges, or object corners that are hidden from view. In rare cases, hidden lines are omitted if they would complicate a drawing or sketch.

Line Type	*Description*
Centerlines Thin Center Line	*Centerlines* are used to show the axes of round or symmetrically shaped holes or solids and reference lines for dimensions. Centerlines are drawn thin, with chain patterns of long and short dashes.
Dimension lines Dimension Line 3.5" Leader Line Extension Line	*Dimension*, *extension*, and *leader lines* are drawn as continuous thin lines. Extension lines are extended from visible lines to indicate the limits of the feature to be dimensioned, and they are usually continuations of the object line. When drawn or sketched, the extension line should not come in contact with the dimensioned feature and should extend approximately ⅛ in. beyond the dimension line.
Extension lines	Dimension lines are drawn between *extension lines*. Arrowheads are placed at the end of the dimension line, and they touch the extension line. The dimension itself is placed in an opening along the dimension line, approximately halfway between the extension lines. It is also permissible to place the dimension above the line if more space is required.

(continued)

Table 6-1 *(continued)*

Line Type	***Description***
Leader lines 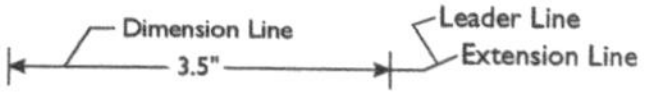 Dimension Line 3.5" Leader Line Extension Line	The leader, or *leader line*, is used to make note of a part or feature. It is drawn or sketched at an angle of 60°, 45°, or 30° to the center of the feature. At the end of the leader, the arrowhead is used to touch circumferences or perimeters.
Section lines Thin Section Lines 	*Section lines* are thin, continuous lines that are usually drawn at a 45° angle. The spacing between these lines will vary, but it will normally have ⅛-in. separation. Section lines are used to show the cut surface in section views. Certain industries use special symbols to signify various materials. These special symbols are sometimes referred to as "cross-hatching" lines.
Cutting and viewing plane lines  Thick Cutting Plane Line	*Cutting and viewing plane lines* are extra-thick lines that can be drawn either continuously or as chains. Cutting plane lines indicate the position of imaginary cutting planes. The cutting plane is found in section-view drawings where portions of an object are "cut away" to show what the interior looks like. At the end of a cutting plane line will be a set of arrowheads that indicate the viewing position.

Line Type	*Description*
Break lines Thick — Short Break Line Thin — Long Break Line	*Break lines* are used to shorten a view of a long part. There are three types of break lines: a thick, wavy line that is used for short breaks; a thin, straight line with a zigzag; and an "S" break used for cylindrical objects.
Phantom lines Thin — Phantom Line	*Phantom lines* are drawn as thin lines with a dash pattern of one long and two shorts. These lines are used to show movable parts in one location (usually the extreme), the path of a moving part, the finished or final machined surfaces on a casting, the outline of a rough casting on finished parts, the position of a part that is next to or fits the part being drawn, or the portion of the part that must be removed.

In Fig. 6-2, an object is placed behind two transparent viewing planes. If the lines of sight were projected perpendicular to the viewing planes, these rays would be parallel to each other. Where they intersect on both viewing planes is where the orthographic views are observed.

An orthographic view that shows the object as viewed from the front is projected onto the front plane. The front plane gives information only about the length and height of the object. The use of only one plane of projection is usually not enough to communicate all the necessary information about an object. Additional information about the object's

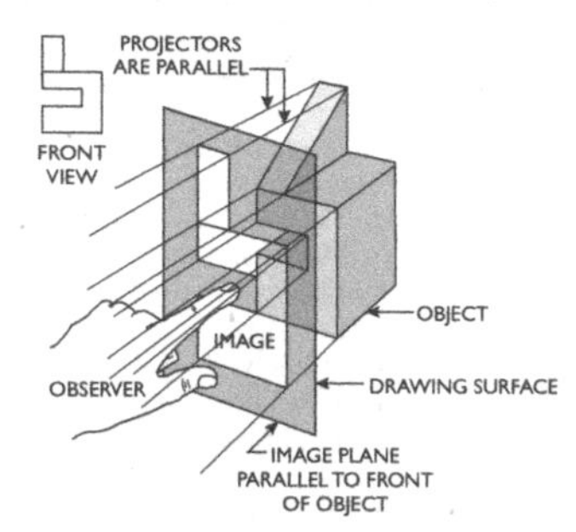

Fig. 6-2 Orthographic projection.

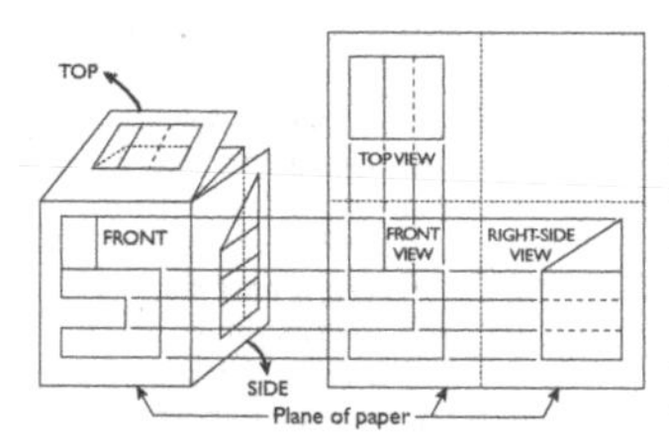

Fig. 6-3 Front, top, and side planes.

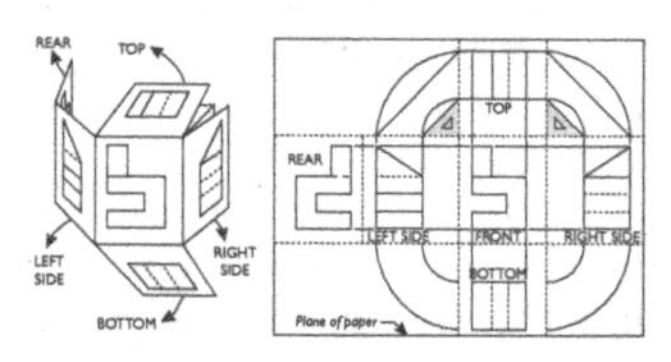

Fig. 6-4 Six planes of projection.

width or details about surfaces perpendicular to the front plane are needed to provide this knowledge. For this, additional picture planes are used.

The top plane is located perpendicular to the front plane and above the object. The top view is obtained by projecting rays of sight 90° from the object to that plane, and it presents both length and width dimensions. The side view (usually a right side) is perpendicular to both top and front planes and gives height and width dimensions. When all three views (front, top, and right side) are given, all the information relative to the three-dimensional shape of the object will be presented (Fig. 6-3).

Six planes of projection can be used to represent all sides of the viewing box: front, top, bottom, back, right side, and left side planes (Fig. 6-4). These six views are called the principal views of projection.

Follow these simple rules when making or reading a drawing with respect to placement of the views:

- The top and bottom views should always be placed above and below, respectively, the front view.
- When the front view is not used, the top and bottom views should be placed above and below, respectively, the rear view.
- The right and left profile views should always be placed to the right and left, respectively, of the front view.
- If the front view is not used, the right and left profile views should be placed to the left and right, respectively, of the rear view.

Often, when viewing the surface of an object, many of the edges and intersections behind the surface of the object are not visible. To be complete, a drawing or sketch must include lines, which represent these edges and intersections. Lines made up of a series of small dashes, called invisible outlines or hidden lines, are used to represent these behind-the-surface edges and intersections. Fig. 6-5 shows the use of a dashed line to show a hidden or invisible edge.

When surfaces of an object are at right angles to each other, regular views are adequate for their representation. When one or more of the views are inclined and slant away from either the horizontal or vertical plane, the regular view will not show the true shape of the inclined surface. To show the true shape, an auxiliary view (Fig. 6-6) is used. In an auxiliary view, the slanted surface is projected to a plane, which is parallel to it. Shapes that would appear in distorted form in the regular view appear in their true shape and size in the auxiliary view.

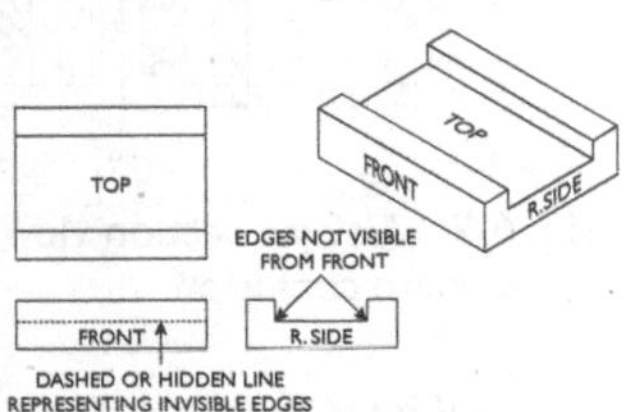

Fig. 6-5 Hidden lines.

Orthographic projection techniques can be used to illustrate the external, or surface, features of complex objects.

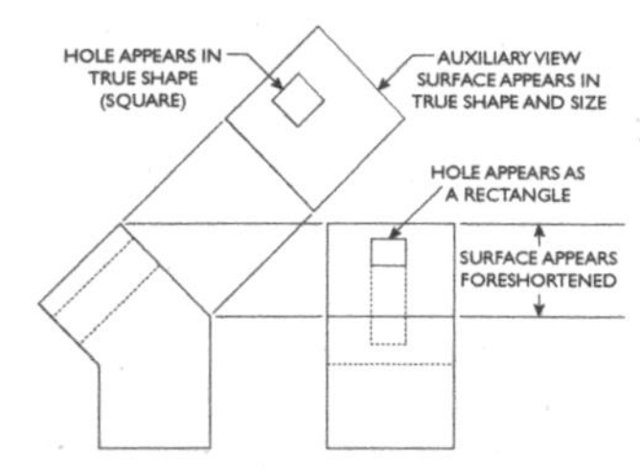

Fig. 6-6 Auxiliary view.

But when interior details must be presented, standard projection techniques are usually ineffective. This is true with parts that have complex cavities or subassemblies located behind an exterior wall or shell. See Fig. 6-7 for an example where the use of hidden lines is confusing and difficult to interpret.

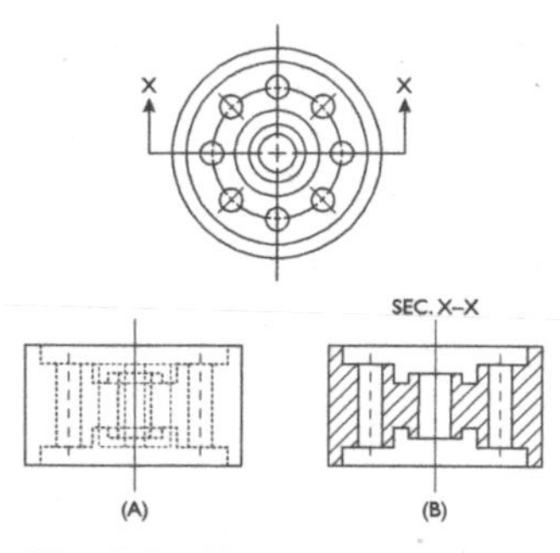

Fig. 6-7 Using a section view to eliminate confusion.

These situations require a technique for viewing that shows the object as it would appear if a portion or section were to be cut away to reveal the internal features. This is known as the section view. On cutaway section views, the invisible edges become visible and may be represented by solid-object lines. You might create a mental picture where a section view appears as if a saw had been used to cut away a section or portion of the object in order to reveal its internal features. This "saw" is referred to as the cutting plane and is presented as a line (straight, curved, or multidirectional) in the drawing. Hence the term *cutting plane line*. See Fig. 6-8.

A cutting plane line shows the path that the imaginary cutting plane takes as it exposes the object's interior surface. As a line, it represents an edge view of the cutting plane. Arrowheads are used and drawn at the ends of these lines to show the viewing direction. The letter notation (e.g., A-A or

Y-Y) is usually placed at each arrowhead to identify the section view that corresponds to the appropriate cutting plane.

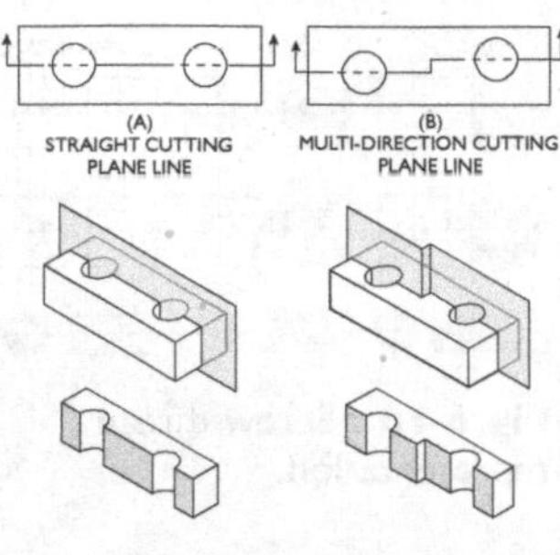

Fig. 6-8 Cutting plane lines.

Symbols are used to illustrate the difference between cut and uncut portions and various other parts in a mechanical drawing. In section views, these symbols are known as section lines. Section lines have been standardized to show the type of material of which a part is made. Fig. 6-9 shows a wide range of standardized crosshatching symbols for common material.

Screw threads are an integral part of most machinery, so there is a repeated need to show and specify threads on a mechanical drawing. Threads can be drawn with a detailed representation showing them as they actually appear, but this would be very time-consuming. It is almost never done. Usually threads are given a symbolic representation suitable for general understanding. Three types of methods are used: detailed representation (approximates the true appearance), schematic representation (nearly as pictorial and much easier to draw), and simplified representation (the easiest to draw and therefore the most commonly used). See Fig. 6-10.

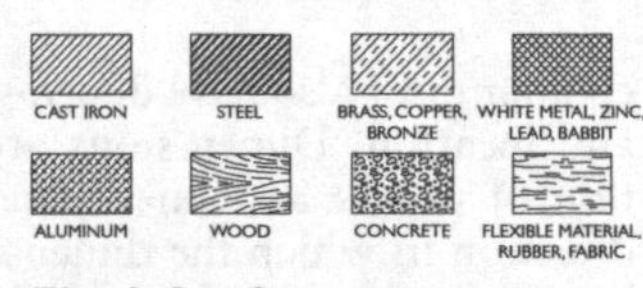

Fig. 6-9 Common crosshatching symbols.

The Unified Standard Thread—which incorporates the earlier National American, SAE, and ASME standards—is the most widely used thread form (Fig. 6-11). It is designated on drawings by a note specifying in sequence the nominal

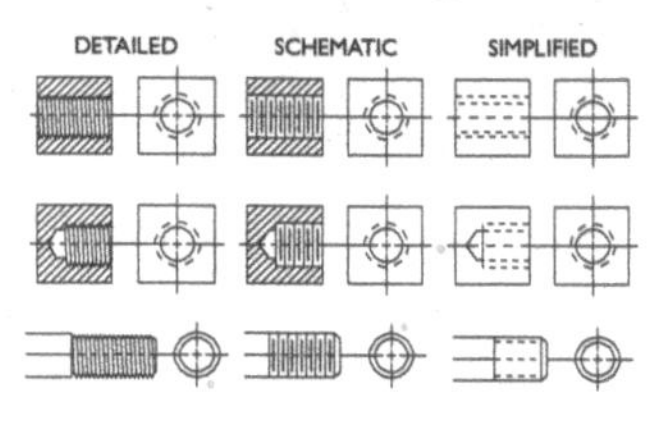

Fig. 6-10 Screw thread representation.

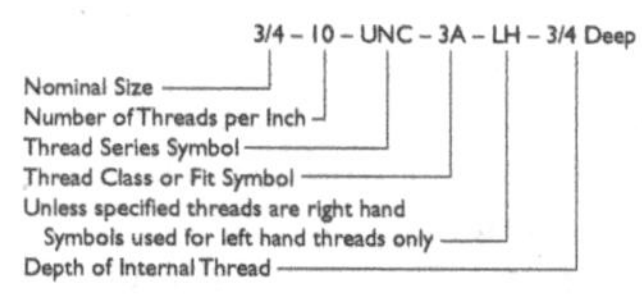

Fig. 6-11 Thread identification.

size, number of threads per inch, thread series symbol, and the thread class symbol. The symbols used are (UNCC), indicating a coarse thread series, (UNF), denoting a fine thread series, and (UNEF), which stands for an extra-fine thread series. Thread class #1 indicates a loose free fit, #2 indicates a free fit with very little looseness, and #3 indicates a very close fit. The #2 class fit is used for most applications. The letter A indicates an external thread and the letter B an internal thread.

Drawings made up of lines to describe shape and contour must also have dimen-sions and notes to supply sizes and location. Dimen-sions are placed between a combination of points and lines. The dimension line indicates the direction in which the dimension applies, and the extension lines refer the dimension to the view. Leaders are used with notes to indicate the feature on the drawing to which the note applies. See Fig. 6-12.

Three systems of writing dimension values are in general use. The first is the "common fraction" system, with all dimensions values written as units and common fractions. The second system uses decimals to express fractions when distances require precision greater than plus or minus 1/64 of an inch. It might be called the "combination fraction and decimal" system. The third system, the "complete decimal system," uses decimal fractions for all dimensional values. Two-place decimals are used where common fractions are used in the other two systems. When large values must be shown with the common fraction system, feet and inch units may be used. The foot (') and inch (") marks may be used to identify the units. Feet and inch dimensions should only be used for distances exceeding 72 in. When dimensions are all in inches, the inch marks are preferably omitted from all dimensions and lines.

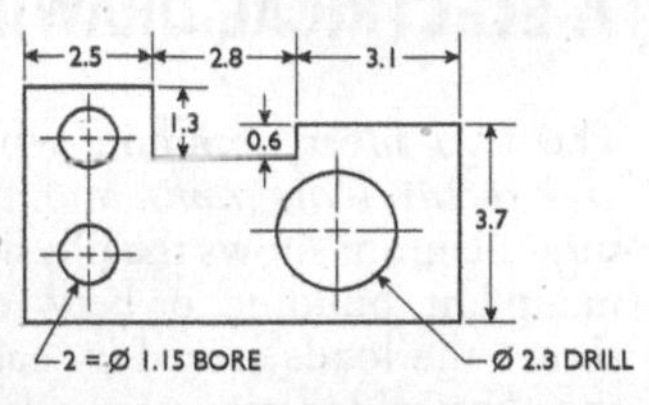

Fig. 6-12 Example of leader, extension, and dimension lines.

7. ELECTRICAL DRAWING

The two most common types of electrical drawings are *architectural diagrams* and *circuit diagrams*. An architectural diagram shows the physical location of the electric lines in a plant, building, or between buildings. A circuit diagram shows the loads served by each circuit. It does not indicate the physical location of any load or circuit.

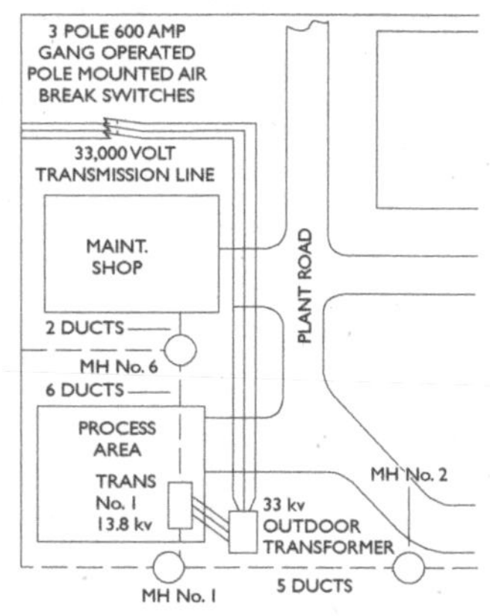

Fig. 7-1 Plot plan.

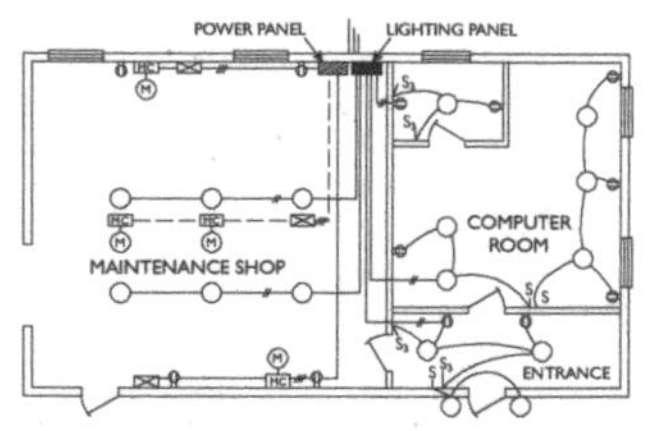

Fig. 7-2 Floor plan.

Architectural Diagrams

A *plot plan* is an architectural diagram. It shows the electrical distribution to all the plant buildings. An example of a plot plan is shown in Fig. 7-1.

Another type of architectural diagram is the *floor plan*, which shows where branch circuits are located in one building or shop (Fig. 7-2).

The last type of architectural diagram is the *riser diagram* used to show how wiring goes to each floor of a building or plant, shown in Fig. 7-3.

Circuit Diagrams

A circuit diagram shows how a single circuit distributes electricity to various loads. As mentioned before, it does not show the location of these loads, as an architectural diagram does. It is common practice to represent various pieces of electrical equipment with symbols on a circuit diagram. A *single-line circuit diagram* is used in many industrial locations to show the use of a three-wire 440-volt circuit. Rather than cluttering the diagram by drawing three wires over and over, a single line is used to represent all three wires. Fig. 7-4 shows a typical single-line diagram.

Control circuits are normally shown as ladder diagrams. This type of schematic diagram shows the parts of the circuit drawn on horizontal lines, like the rungs of a ladder. Fig. 7-5 shows an example of a ladder diagram.

Electrical diagrams use special symbols to represent switches, transformers, motors, relays, and more. Table 7-1 shows a list of standard symbols used throughout the electrical industry.

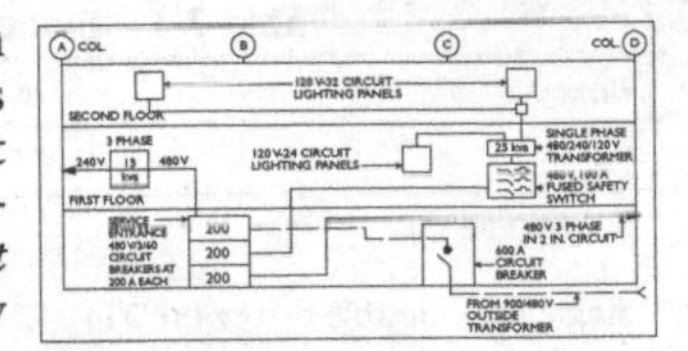

Fig. 7-3 Riser diagram.

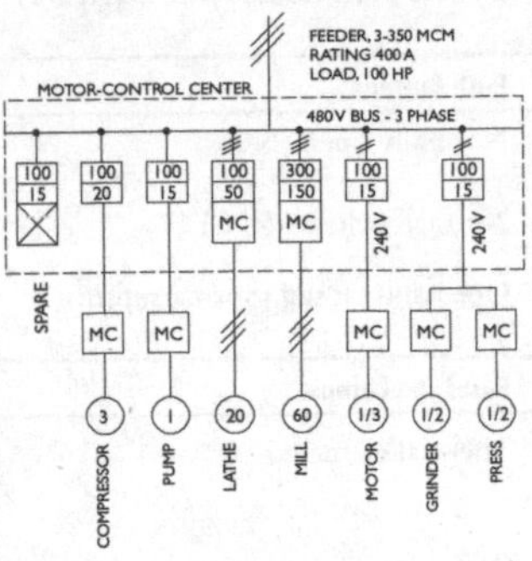

Fig. 7-4 Single-line diagram.

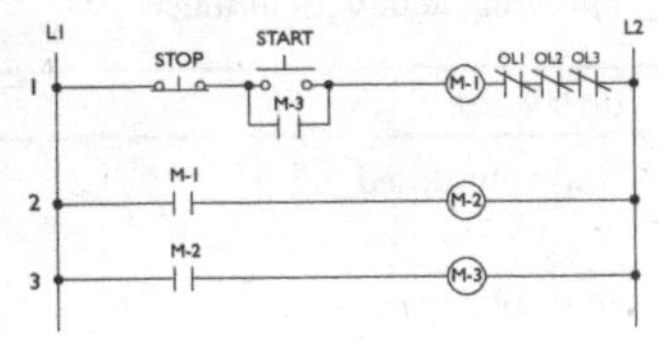

Fig. 7-5 Ladder diagram.

Table 7-1 Electrical Symbols

Name	*Symbol*
Switches	
Single pole (SP)	
Single pole, double throw (SPDT)	
Double pole, single throw (DPST)	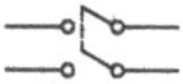
Double pole, double throw (DPDT)	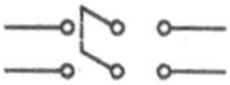
Push Buttons	
Normally open (NO)	
Normally closed (NC)	
Open and closed (Spring return)	
Fuses and Lamps	
Thermal element	
Fuse	
Indicating light with leads	
Indicating light with terminal	
Limit Switch	
Normally closed	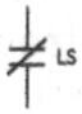 LS
Normally open	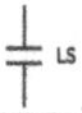 LS
Conductors	
Crossing—not connected	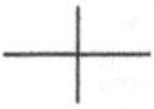

Name	*Symbol*
Conductors	
Crossing and connected	
Joining	
Circuit Breaker	
Single-pole	CIRCUIT BREAKER ELEMENT
Two-pole	
Thermal Overloads	
Three-phase thermal element heater	THERMAL ELEMENT (HEATER)
Resistor	
Fixed with leads	
Fixed with terminals	
Coils	
Non-magnetic — fixed	
Magnetic core — fixed	
Misc.	
Battery	OR
Capacitor	
Ground	
Motor	

8. ISOMETRIC DRAWING

The isometric system of drawing is three-dimensional and more pictorial than the orthographic system. The word isometric means equal measure and refers to the isometric position that is the basis for isometric drawing.

Rotating a cube around its vertical axis and tilting it forward until all faces are foreshortened equally, as shown in Fig. 8-1a, shows how the isometric position is developed. The overall outline is a regular hexagon (Fig. 8-1b). The three lines of the front corner of the cube in isometric position make equal angles with each other and are called the isometric axes (Fig. 8-1c).

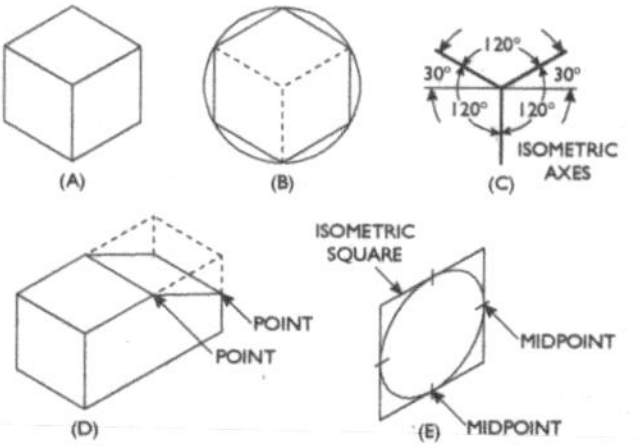

On isometric drawings, horizontal object lines are drawn parallel to the 30° axis. Actual lengths are shown to scale, frequently resulting in a distorted appearance due to the foreshortening effect. Lines that are not parallel to the isometric axes do not appear in their true length. To draw such lines, their ends are located on isometric lines and the points connected (Fig. 8-1d). A circle shows in isometry as an ellipse, as shown in Fig. 8-1e. The midpoint of each side is located and the arcs of circles drawn to be tangent at the midpoints, as shown in Fig. 8-1e.

Fig. 8-1

When making sketches of objects having non-isometric surfaces, it's helpful to imagine the object as contained in rectangular boxes. By sketching the rectangular boxes in isometric position, points may be accurately located and constructed in simplified form, as shown in Fig. 8-2.

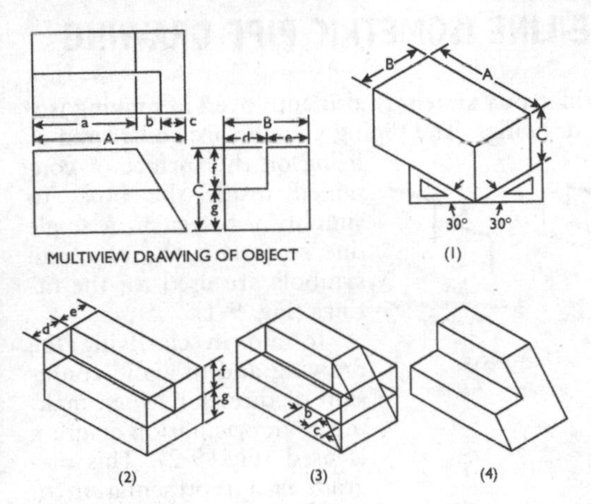

Fig. 8-2

There are two types of scale used in isometric drawing. The first is full size, where all measures are made as if the object were projected full size. For example, if a 2-in. isometric cube were drawn, all sides would be measured to the length of 2 in. In reality, the length of the sides of an isometric object is actually $\sqrt{2/3}$ of true size, equal to 81.65 percent of actual size. Thus, the 2-in. cube would be drawn with sides equal to 1.633 in. To approximate this proportion, most drawings use a ¾-in. size scale. The two methods are shown in Fig. 8-3.

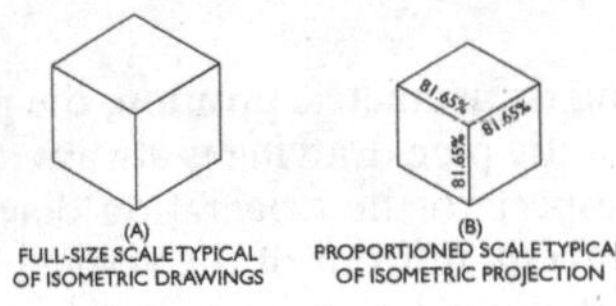

Fig. 8-3

9. SINGLE-LINE ISOMETRIC PIPE DRAWING

The rectangular box system is also employed in making isometric pipe drawings. The piping systems are considered as being on the surface or contained inside the box. To simplify the system, a single line represents the pipe and symbols are used for the fittings (Fig. 9-1).

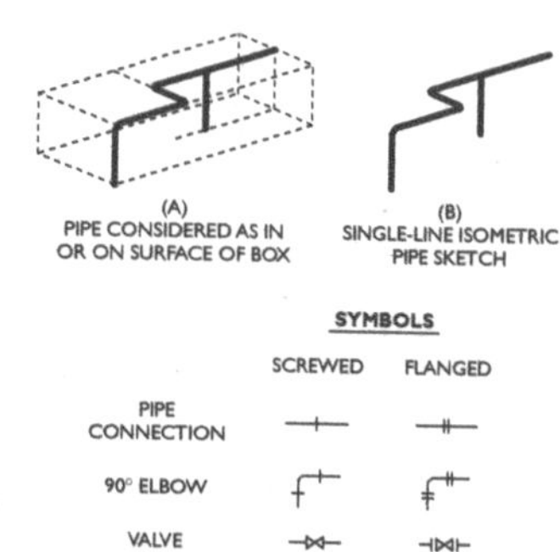

Fig. 9-1

To aid in clarifying the drawing and to avoid confusion of direction when making it, an orientation diagram is used (Fig. 9-2). This diagram is a representation of the isometric axes, each labeled to indicate relative position and direction. The vertical axis is always labeled "UP" and "DOWN." The two horizontal axes are given appropriate direction labels such as "FRONT," "BACK," "NORTH," and "SOUTH."

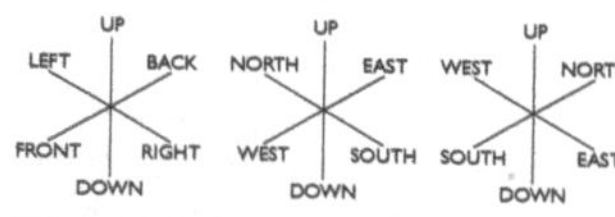

Fig. 9-2

As was the case in developing the isometric position, the point of observation for an isometric pipe drawing is always directly in front of the object. In respect to the orientation diagram, the observation point is directly in front of the vertical "up" and "down" axis. When the point has been selected from which pipe system will be viewed, the diagram is labeled to conform to this point. The lines on the drawing are then made to indicate the actual direction of the pipe.

It is very important when selecting a viewing position to consider which position will result in a drawing with the most clarity. Two drawings of the same system when viewed from different points will result in one being much easier to read than the other, as illustrated in Fig. 9-3.

Symbols are used for standard fittings, valves, and other piping components. A complete list of all the symbols is shown in Table 9-1 and Table 9-2.

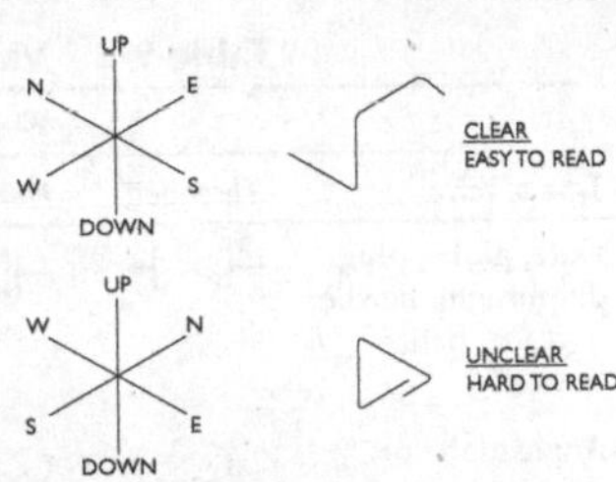

Fig. 9-3

Table 9-1 Pipe Fitting Symbols

For Use on Pipe Drawings and Isometrics

Type of Fitting	*Threaded*	*Buttweld Socket-Weld*		*Flanged*	*Other Fabrication*
Eccentric					Swage
Swaged nipple	Fitting				
Elbows 45°					Bend
90°		Full	Red.		Bend
90° turned away					

Table 9-2 Valve Symbols

	For Pipe Drawings			
Type of Valve	*Threaded*	*Flanged*	*Buttwelded*	*Socket-Welded*
Gate, globe, plug, diaphragm, needle, Y-globe, ball, butterfly				
Angle globe or non-return plan				
Elevation				
Swing, lift, tilt, or wafer-type check				
Three-way plug valve				

The direction of the crossing lines or symbol marks depends on the direction of the pipe. Flange faces on horizontal pipe runs are vertical; therefore, fitting marks on horizontal runs should be "Up" and "Down." Vertical pipe run flanges are horizontal and may be drawn on either horizontal axis (Fig. 9-4).

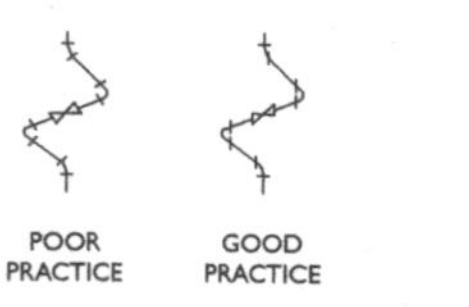

Fig. 9-4

Isometric extension and dimension lines are drawn parallel to the isometric axis. Dimensions may extend to object lines, but the preferred practice is to have arrowheads end on extension lines. If possible, all extension and dimension lines should be placed outside the object (Fig. 9-5).

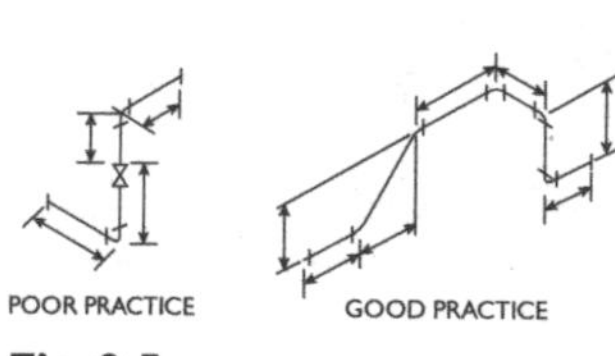

Fig. 9-5

10. SKETCHING

Regardless of the skill and proficiency that a craftsperson has in building, repairing, or maintaining equipment, the ability to make a sketch to convey ideas to others is a necessity. Field sketches are critical to convey the installation instructions and show how something was disassembled.

A popular misconception about sketching is that a person must be artistic or able to "draw." Not so at all. Mechanical, electrical, and piping sketches use simple lines and symbols to depict objects. No artistic shading or other embellishments are needed to show a truthful representation of the object.

Sketching an Object Freehand

All that is needed to produce a freehand sketch is a few good soft lead pencils, an eraser, and some coordinate paper with four or eight spaces to the inch (often called a quadrille pad). Sketching is just making lines and curves on a piece of paper. It's difficult to draw long straight lines on a paper. Short light strokes are better. Or one trick is to place the pencil in a normal writing position and use the third and fourth fingers to ride along the edge of the paper pad to keep a line parallel and straight. Rotating the pad 90° allows the person to use the same technique to make lines perpendicular to the first line that was drawn.

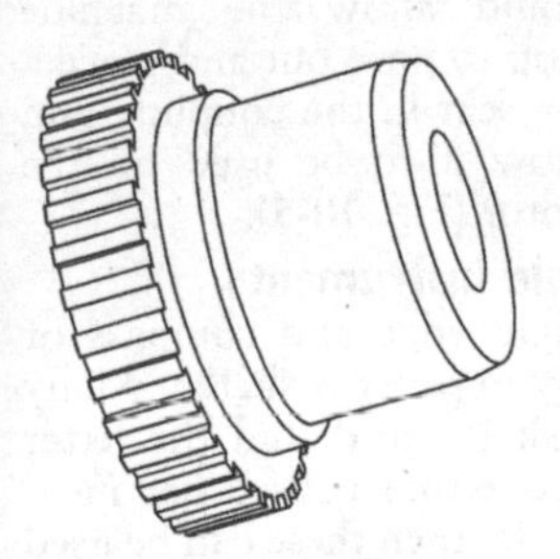

Fig. 10-1 Coupling half.

Fig. 10-1 shows a picture of a coupling half that was purchased at minimum bore and needs to be bored out in order to be installed on a motor shaft.

Using the sketching techniques just described, use the edges of the paper to draw a

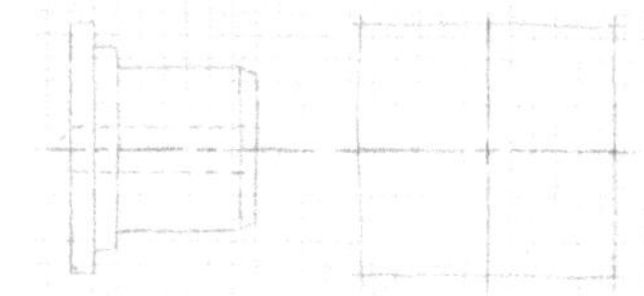

Fig. 10-2 Boxing in the shape.

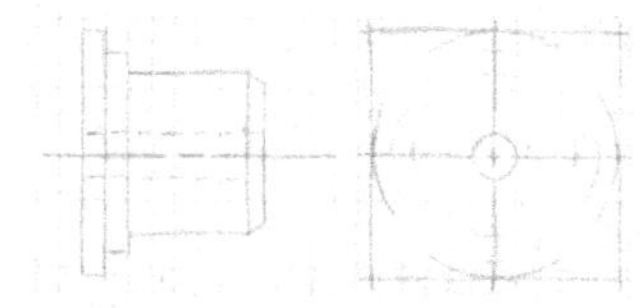

Fig. 10-3 Sketching the arcs of the circle.

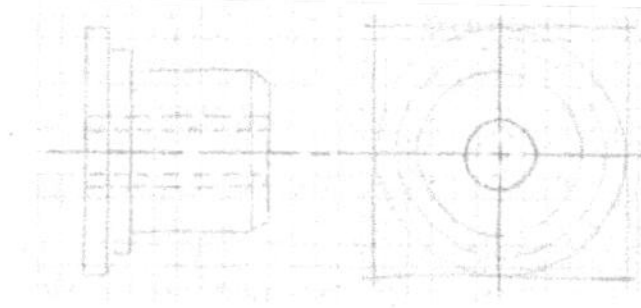

Fig. 10-4 Completed sketch without dimensions.

series of straight lines to define some of the borders of the object in the front view. Use very light lines to "box" in the circular shape of the coupling in the side view. An example of this technique is shown in Fig. 10-2.

Next, block the circular parts of the coupling by sketching four circular arcs at four points where the circles are tangent to the light lines of the square (Fig. 10-3).

Finally, using heavier strokes and a good eraser, complete the sketch—without instructions or dimensions as shown in Fig. 10-4.

Finally, the sketch is dimensioned. Specific information is required that would allow the machine shop to bore out and place a key seat in the coupling and allow it to be used on the motor (Fig. 10-5).

Sketching an Object Using Simple Instruments

Adding a few items, like a straight edge and compass or even a circle template, will greatly improve a sketch. Many people have trouble lettering a sketch and find the letter template helpful. If the straight edge that is used has measured increments, like a six-inch rule, then these can be used to proportion the lines more adequately to make the sketch true to life. The compass or circle template makes creations of these elements look much more professional. Fig. 10-6 shows the same coupling half drawn with the use of a straight edge and compass. The sketch conveys the same information as Fig. 10-5, but it has a more professional look and probably took less time to draw.

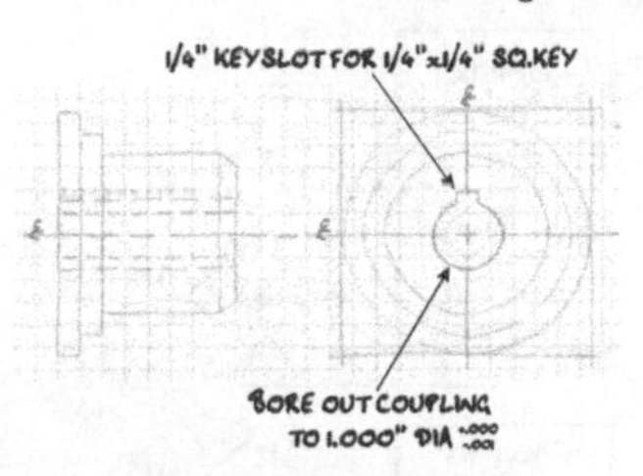

Fig. 10-5 Showing dimensions and instructions.

Troubleshooting trees are very useful sketches to show the logic of determining the extent or root of a problem. Often a millwright, electrician, or mechanic—after figuring out what was wrong with a machine—can aid the next person by sketching a troubleshooting tree. There are plastic templates that depict each element of logic. Fig. 10-7 shows a typical tree used to troubleshoot a simple "light is out" work order. While a craftsman would probably not develop a troubleshooting tree for this problem, the diagram does depict the proper use of this type of template in the sketching process.

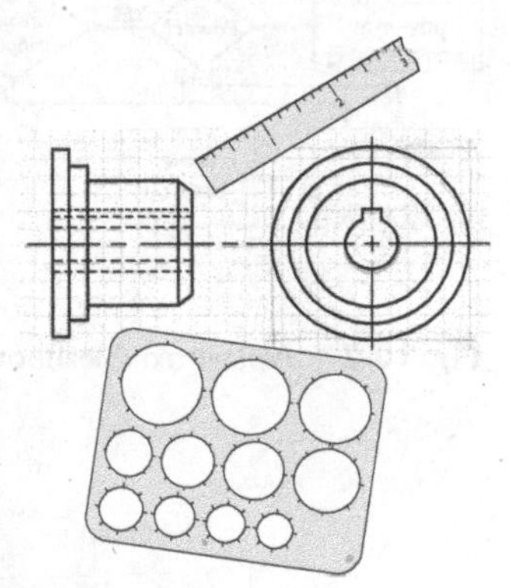

Fig. 10-6 Better sketch using instruments.

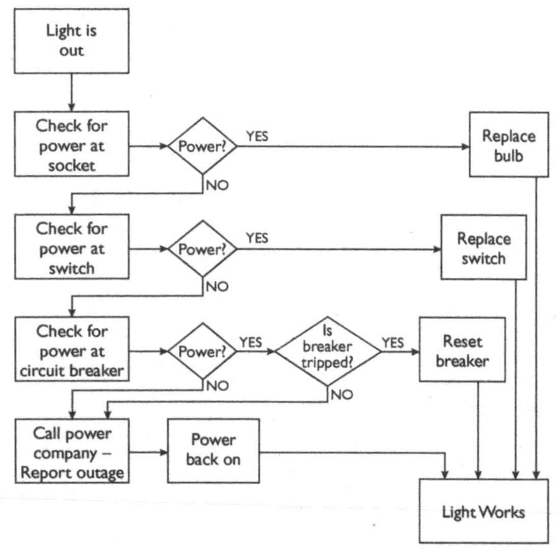

Fig. 10-7 Typical troubleshooting tree diagram.

11. FIELD LAYOUT

Installing machinery and equipment usually requires its location in respect to given points, objects, and surfaces. These may be building columns, walls, machinery, or other equipment.

To accomplish this in field layout work, *baselines* are located with respect to these references and the layout developed from the baselines. For plane surface layout, two baselines at right angles are usually sufficient.

Specific lines and points are then located by laying out *centerlines* parallel to these base lines. A common field layout problem is the layout of right-angle baselines.

Constructing Perpendicular Lines

In Fig. 11-1, baseline (a) is laid out by measurement parallel to the building columns, and baseline (b) is laid out at right angles to baseline (a). The centerlines for machinery installation are then laid out parallel to the baselines.

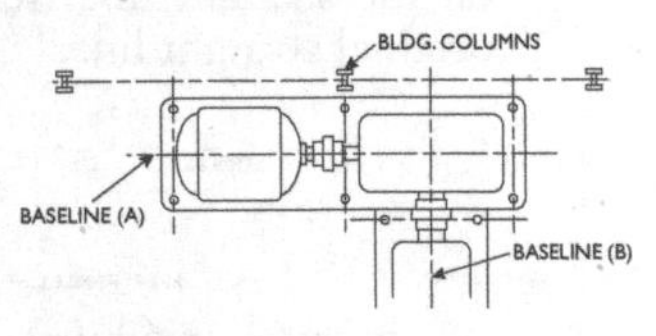

Fig. 11-1

The large steel square (or carpenter's framing square) is suitable for relatively small layout work, but not for large-scale work. A slight error of the square may be magnified to unacceptable proportions when lines are extended. Developing the layout to suit the size of the job is a more dependable and accurate method.

Several right-angle layout methods will give accurate results when used for large-scale field layout. The two most suitable methods are development by swinging arcs and the 3-4-5 triangle layout. In both cases, only simple measurements made with care are required. Accuracy can be achieved by using a larger scale. Using a 6-8-10 triangle will give greater accuracy than the 3-4-5, and a 9-12-15 triangle will improve the accuracy even more.

Arc Method

Use a steel tape, wire, or similar device that cannot stretch. Do not use a string line. You must follow these steps to obtain accurate results, as illustrated in Fig. 11-2:

1. Locate two points, (a) and (b), on the straight line at equal distances from the given point. This may be done by measurement, using a marked stick, or swinging a steel tape.
2. Swing arcs from points (a) and (b), using a radius length about 1½ times the measurement used to locate these points. Exactly the same radius must be used for both arcs. Locate point (c) where the arcs intersect.
3. Construct a line from point (c) through the given point on the straight line. It will be at right angles to the original straight line.

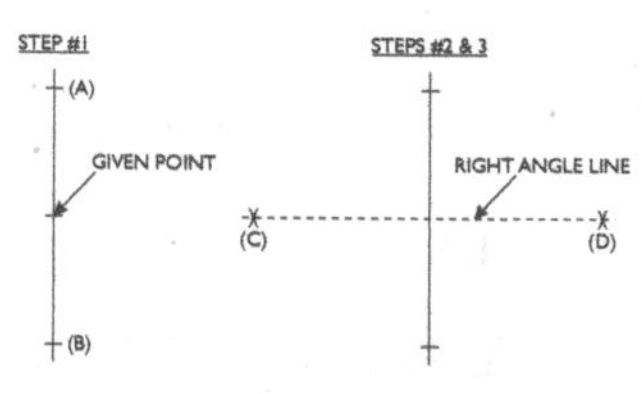

Fig. 11-2 Arc method.

If conditions permit, when swinging arcs to locate point (c) arcs locating point (d) may also be swung. This will provide a double check on the accuracy of the layout, as the three points—(c), given point, and (d)—should form a straight line.

3-4-5 Triangle Method

The 3-4-5 layout method is based on the fact that any triangle having sides with a 3-4-5 length ratio is a right triangle (Fig. 11-3):

1. Select a suitable measuring unit. Use the largest one practical, as the larger the layout the less effect minor errors will have. Measure three units from the given

point at approximately a right angle from the original line and swing arc (a).

2. Measure four units from the given point along the original line and locate point (b).
3. Measure five units from point (b) and locate point (c) on the arc line (a).
4. Construct a line from point (c) through the given point. It will be at right angles to the original line.

If conditions permit, a similar triangle may be constructed on the opposite side of the original line. This will provide a double check on the accuracy of the layout, as the three points—(d), given point, and (c)—should form a straight line.

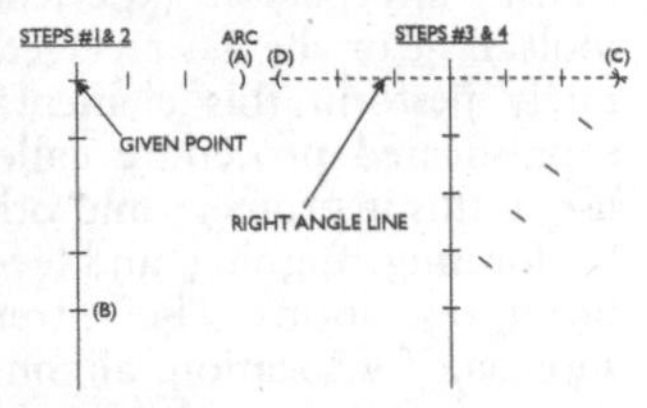

Fig. 11-3 3-4-5 method.

Builder's or Engineer's Transit

In many cases, establishing lines in respect to reference points and locating centerlines and intersecting lines may be much more easily accomplished by instrument. The engineer's transit, in common use by surveyors, is an excellent instrument to use for this purpose. Although the transit is designed for general surveying work, of which simple right-angle layout is an elementary operation, it can accomplish this operation with great accuracy.

To lay out a right-angle line with an engineer's transit, select a point along the base line as the reference location from which sightings are to be made. Place the transit approximately over the point, and move the tripod legs as required to bring the plumb bob into approximate position. Then level the instrument approximately and shift laterally on top of the tripod until the plumb bob is exactly over the reference point. Then level the instrument using the leveling screws and level tubes. Bring each level tube approximately to center, then center each bubble carefully. Then make a sighting through the telescope at a second point on the base line, to align the instrument with the base line. Then set the horizontal circle graduations on the instrument to exactly zero with the aid of the vernier adjustment. The establishment of a right-angle line now merely requires the rotation of the instrument 90° and a sighting to locate a point on the desired right-angle line.

Although surveyors and/or engineers normally handle the transit, any person experienced with precision measuring tools, if he or she has received proper instruction, may accurately perform this elementary operation. A much more sophisticated procedure called industrial surveying makes use of this instrument and other similar, related instruments for locating, aligning, and leveling in the course of manufacturing operations. The extremely close tolerances of measurement for location, alignment, and elevation required in the manufacture and assembly of large machines, aircraft, and space vehicles necessitated the development of this skill. The ordinary tools, jigs, and measuring devices are not adequate, and methods and apparatus similar to those of surveying have been substituted.

Piano Wire Lines

In cases where layout lines are also used as reference or alignment lines for machinery and equipment installation, piano wire lines are recommended. High-tensile-strength piano wire line is superior to fiber line because it will not stretch, loosen, or sag. When properly tightened, it will stay taut and relatively stationary in space, retaining its setting and providing a high degree of accuracy.

You can locate other lines and points precisely from such a line by measurement, use of plumb lines, or other methods. You can easily mark a fixed-point location on a wire line by crimping a small particle of lead or other soft material to the wire.

To make a wire line taut, draw it up very tightly and place it under high tensile load. Accomplishing this, and holding the line in this condition, requires rigid fastenings at the line ends and a means of tightening and securing the line. You must also make provisions for adjustment if the accuracy of setting, which is the principal advantage of a wire line, is to be accomplished.

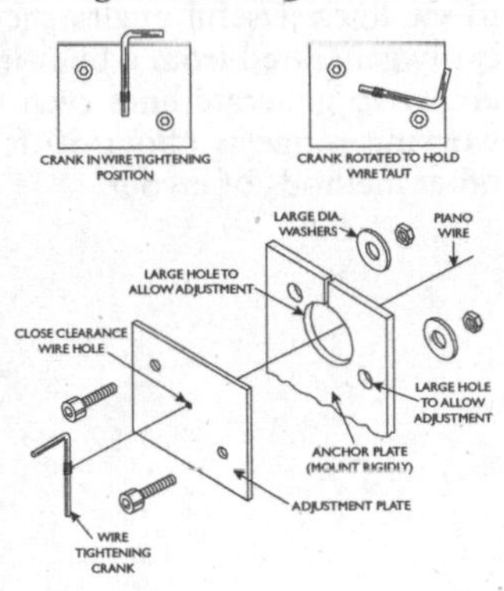

Fig. 11-4

Fig. 11-4 shows a device that holds, tightens, adjusts, and secures a piano wire line. The *anchor plate* must be rigidly mounted to maintain the high tensile load on the line. Adjust it by lightly tapping on the *adjusting plate* to move it in the direction required. When making adjustments, the *clamping screws* should be only lightly tightened, as the adjusting plate must move on the surface of the anchor plate. When the line is accurately positioned, the clamping screws should be securely tightened.

Laser Layout

An important new tool for layout work is the laser layout tool. A transit level similar to a builder's level can now incorporate a laser. These units are fairly expensive, but they can be rented for the few days that might be required for field layout.

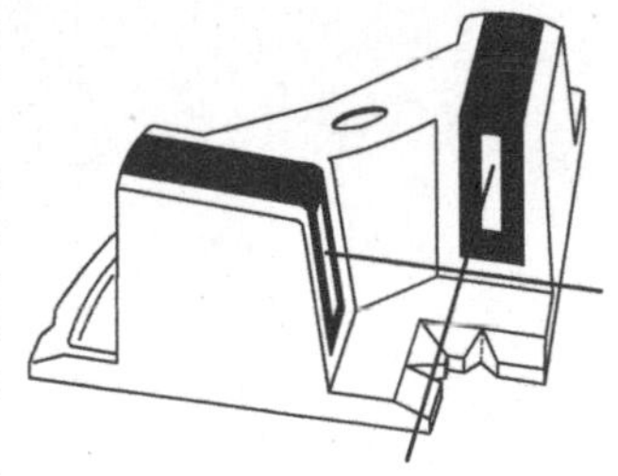

Fig. 11-5

Another tool is the laser square or laser protractor,

which is usually far less expensive (Fig. 11-5). This device generates precise 90° angles with speed and accuracy. The device allows for a one-person operation and projects easy-to-see lines. Useful angles such as 90°, 45°, and 22.5° are easily generated from a built-in protractor. The laser has the ability to generate lines even over wet concrete. The usual working range is about 50 ft. Accuracy is far superior to other methods of layout.

12. MACHINERY INSTALLATION

Foundations

The first step in machinery installation is to provide a suitable base or support, or a "foundation." It must be capable of carrying the applied load without settlement or crushing. Heavy machinery foundations are usually concrete-type structures, structural steel being used for lighter applications and where space and economy are determining factors.

Anchoring—Using Anchor Bolts

Anchor bolts are used to secure machinery rigidly to concrete foundations. The anchor bolts are usually equipped with a hook or some other form of fastening device to ensure unity with the foundation concrete. The most common method of locating anchor bolts is making a template with holes corresponding to those in the machine to be fastened, as shown in Fig. 12-1.

Machine units should not be mounted directly on a concrete foundation and secured with anchor bolts. Such a practice presents many difficult problems with units requiring alignment. Good design incorporates a bedplate, secured to the foundation with anchor bolts, on which machine units are mounted and fastened. The bedplate provides a solid, level, supporting surface, allowing ease and accuracy of shimming and alignment. As the mounting bolts for the machinery are separate from the bedplate anchor bolts, the bedplate is not disturbed as machine mounting bolts are loosened and tightened during alignment.

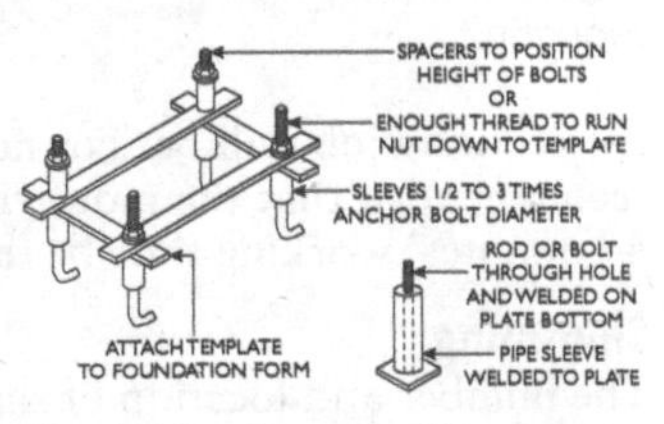

Fig. 12-1 Anchor bolting.

Grout Non-Shrink Characteristics

Because it is impractical to mount a bedplate directly on a concrete foundation, shims and grout are used. The shims provide a means of leveling the bedplate at the required elevation; the grout provides support and securely holds all parts in position. This method of foundation mounting is shown in Fig. 12-2.

Grout is a fluid mixture of mortar-like concrete that is poured between the foundation and bedplate. It secures and holds the leveling shims and provides an intimate support surface for the bedplate. The foundation should be constructed with an elevation allowance of ¾ to 1½ in. for grout. Less than ¾ in. grout may crack and break up in service, while 1½ in. is a practical limit for shim thickness.

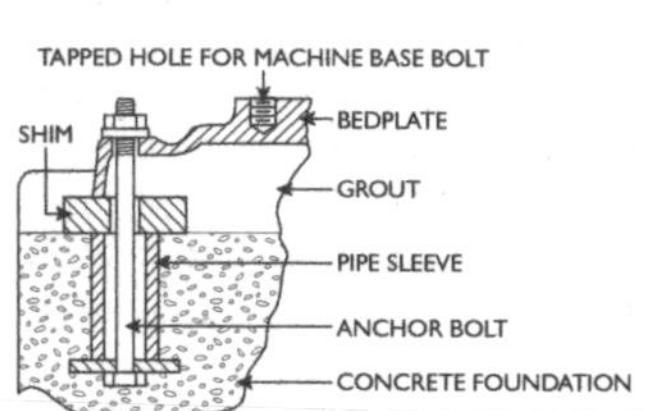

Fig. 12-2 Bedplate mounting.

A key factor in selecting a grout is the working time. There must be adequate time for the grout to be properly placed. The longer working time a grout has, the better the grouting job will be. For normal grouting, it is generally recommended that the industrial standard of a minimum of 45 minutes' working time be required.

Shimming

The number and location of shims will be determined by the design of the bedplate. Provide firm support at points where weight will be concentrated and at anchor bolt locations. Flat shims of the heaviest possible thickness are recommended. The use of wedge-type shims simplifies the leveling of the bedplate; however, this must be done properly to give adequate support. Wedge shims should be doubled and placed in line with the bottom surface of the bedplate edge, as shown in Fig. 12-3a. The improper practice of using single wedges at right angles to the bedplate edge is shown in Fig. 12-3b. As only a relatively small area of the single shim is supporting the bedplate, unit stress may be excessive and crushing may occur.

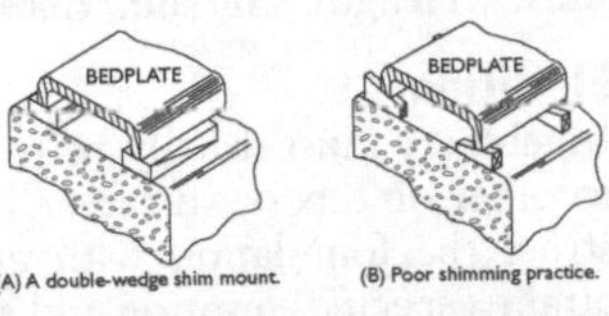

(A) A double-wedge shim mount. (B) Poor shimming practice.

Fig. 12-3 Bedplate shimming.

The term *leveling*, with respect to machinery installation, means the operation of placing machinery or equipment on a true horizontal plane. The tool used, called a *level*, is an instrument incorporating a glass tube containing spirits. A bubble forms in the tube when it is slightly less than completely filled with the spirits. The tube is then mounted in the body of the level and calibrated to indicate true horizontal position when the bubble is centered. The accuracy of a level may be checked by comparing the bubble readings when the level is placed on a horizontal plane and then reversed end-for-end. The bubble readings should be identical in both positions.

When leveling machinery, the level should be used as a measuring instrument, not as a checking device for trial-and-error adjustments. The level should be used in conjunction with a feeler or thickness gauge to measure the amount of error or off level. The leveling shim thickness can then be calculated from this measurement.

Determining Shim Thickness

1. Insert thickness gauge as needed at the low end of the level to get a zero bubble reading.
2. The shim thickness required will be as many times thicker than the gauge thickness as the distance between bearing points is greater than the level length (See Fig. 12-4).

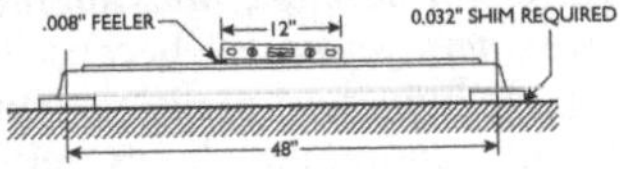

.008" FEELER — 12" — 0.032" SHIM REQUIRED — 48"

Fig. 12-4

The shim thickness is four times the gauge thickness because the distance between bearing points is four times greater than the level length. The shim thickness is 4 × .008, or .032.

Elevation

Machinery must usually be placed at a given height in respect to other objects or surfaces. The normal procedure is to construct the foundation with an allowance for shimming. To attain a specific elevation and a level position, roughly shim the machine to the approximate level, at slightly below the desired elevation. One bearing point can then be shimmed to the given elevation. As this point will be high, all other bearing points can then be raised to correct elevation by shimming them level.

Machinery Mounts—Vibration Mounts

In many manufacturing plants, machinery is no longer permanently fastened in place with anchor bolts. This is especially true in metal manufacturing industries, where frequent model changes or design improvements of the product are made. Flexibility and mobility are a necessity because production lines are constantly being rearranged and new or improved machines replace obsolete units.

In some cases where a line of many machines may be shut down if a single unit fails, the mobility capability may be carried even further. Disabled units may be removed from the line and replaced by new or rebuilt replacement units to minimize production down time. To accomplish this, the machine must be equipped with flexible, quick-disconnecting coupling service lines for power, air, water, and so on.

A prime requirement for this kind of mobility is devices that support the machine and the ability to make quick adjustments for leveling and alignment. These devices are called *machinery mounts,* because they eliminate the need for anchor bolts and make machines free-standing. This type of machine mounting allows units to be picked up and moved to another location with a minimum of installation work. Using this system of mounting eliminates the need to remove anchor bolts or other fasteners or to break floors. The leveling features built into such mounts allow leveling of the machine in a minimum amount of time to precise limits. Because the weight of the machine is the primary anchor for this system of mounting, it is very important that the weight be solidly transferred to the floor. Because a floor surface is seldom smooth and flat, mounts with a leveling feature are usually used.

Because of the ever-increasing variety of industrial machinery, the problems of vibration, shock, and noise, in addition to that of mobility, have become important considerations. To overcome these problems, various types of special machine mounts have been developed. One widely used style of machine mount, with vibration and shock damping capability as well as adjustment and leveling features, is shown in Fig. 12-5. This style of mount may be used on machinery ranging from office and laboratory installations to heavy metal-working machines and similar equipment subjected to high-impact shock loading and severe vibrations. This style of mount eliminates the need for anchor bolts or floor lag bolts, yet it keeps the machinery firmly in place.

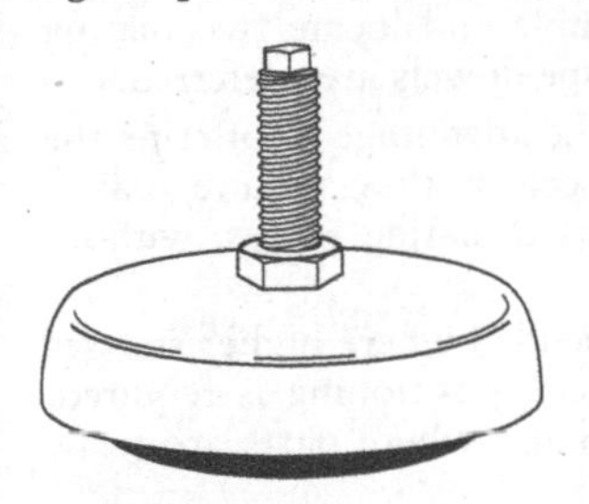

Fig. 12-5 Shock and vibration damping machinery mount.

Use of this style of mount makes possible quick and easy relocation of machinery while eliminating the problem of shock and vibration from one machine disturbing another. Sensitive, high-precision machinery can be located for best workflow without danger of impact or vibrations from other machinery nearby. Noise, as well as vibrations transmitted from the base of the machine through the floor and building, is also reduced. Another very important benefit resulting from the use of vibration-damping mounts is the reduction of internal stresses in the machine. When vibrating equipment is rigidly bolted to the floor, an amplification of internal stresses occurs. The results often are misalignment of machine frames and undue wear on bearings and related parts.

Doweling

After setting the machine on the bedplate and making all the necessary alignment moves, it is often useful to dowel the machine to do the following:

1. Preserve alignment between two or more mating parts
2. Resist lateral forces tending to separate the parts

There are three basic types of dowel pin: straight, tapered, and spring-type.

Selecting a Dowel Pin

1. Selection of the type of pin to use—straight, tapered, or spring-type—depends on the application.
2. Where it is used to assemble and locate two mating parts, tapered or spring-type dowels are preferred.
3. Spring-type dowels have the advantage of offering the lowest installation cost because they require only a drilling operation in the two mating pieces, without any subsequent reaming.
4. Taper dowels are recommended where higher stresses are involved or more precise positioning is required. They are not intended for use where parts are to be interchangeably assembled.
5. Straight dowels should be used where parts are frequently disassembled or where more precise positioning is required than is obtainable with spring-type dowels. The original cost of machining the two mating parts for straight dowels is higher because very close tolerances are required on hole diameters, hole locations, and squareness of the holes with the mating surfaces.

6. Taper pins, large end threaded, instead of plain taper pins, are for use in applications where the small end of the pin is not accessible for removal by striking with a drift pin.

Installing a Dowel Pin

1. When the dowel pin-hole extends completely through the two mating parts, the pin installation is referred to as a through pin. This type of installation is preferred and shown in Fig. 12-6.
2. Fig. 12-7 shows a semi-blind type. The smaller hole serves as a vent and also as a knockout hole.
3. Fig. 12-8 shows a blind type of hole; with this arrangement a dowel pin should always be a sliding fit in the blind hole and a force fit in the mating part. A blind-type dowel should never be used when it is feasible to employ either the through type or the semi-blind type.

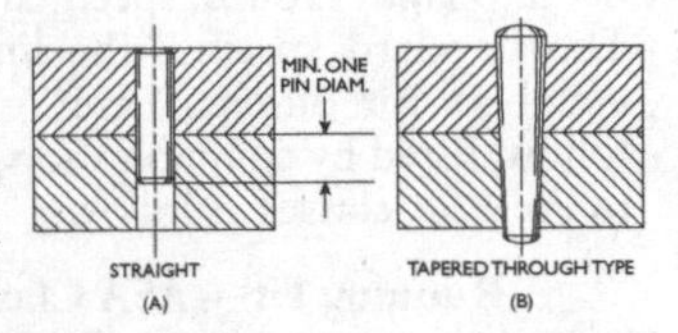

Fig. 12-6 Through pin type.

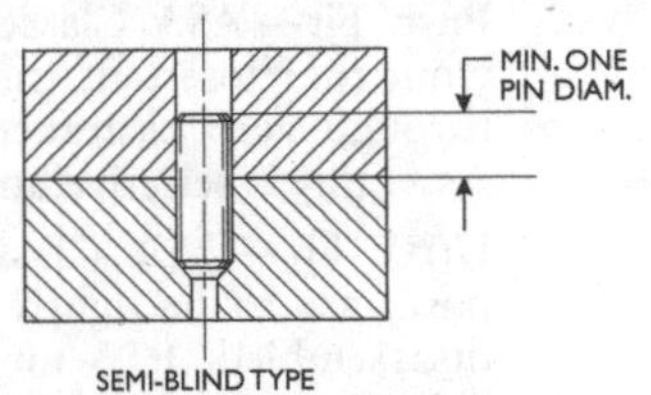

Fig. 12-7 Semi-blind type.

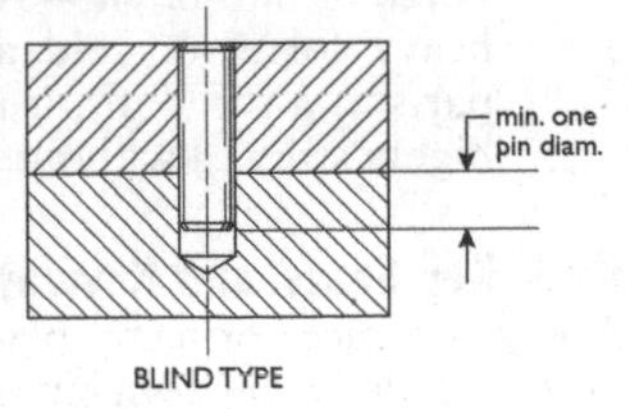

Fig. 12-8 Blind type.

13. MACHINE ASSEMBLY

Allowance for Fits

The American Standards Association (ASA) classifies machine fits into eight groups, specified as Class #1 through Class #8. The standard specifies the limits for internal and external members for different sizes in each class. The groups that follow, listed by common shop terms, compare approximately to the ASA classes as follows:

Running Fit—ASA Class #2. This fit is for assemblies where one part will run in another under load with lubrication.

Push Fit—ASA Classes #4 and #5. This fit ranges from the closest fit that can be assembled by hand, through zero clearance, to very slight interference. Assembly is selective and not interchangeable.

Drive Fit—ASA Class #6. This fit is used where parts are to be tightly assembled and not normally disassembled. It is an interference fit and requires light pressure. It is also used as a shrink fit on light sections.

Force or Shrink Fit—ASA Class #8. This fit requires heavy force for cold assembly or heat to assemble parts as a shrink fit. It is used where the metal can be highly stressed without exceeding its elastic limit.

Keys, Key Seats, and Keyways

A *key* is a piece of metal placed so that part of it lies in a groove, called a *key seat,* cut in a shaft. The key then extends somewhat above the shaft and fits into a *keyway* cut in a hub. See Fig. 13-1.

Table 13-1 Recommended Allowances

	Running Fit		Push Fit	
Diameter (Inches)	Ordinary Loads	Severe Loads	Light Service	No Play
	(Clear)	(Clear)	(Clear)	(Inter)
Up to ½	.0005	.001	.00025	.0000
½ to 1	.001	.0015	.0003	.00025
1 to 2	.002	.0025	.0003	.00025
2 to 3½	.0025	.0035	.0003	.0003
3½ to 6	.0035	.0045	.0005	.0005

	Drive Fit		Force or Shrink Fit	
Diameter (Inches)	Field Assembly (Inter)	Shop Assembly (Inter)	Force (Inter)	Shrink (Inter)
Up to ½	.0002	.0005	.00075	.001
½ to 1	.0002	.0005	.001	.002
1 to 2	.0005	.0008	.002	.003
2 to 3½	.0005	.001	.003	.004
3½ to 6	.0005	.001	.004	.005

(Clear) Indicates clearance between members
(Inter) Indicates interference between members

The simplest key is the square key, placed half in the shaft and half in the hub. A flat key is rectangular in cross-section and is used in the same manner as the square key for members with light sections. The gib head key is tapered on its upper surface and is driven in to form a very secure fastening.

A variation on the square key is the *Woodruff key.* It is a flat disc made in the shape of a segment of a circle. It is flat on top with a round bottom to match a semi-cylindrical key seat.

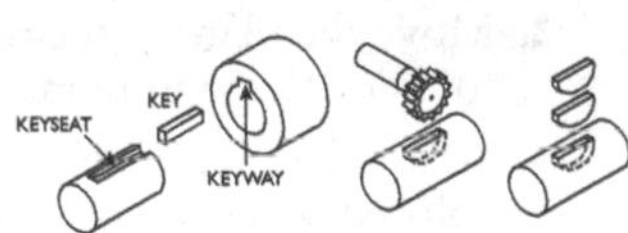

Fig. 13-1

Scraping

In metalworking, slight errors in plane or curved surfaces are often corrected by hand scraping. Most machine surfaces that slide on one another, as in machine tools, are finished in this manner. Also plane-bearing boxes are scraped to fit their shafts after having been bored or, in the case of babbitt bearings, after having been poured.

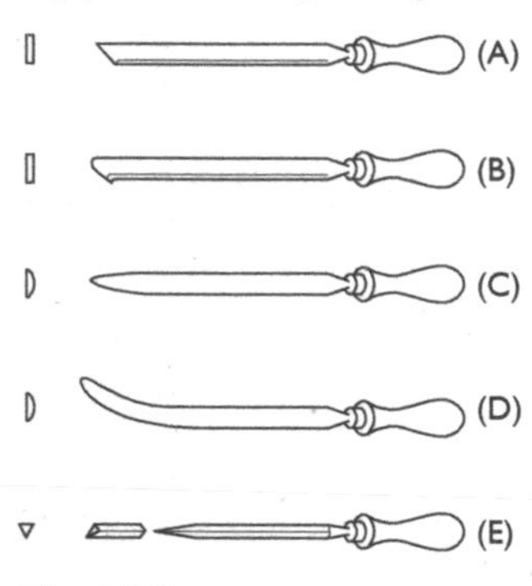

Fig. 13-2

Various styles of hand scrapers are shown in Fig. 13-2. The flat scraper (A) is used for flat scraping. Some people prefer the hook scraper (B), which is also used on flat surfaces. Flat and curved scrapers with a half-round cross-section, (C) and (D), are used for scraping bearings. The three-cornered scraper (E) is used to some extent on curved surfaces and to remove burrs and round the corners of holes.

Flat Surface Scraping

1. Coat the entire surface of a true surface plate with a suitable scraping dye such as Prussian blue.
2. Place the surface plate on the surface to be scraped or, if the work piece is small, place it on the surface plate.
3. Move the plate or the piece back and forth a few times to color the high points on the work piece.
4. Scrape the high spots on the work piece, which are colored with the dye where they touched the surface plate (Fig. 13-3).

Bearing Scraping

1. Coat the journal of the shaft with a thin layer of Prussian blue, spreading it with the forefinger.
2. Place the shaft in the bearing, or vice versa, tighten it, and turn one or the other several times through a small angle.
3. Scrape the high spots in the bearing, which are colored where they touched the shaft journal (Fig. 13-3).

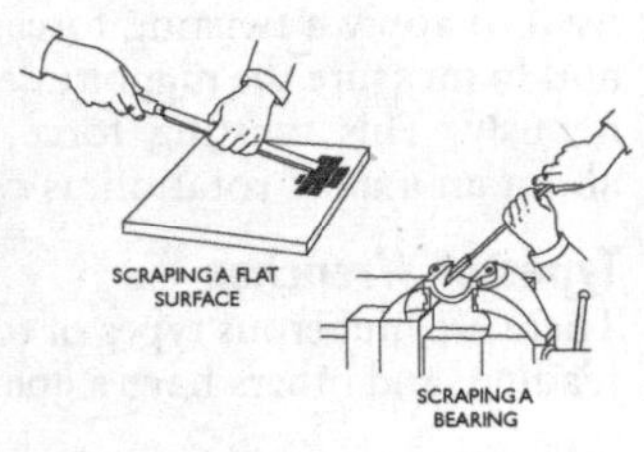

Fig. 13-3

14. USE OF THE TORQUE WRENCH

The words *torque wrench* are commonly used to describe a tool that is a combination wrench and measuring tool. It is used to apply a twisting force, as do conventional wrenches, and to measure the magnitude (amount) of the force simultaneously. This twisting force, which tends to turn a body about an axis of rotation, is called *torque.*

Types of Wrenches

There are numerous types of torque wrenches; some are direct reading, and others have signaling mechanisms to warn when the predetermined torque value is reached. Fig. 14-1 shows a round beam torque wrench with a scale and a dial indicator style of wrench. Both are direct-reading wrenches.

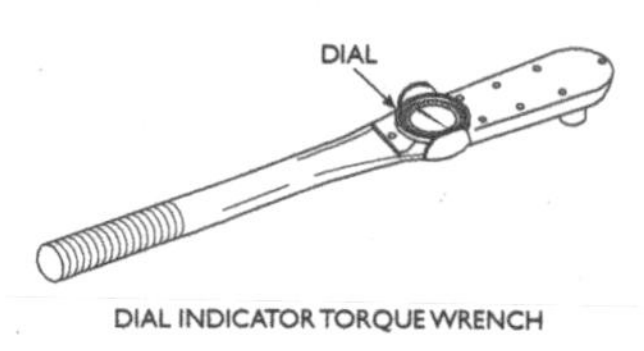

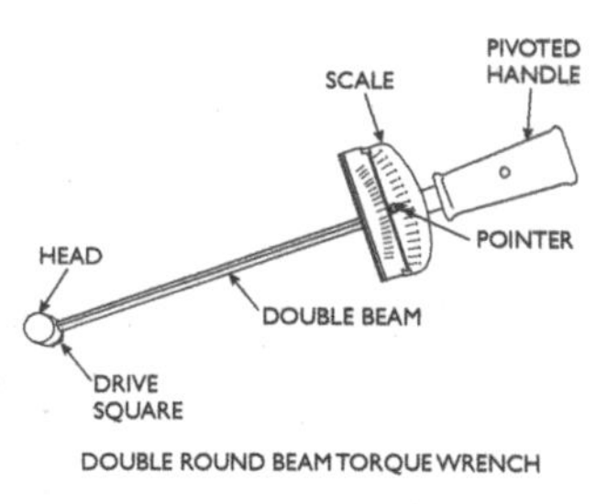

Fig. 14-1 Torque wrench types.

Torque Calculations

All torque measurements are based on the fundamental law that torque equals force multiplied by distance (Fig. 14-2).

Torque units of measure (inch-pounds and foot-pounds) are the product of the force measured in pound units and a lever length in either inch units or foot units. Fig. 14-3 shows examples of the calculation of torque for both inch-pounds and foot-pounds and illustrates how one measurement can be converted to the other.

Torque Wrench Terms

Push or Pull

Force should be applied to a torque wrench by pulling whenever possible, primarily because there is a greater hazard to fingers or knuckles when pushing, should some part fail unexpectedly. Either can produce accurate results.

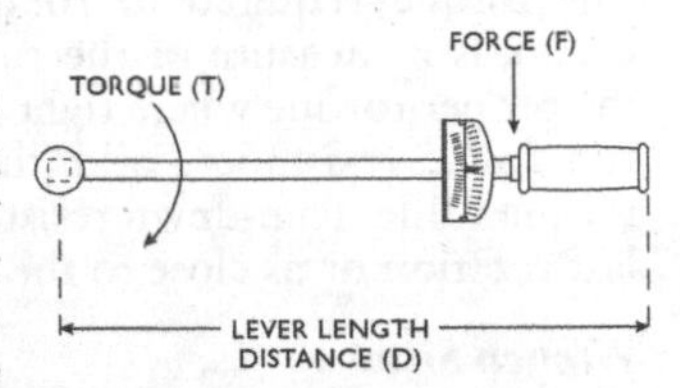

Fig. 14-2 Torque wrench fundamentals.

Breakaway Torque

The torque required to loosen a fastener is generally some value lower than that to which it has been tightened. For a given size and type of fastener, there is a direct relationship between tightening torque and breakaway torque. When this relationship has been determined by actual text, you can check tightening torque by loosening and checking breakaway torque.

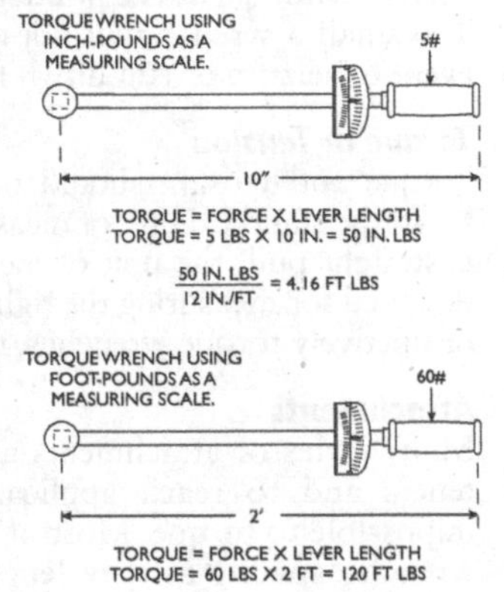

Fig. 14-3

Set or Seizure

In the last stages of rotation in reaching a final torque reading, seizing or set of the fastener may occur. When this occurs there is a noticeable popping effect. To break the set, back off and then again apply the tightening torque. Accurate torque setting cannot be made if the fastener is seized.

Run-Down Resistance

The torque required to rotate a fastener before makeup occurs is a measure of the run-down resistance. To obtain the proper torque where tight threads on locknuts produce a run-down resistance, add the resistance to the required torque value. Run-down resistance must be measured on the last rotation or as close to the makeup point as possible.

Wrench Sizes

The correct size wrench for a job is one that will read 25 to 75 percent of the scale when the required torque is applied. This will allow adequate capacity and provide satisfactory accuracy. Avoid using an oversize torque wrench; obtaining correct readings as the pointer starts up the scale is difficult. Too small a wrench will not allow for extra capacity in the event of seizure or run-down resistance.

Torque or Tension

Torque and tension should not be confused. Torque is the twist, the standard unit of measure being foot pounds. Tension is straight pull, the unit of measure being pounds. Wrenches designed for measuring the tightness of a threaded fastener are distinctively torque wrenches, not tension wrenches.

Attachments

Many styles of attachments are available to fit various fasteners and to reach applications that may otherwise be impossible to torque. Most of these attachments increase the wrench capacity as they lengthen the lever arm. Fig. 14-4 shows an example of a torque wrench using an attachment.

Many mechanics are unaware of the various attachments that exist for use with a torque wrench. A complete torque wrench set includes crowfoot, open-end, spanner, and flare-nut attachments.

When using such attachments, scale-reading corrections must be made. The scale correction will be in reverse ratio to the increase in the length of the lever arm. If adding an attachment to the wrench doubles the arm, its capacity is doubled and the scale shows only one-half of the actual torque applied. Use the formula in Fig. 14-4 to determine correct scale readings when using an attachment.

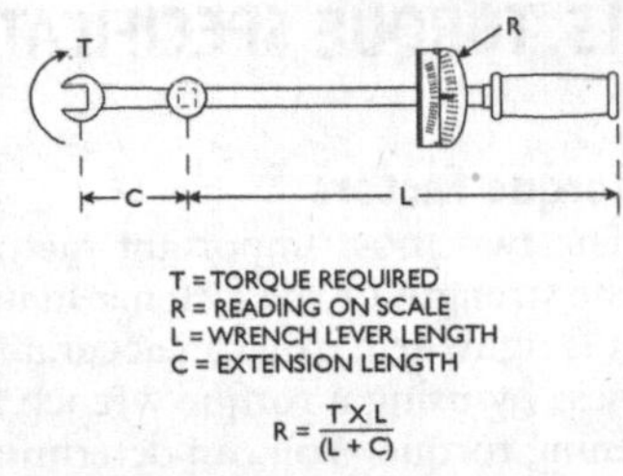

Fig. 14-4

15. TORQUE SPECIFICATIONS

Torque Factors

The two most important factors in fastener tightening are the strength of the fastener material and the degree to which it is tightened. You can accurately control the degree of tightness by using a torque wrench to measure the applied tightening torque. You can determine the strength of the fastener by the markings on the head of the bolt or screw. SAE and ASTM developed these head markings to denote relative strength, and they have been incorporated into standards for threaded fastener quality. They are called *grade markings* or *line markings* and are listed in Table 15-1.

Table 15-1 Bolt Grade Markings

Grade Marking	*Specification*	*Material*
	SAE—Grade 0 SAE—Grade 1 ASTM—A 307 SAE—Grade 2	Steel Low carbon steel Low carbon steel
	SAE—Grade 3	Medium carbon steel, cold worked
	SAE—Grade 5 ASTM—A 49	Medium carbon steel, quenched and tempered
	ASTM—A 325	Medium carbon steel, quenched and tempered

Grade Marking	Specification	Material
BB	ASTM—A 354 Grade BB	Low alloy steel, quenched and tempered
BC	ASTM—A 354 Grade BC	Medium carbon alloy steel, quenched and tempered, roll threaded after heat treatment
	SAE—Grade 8	Medium carbon alloy steel, quenched and tempered
	ASTM—A 354 Grade BD	Alloy steel, quenched and tempered
A 490	ASTM—A 490	Alloy steel, quenched and tempered

ASTM Specifications:
A 307—Low carbon steel externally and internally threaded fasteners
A 325—High-strength steel bolts for structural steel joints, including suitable nuts and plain hardened washers
A 449—Quenched and tempered steel bolts and studs
A 354—Quenched and tempered alloy steel bolts and studs with suitable nuts

As shown in Table 15-1, unmarked bolt heads or cap screws are generally considered to be mild steel. The greater the number of marks on the head, the higher the quality. Thus, bolts of the same diameter vary in strength and require a correspondingly different tightening torque or preload.

Torque Tables

The suggested maximum torque values in Table 15-2 for fasteners of various materials should be used as a guide only. Be sure to follow manufacturers' specifications on specific torque applications.

Table 15-2 Torque in Foot-Pounds

Fastener Diameter, in.	*Threads per Inch*	*Mild Steel*	*Stainless Steel 18-8*	*Alloy Steel*
1/4	20	4	6	8
5/16	18	8	11	16
3/8	16	12	18	24
7/16	14	20	32	40
1/2	13	30	43	60
5/8	11	60	92	120
3/4	10	100	128	200
7/8	9	160	180	320
1	8	245	285	490

Torque Values for Steel Fasteners

The strength of a bolted connection depends on the clamping force developed by the bolts. The tighter the bolt, the stronger the connection. The two principal factors limiting the clamping force that the bolt may develop are the bolt size and its tensile strength.

The tensile strength of a bolt depends principally on the material from which it is made. Bolt manufacturers identify bolt materials by head markings that conform to SAE and ASTM specifications. These marks are known as grade markings. The most commonly specified grades are listed in Table 15-3.

The values in Table 15-3 do not apply if special lubricants such as colloidal copper or molybdenum disulphide are used. Use of special lubricants can reduce the amount of friction in the fastener assembly so that the torque applied may produce far greater tension than desired.

Table 15-3 Torque Values for Graded Steel Bolts

Grade		SAE 1 or 2	SAE 5	SAE 6	SAE 8
Tensile Strength		64000 psi	105000 psi	130000 psi	150000 psi
Grade Mark					
Bolt Diameter (in Inches)	Threads per Inch	Torque in Foot-Pounds			
1/4	20	5	7	10	10
5/16	18	9	14	19	22
3/8	16	15	25	34	37
7/16	14	24	40	55	60
1/2	13	37	60	85	92
9/16	12	53	88	120	132
5/8	11	74	120	169	180
3/4	10	120	200	280	296
7/8	9	190	302	440	473
1	8	282	466	660	714

Hydraulic Torquing for Large Machinery

On very large machinery, especially equipment that is subjected to high-impact loads or that can shake and vibrate, hydraulic bolt torquing equipment must be used to ensure secure fastening of the machinery components under load. The term *torquing* is a bit of a misnomer because many of these devices work by using a high-pressure fluid from a pumping unit used to pressurize a hydraulic load cell. This load cell creates a force that allows a tensioner component to actually stretch the stud. When this happens, the nut lifts off the flange or washer on which it resting. You can return the nut to its seated position by turning the nut with a wrench or special tool. When the nut is tight against the washer again, the pressure in the load cell is released and the stud remains stretched with the load locked in. In a normal torque wrench situation, the nut is turned to stretch the bolt, while this hydraulic unit actually stretches the bolt and uses the nut as a fixed stop to retain the bolt in a stretched condition. These tools are usually very expensive, and renting them or hiring an outside contractor to provide the service often makes good economic sense.

16. MEASUREMENT

The most useful tools for machinery measurement are the *steel rule*, *vernier caliper*, *micrometer*, *dial indicator*, *screw-pitch gauge*, and *taper gauge*.

Steel Rule

A plain flat 6-in. *steel rule* is a type commonly used. The inches are graduated into eighths and sixteenths of an inch on one side of the rule, and thirty-seconds and sixty-fourths on the other (Fig. 16-1).

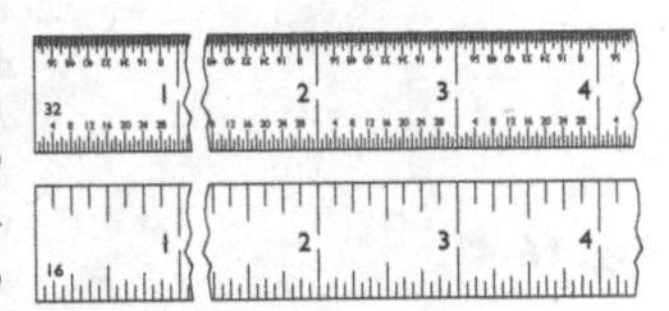

Fig. 16-1 Steel rule.

When taking measurements with the steel rule, place the 1-in. index mark—instead of the end of the rule—at one edge of the piece to be measured. Lay the steel rule parallel to the edge of the piece (Fig. 16-2). Remember to deduct 1 in. from the reading. Be careful when you set the 1-in. mark at the edge of the piece. The edge of the piece should coincide with the center of the index mark on the steel rule (Fig. 16-3).

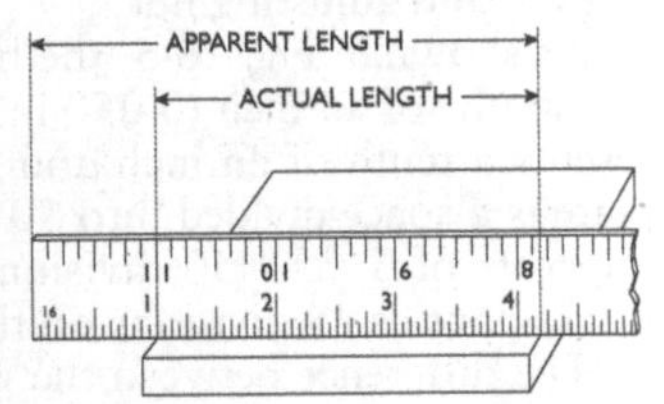

Fig. 16-2

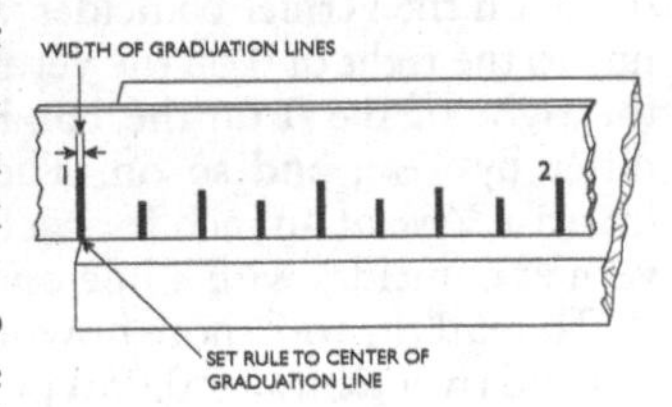

Fig. 16-3

Use two steel rules to measure the depth of a hole (Fig. 16-4). Place one steel rule, or a straightedge, across the end of the piece with the index mark of the measuring rule registering accurately with the straightedge. You can

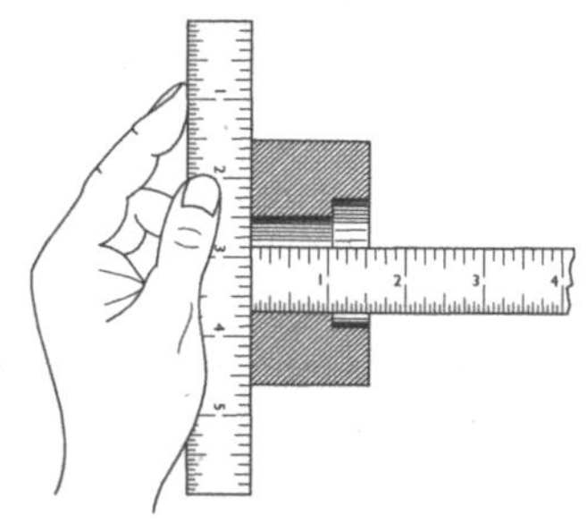

Fig. 16-4

read the depth of the hole accurately from the measuring rule.

Vernier Caliper

The *vernier caliper* was invented by French mathematician Pierre Vernier. It consists of a stationary bar and a movable vernier slide assembly. The stationary rule is a hardened graduated bar with a fixed measuring jaw. The movable vernier slide assembly combines a movable jaw, vernier plate, clamp screws, and adjusting nut.

As shown in Fig. 16-5, the bar of the tool is graduated in twentieths of an inch (0.050 in.). Every second division represents a tenth of an inch and is numbered. On the vernier plate is a space divided into 50 parts and numbered 0–50 at intervals of 5. The 50 divisions on the vernier occupy the same space as 49 divisions on the bar.

The difference between the width of one of the 50 spaces on the vernier and one of the 49 spaces on the bar is therefore $\frac{1}{50}$ of $\frac{1}{20}$, or $\frac{1}{1000}$ of an inch. If the tool is set so that the 0 line on the vernier coincides with the 0 line on the bar, the line to the right of 0 on the vernier will differ from the line to the right of the 0 on the bar by $\frac{1}{1000}$; the second line will differ by $\frac{2}{1000}$, and so on. The difference will continue to increase $\frac{1}{1000}$ of an inch for each division until line 50 on the vernier coincides with a line on the bar.

To read the tool, note how many inches, tenths (or 0.100 in.) and twentieths (or 0.050 in.) the 0 mark on the vernier is from the 0 mark on the bar. Then note the number of divisions on the vernier from 0 to a line, which exactly coincides with a line on the bar.

Example

In Fig. 16-6, the vernier has been moved to the right $1\frac{4}{10}$ and $\frac{1}{20}$ in. (1.450 in.), as shown on the bar (b), and the fourteenth line on the vernier coincides with a line, as indicated by the stars, on the bar (c). Therefore, fourteen thousandths of an inch should be added to the reading on the bar, and the total reading is $1\frac{464}{1000}$ in. (1.464 in.).

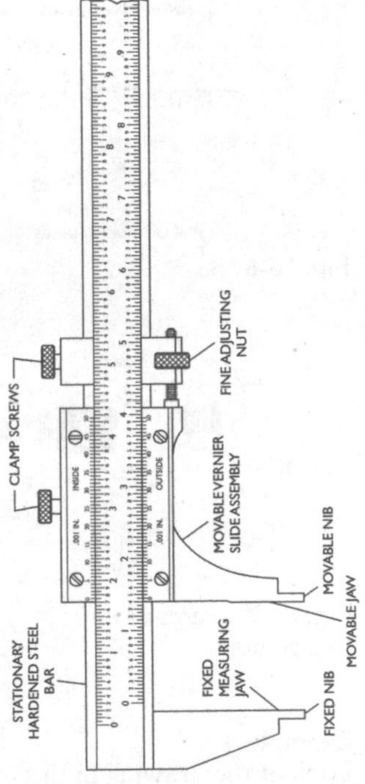

Fig. 16-5 Parts of a vernier caliper.

Micrometer

Reading a Micrometer Graduated in Thousandths of an Inch (0.001 in.)

The pitch of the screw thread on the spindle of a micrometer is 40 threads per inch. One revolution of the thimble advances the spindle face toward or away from the anvil face exactly $\frac{1}{40}$ in. (0.025 in.).

The reading line on the sleeve is divided into 40 equal parts by vertical lines that correspond to the number of threads on the spindle. Therefore, each vertical line designates $\frac{1}{40}$ in. (0.025 in.), and every fourth line, which is longer than the others, designates hundreds of thousandths. For example, the line marked "1" represents 0.100 in., the line marked "2" represents 0.200 in., and the line marked "3" represents 0.300 in.

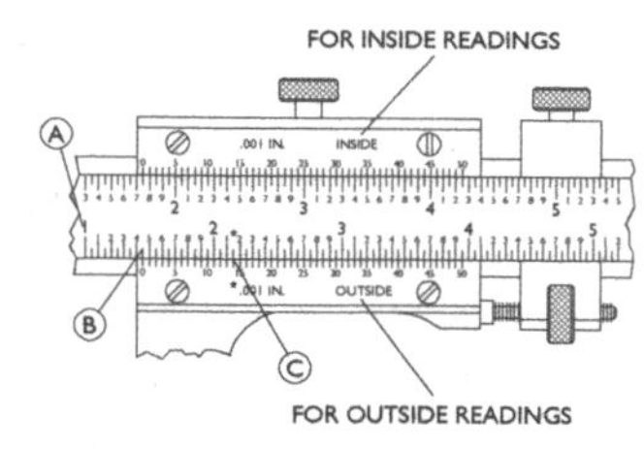

Fig. 16-6

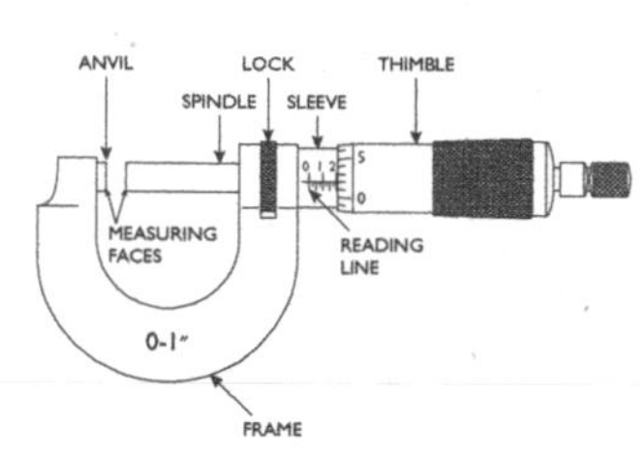

Fig. 16-7 Parts of a micrometer.

The beveled edge of the thimble is divided into 25 equal parts with each line representing 0.001 in. and every line numbered consecutively. Rotating the thimble from one of these lines to the next moves the spindle longitudinally 1/25 of 0.025 in., or 0.001 in.; rotating two divisions represents 0.002 in. Twenty-five divisions indicate a complete revolution, 0.025 in. (1/40 of an inch).

To read a micrometer in thousandths, multiply the number of vertical divisions visible on the sleeve by 0.025 and to this add the number of thousandths indicated by the line on the thimble that coincides with the reading line on the sleeve.

Example

Look at the drawing in this example.

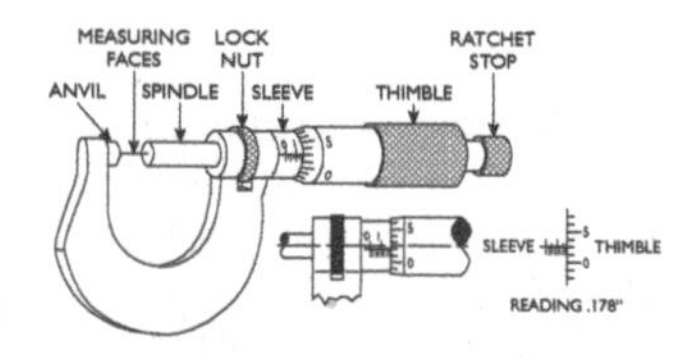

The "1" line on the sleeve is visible, representing 0.100 in.

There are three additional lines visible, each representing 0.025 in.	3 × 0.025 in. = 0.075 in.
Line "3" on the thimble coincides with the reading line on the sleeve, with each line representing 0.001	3 × 0.001 in. = 0.003 in.
The micrometer reading is	0.178 in.

Reading a Micrometer Graduated in Ten-Thousandths of an Inch (0.0001 in.)

When measuring shafts for bearing installation the measurements are usually given to the ten-thousandth of an inch. A micrometer with a vernier scale capable of measuring to the ten-thousandth must be used.

The micrometer is read to the thousandth, but now the vernier micrometer shown has 10 divisions marked on the sleeve occupying the same space as 9 divisions on the beveled edge of the thimble. The difference between the width of one of the 10 spaces on the sleeve and one of the 9 spaces on the thimble is one-tenth of a division on the thimble. Because the thimble is graduated to read in thousandths, one-tenth of a division would be one ten-thousandth. To make the reading, first read to the thousandth as with a regular micrometer, then see which of the horizontal lines on the sleeve coincides with a line on the thimble. Add to the previous reading the number of ten-thousandths indicated by the line on the sleeve that exactly coincides with the line on the thimble.

Example

In Fig. 16-8 (a) and (b), the 0 (zero) on the thimble coincides exactly with the axial line on the sleeve and the vernier 0 (zero) on the sleeve is the one that coincides with a line on the thimble. The reading is, therefore, an even 0.2500 in. Fig. 16-8(c) shows the 0 (zero) line on the thimble has gone beyond the axial line on the sleeve, indicating a reading of more than 0.2500 in. Checking the vernier shows that the

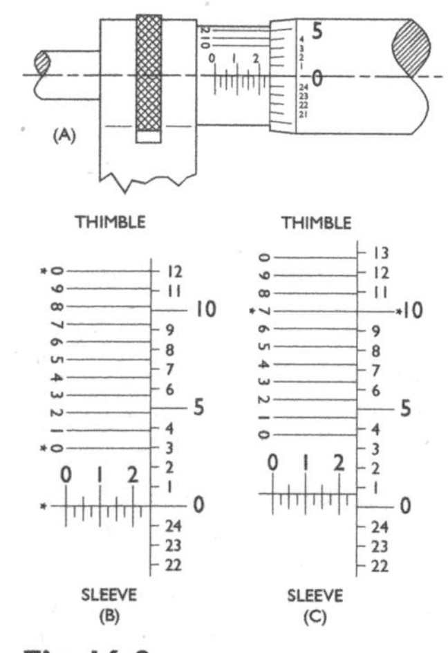

Fig. 16-8

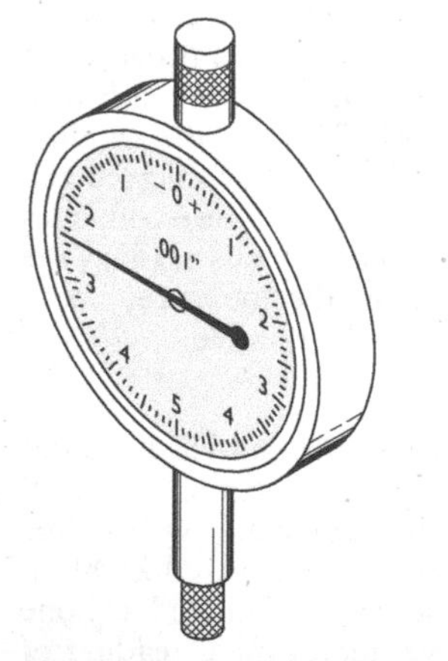

Fig. 16-9 Dial indicator.

seventh vernier line on the sleeve is the one that exactly coincides with a line on the thimble; therefore, the reading is 0.2507 in.

Dial Indicator

A *dial indicator* is an instrument for indicating size differences, rather than for making measurements. The dial indicator ordinarily is not used to indicate distance, but it can be used in combination with a micrometer to measure the exact distance.

The movement of a hand on the dial indicator shows variations in measurements. The dial is graduated in thousandths of an inch; each division on the dial represents the contact point movement of 0.001 of an inch.

The dial indicator is useful for checking shafts for alignment and straightness, checking cylinder bores for roundness and taper, and testing bearing bores (Fig. 16-9).

Screw-Pitch Gauge

The number of threads per inch, or pitch, of a screw or nut can be determined by the use of the *screw-pitch gauge* (Fig. 16-10). This device consists of a holder with a number of thin blades that have notches cut on them to represent different numbers of threads per inch.

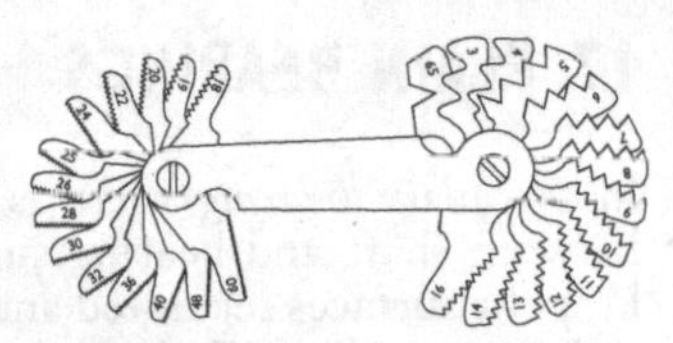

Fig. 16-10 Screw-pitch gauge.

Taper Gauge

This type of gauge is made of metal and has a graduated taper (Fig. 16-11). It is used for bearing and alignment work.

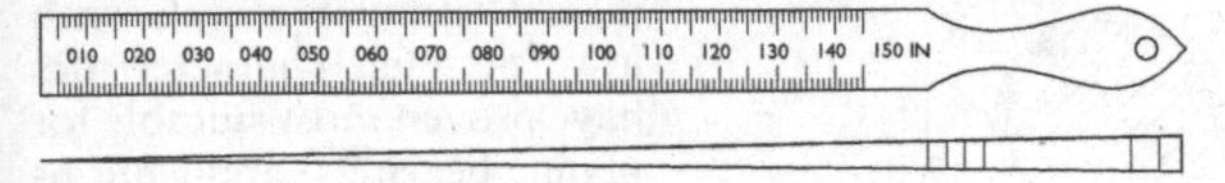

Fig. 16-11 Taper gauge.

17. PLAIN BEARINGS

In the *plain bearing* there is a relative sliding movement between shaft and bearing surfaces. A lubricant is used to keep the surfaces separated and overcome friction. If the film of lubricant (Fig. 17-1) can be maintained, the surfaces will be prevented from making contact and long bearing life will result. This is extremely difficult to accomplish, and usually some surface contact occurs when operating plain bearings. Dissimilar metals with low frictional characteristics have proven most suitable for plain bearing applications because they are less susceptible to seizure on contact. The common practice is to use a steel shaft and to make plain bearings of bronze, babbitt, or some other material that is softer than steel. A bearing metal that is softer than steel will wear before the shaft and usually can be replaced more easily, quickly, and cheaply.

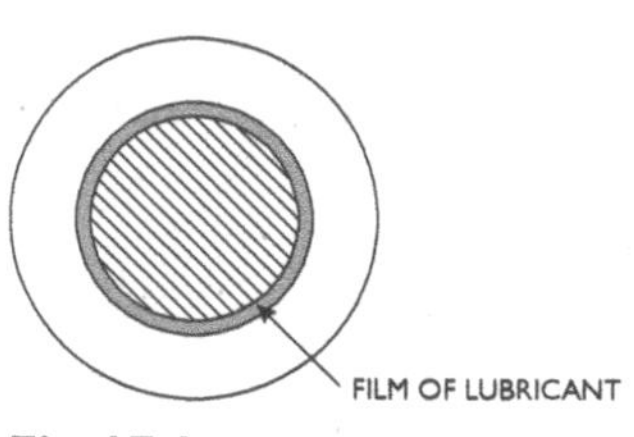

Fig. 17-1

A bearing, in the mechanical terms used by millwrights and mechanics, is a support for a revolving shaft. The word *bearing* is also used in some cases to describe a complete assembly rather than just the supporting member. Assembling a plain bearing generally requires the following parts: the shaft or the surface of the shaft, which is called the journal; the bearing liner, box, sleeve, bushing or insert (depending on the construction and design); and the box block or pedestal, which contains and supports the liner. If the box or block is of the split type, there would also be a cap fastened by bolts or studs. The meanings of some of the names and terms commonly used in connection with bearings and associated parts are given next.

Plain Bearing Nomenclature

- A *journal* is that part of a shaft, axle, spindle, etc., that turns in a bearing.
- The *axis* is a straight line (imaginary) passing through a shaft on which the shaft revolves or may be supposed to revolve.
- *Radial* means extending from a point or center in the manner of rays (as the spokes of a wheel are radial).
- *Thrust* is pressure of one part against another part, or force exerted endwise or axially through a shaft.
- *Friction* is resistance to motion between two surfaces in contact.
- *Sliding motion* is where two parallel surfaces move in relation to each other.
- *Rolling motion* takes place with round objects rolling on mating surfaces with theoretically no sliding motion.

Plain Bearing Theory

Plain bearings are generally very simple in construction and operation, although they vary widely in design and materials. They are amazingly efficient and can support extremely heavy rotating loads. The secret of this tremendous load-carrying ability lies in the tapered oil film developed between the journal and the bearing surface. To understand the formation of this film, a detailed discussion of the factors involved is required.

Friction has been defined as *the resistance to motion between two surfaces in contact*. It is present at all times where any form of relative motion occurs. It is the governing factor in machine design and operation, and in most cases friction is the major limiting factor in machine speed, capabilities, etc. It is overcome in machinery by separating the moving surfaces with a fluid film or lubricant, usually an oil.

Fig. 17-2 shows enlarged views of mating surfaces with and without an oil film. Microscopic examination of bearing surfaces shows minute peaks and valleys similar to those shown. The smoother or more highly finished the surfaces, the thinner the oil film required to keep them separated. The *wearing-in* or *running-in* of a plain bearing is a process of wearing off the peaks and filling in the valleys so that extreme high points do not penetrate the film and make contact.

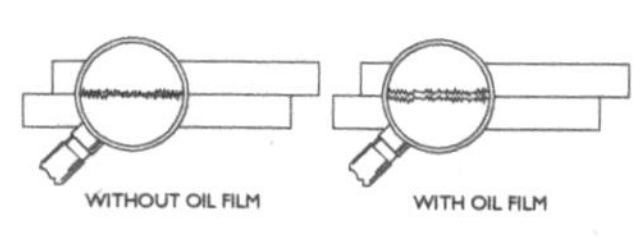

Fig. 17-2

When the surfaces are separated with an oil film, the only friction then is within the fluid, and generally this is of small consequence. If an oil film can be formed and complete separation of the surfaces maintained, the condition is described as thick-film lubrication. In plain bearings, the shaft's rotation generates this essential film between the journal and bearing surfaces. While stationary, the weight of the shaft squeezes out the oil, causing the journal to rest on the bearing surface, as shown in Fig. 17-3a. When rotation starts, a wedge-shaped film develops under the journal. In a properly designed bearing, the pressure that develops in this wedge lifts the journal away from contact with the bearing, as shown in Fig. 17-3b.

While plain bearings are operating at normally rated load and speed, practically no wear occurs because the oil film keeps the bearing surfaces separated. This oil film exists because of *hydromatic pressure*. This occurs because the lubricant has a tendency to adhere to the rotating journal and is drawn into the space between the journal and the bearing. Because of the viscosity, or the internal resistance to motion of the

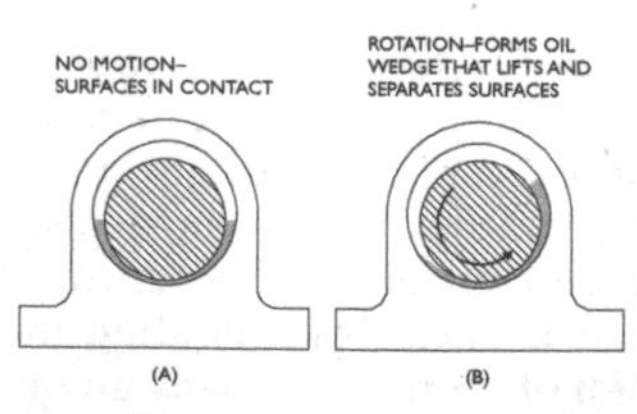

Fig. 17-3

lubricant, hydraulic pressure is built up in the lubricant film. Pressures as high as 600 psi are not unusual in well-designed bearings. Hydrodynamic action occurs in a plain bearing as the shaft rotates.

Plain Bearing Types

A *solid bearing* is the most common plain bearing. Often called a sleeve or bushing, it is usually pressure fitted into a supporting member (Fig. 17-4).

The *split bearing* is divided in two pieces; the split bearing allows easy removal of the shaft (Fig. 17-5).

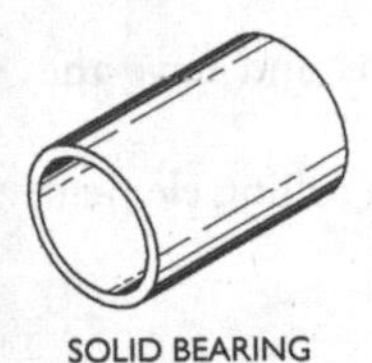

SOLID BEARING

Fig. 17-4

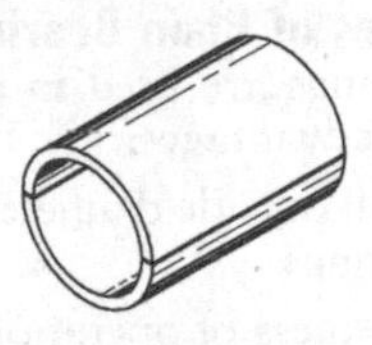

SPLIT BEARING

Fig. 17-5

The *journal bearing* is used for support of radial loads. It takes its name from the portion of the shaft or axle that operates within the bearing (Fig. 17-6).

The *part bearing* and/or half bearing encircling only part of the journal are used when the principal load presses in the direction of the bearing. Their advantages are low material cost and ease of replacement (Fig. 17-7).

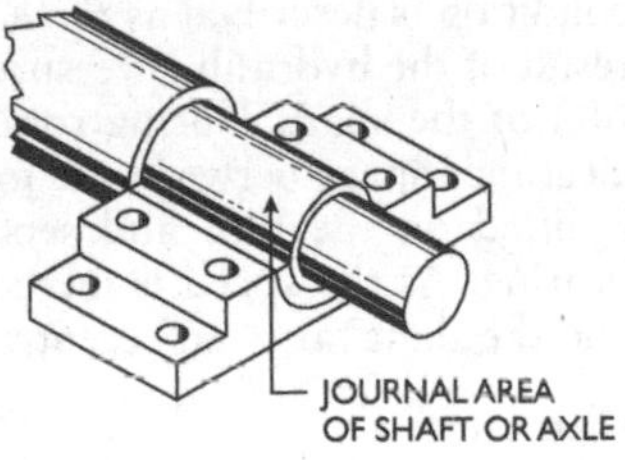

JOURNAL BEARING

Fig. 17-6

The *thrust bearing* supports axial loads and/or restrains endwise movement. A widely used style of thrust bearing is the simple annular

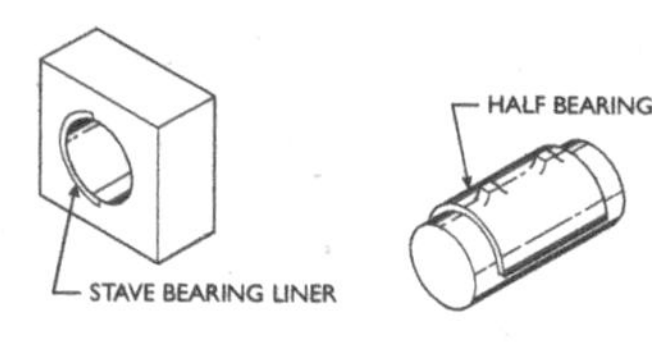

Fig. 17-7

ring or washer. Two or more such thrust rings may be combined to mate a hard steel surface with a softer low-friction material. Another common practice is to support thrust loads on the end surface of journal bearings. When the area is too small for the applied load, a flange may be provided. A shoulder is the usual method of supporting the thrust load on the shaft (Fig. 17-8).

Advantages of Plain Bearings

Plain bearings are used in many applications and have the following advantages:

- Small outside diameter as compared to rolling element bearings
- Quietness of operation
- Good capacity to absorb shock loading
- Ability to take oscillating motion
- Low cost

Boundary Lubrication

When bearing surfaces are separated with an oil film, the condition is described as *thick-film lubrication*. This film is a result of the hydraulic pressure that is generated by the rotation of the shaft. During rotation the oil is drawn into the clearance space between the journal and the shaft. The shaft is lifted on the film and separated from contact with the bearing. As the speed is increased, higher pressure develops and the shaft takes an eccentric position.

Clearance

To allow the formation of an oil film in a plain bearing, there must be clearance between the journal and the bearing. This clearance varies with the size of the shaft and the bearing material, the load carried, and the accuracy of shaft position desired. In industrial design of rotating equipment a diametral clearance of 0.001 in. per inch of shaft diameter is often used. This is a general figure and requires adjustment for high speeds or heavy loading.

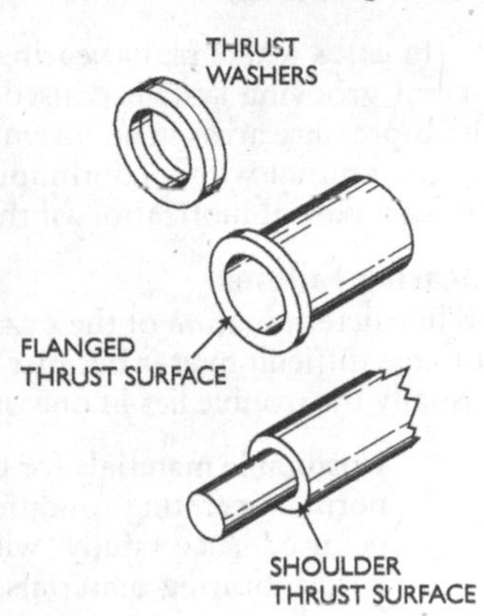

Fig. 17-8

Lubrication Holes and Grooves

The simplest plain bearings have a hole in the top through which the lubricant travels to the journal. For longer bearings, a groove or combination of grooves extending in either direction from the oil hole, as shown in Fig. 17-9, will distribute the oil. Recommended practice is to locate the inlet hole in the center of the bearing in the low-pressure region. On long bearings, two or more inlets may be required. Proper grooving is the most dependable means of distributing the oil so the moving surfaces, as they pass the grooves, will take up a film of oil. In bearings not fed under pressure, this is necessary if a taper oil film is to be developed to provide thick-film lubrication. Frequently, a pocket or reservoir is provided in the low-pressure region to place a volume of oil at the oil-film formation point. Regardless of the system of grooving, it is important that such grooving be confined to the unloaded portion of the bearing surface. If the grooving extends into the loaded or pressure region, the oil film will be disrupted, and boundary lubrication conditions will exist.

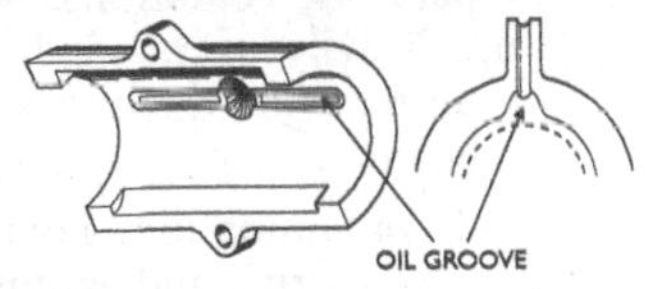

Fig. 17-9

In cases where plain bearings are grease lubricated, a different grooving system is used. Grooves are inserted at the high-pressure areas in an attempt to supply the entire surface of the journal with a continuous coating of grease, ensuring at least partial lubrication of the bearing.

Bearing Failures

While determination of the exact cause of a bearing failure is often a difficult matter because of the many factors involved, usually the trouble lies in one or more of the following areas:

Unsuitable materials for either shaft or bearing. Under normal operating conditions, tin and lead babbitts may be used successfully with soft steel journals. With harder bearing materials, it is frequently necessary to have harder shaft surfaces for satisfactory operation.

Incorrect grooving. If the grooves are incorrectly located, bearing failures can result. If the oil grooves are located in the high-pressure areas, they act as pressure-relief passages. Thus the formation of the hydrodynamic film is interfered with, and the bearing's load-carrying capacity is reduced.

Unsuitable surface finish. The smoother the finish, the more closely the shaft may approach the bearing without danger of metallic contact. Surface finish is important in any plain bearing application.

Insufficient clearance. There must be sufficient clearance between journal and bearing to allow the oil film to exist.

Operating conditions. When machines are speeded up, heavily overloaded, or used for a purpose other than that for which they were designed, bearings take a beating.

Oil contamination. Foreign material in the lubricant causes scoring and galling of the bearing surfaces.

Dry Bearings

Dry bearings operate without a significant fluid film to separate the moving surfaces; therefore, low-friction materials, or materials impregnated with a lubricant, must be used.

Sintered bronze. Sintered bronze is a porous material that can be impregnated with oil, graphite, or PTFE. It is not suitable for heavily loaded applications, but it can be useful where lubrication is inconvenient.

PTFE. This is suitable only in very light applications. PTFE is mechanically weak material that has a tendency to flow and is seriously affected by high temperatures.

Nylon. Similar to PTFE but slightly harder, nylon is used only in very light applications. It is less likely to cold flow than PTFE.

18. ANTI-FRICTION BEARINGS

Anti-friction bearings, so called because they are designed to overcome friction, are of two types: *ball bearings* and *roller bearings*. In the plain bearing, the frictional resistance to sliding motion is overcome by separating the surfaces with a fluid film. The anti-friction bearing substitutes rolling motion for sliding motion by the use of rolling elements between the rotating and stationary surfaces and thereby reduces friction to a fraction of that in plain bearings.

Classified by function, ball bearings may be divided into three main groups: *radial*, *thrust*, and *angular-contact* bearings. Radial bearings are designed primarily to carry a load in a direction perpendicular to the axis of rotation (Fig. 18-1a). Thrust bearings can carry only thrust loads—that is, a force parallel to the axis of rotation tending to cause endwise motion of the shaft (Fig. 18-1b). Angular-contact bearings can support combined radial and thrust loads (Fig. 18-1c).

Roller bearings may also be classified by their ability to support radial, thrust, and combination loads. In addition, they are further divided into styles according to the shape of their rollers: *cylindrical* (also called *straight*), *taper*, *spherical*, and *needle*. Combination load-supporting roller bearings are not called angular-contact bearings, as they are quite different in design. The taper-roller bearing, for example, is a combination load-carrying bearing by virtue of the shape of its rollers (Fig. 18-2).

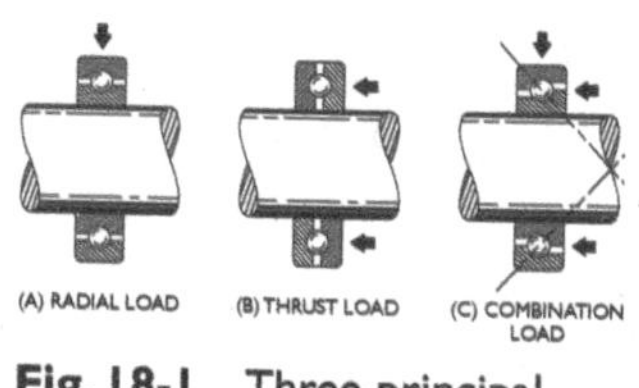

Fig. 18-1 Three principal types of ball bearings.

Nomenclature

All anti-friction bearings consist of two hardened rings called the inner and outer rings, the hardened rolling elements that may be either balls or rollers, and a separator. Bearing size is usually given in terms of what are called boundary dimensions. These are the outside diameter, the bore, and the width. The inner and outer rings provide continuous tracks or races for the rollers or balls to roll in. The separator or retainer properly spaces the rolling elements around the track and guides them through the load zone. Other words and terms used in describing anti-friction ball bearings are the face, shoulders, corners, etc. All are illustrated in Fig. 18-3. The terms used to describe taper-roller bearings are a little different in that what is normally the outer ring is called the cup, and the inner ring is called the cone. The word "cage" is standard for taper-roller bearings rather than "separator" or "retainer." These terms are illustrated in Fig. 18-4. Nomenclature for the straight roller bearing is illustrated in Fig. 18-5.

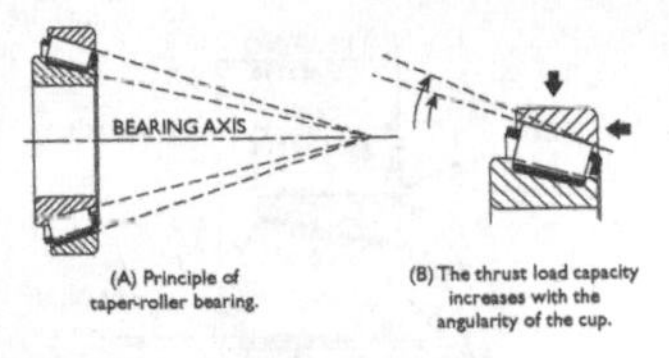

Fig. 18-2 Taper-roller bearing.

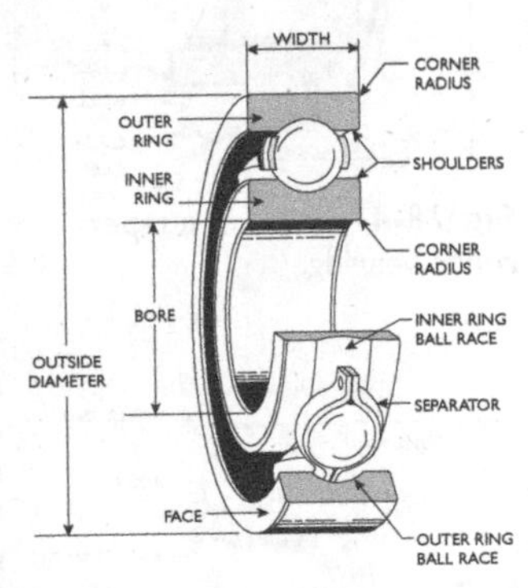

Fig. 18-3 Parts of a ball bearing.

Self-Aligning Bearings

The *self-aligning* bearing is a specialized style of anti-friction bearing that has the capacity of angular self-alignment. This is accomplished by the use of a spherical raceway inside the outer ring of the bearing. The inner ring and the rolling elements rotate at right angles to the shaft centerline in a fixed raceway. The position of the outer ring, because of its

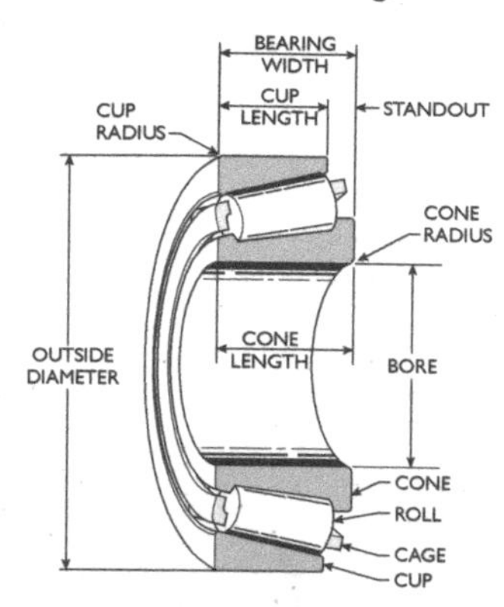

Fig. 18-4 Parts of a taper-roller bearing.

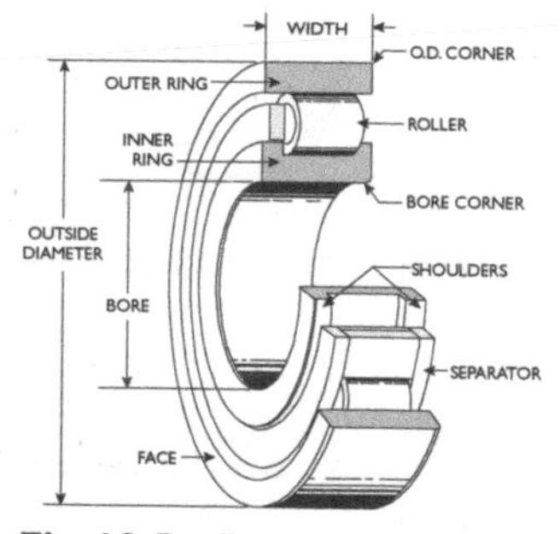

Fig. 18-5 Parts of a straight-roller bearing.

spherical raceway, may be misaligned within the limits of its width and still provide a true path for the rolling elements to follow. Bearings of this type are called *internal* self-aligning.

The *spherical double-row* roller bearing (Fig. 18-6a) is a common self-aligning style of roller bearing. Another widely used style is the *self-aligning* ball bearing (Fig. 18-6b). The angular movement this bearing allows is possible because the two rows of balls are rolling on the spherical inner surface of the outer ring. Another self-aligning ball bearing (single row) incorporates an additional outside ring with a spherical inner surface (Fig. 18-6c). The outside of the regular outer ring is made spherical to match the extra ring. This style of construction is used for single-row self-aligning ball bearings. It is called an *external style* self-aligning ball bearing.

Another specialized style of bearing is the *wide inner ring* bearing. It is used primarily in pillow blocks, flange units, etc., commonly termed transmission units. Wide inner ring bearings are commonly made in two types, rigid (Fig. 18-7a) and self-aligning (Fig. 18-7b). The rigid type has a straight cylindrical outside surface; the self-aligning type has a spherical outside surface. They are made to millimeter outside dimensions and inch bore dimensions.

They are an assembly of standard metric-size ball bearing outer rings with special wide inner rings. The bore of the inner ring is made to inch dimensions to fit standard fractional-inch dimension shafting. As they are usually contained in an assembled unit, they are specified by their nominal fractional-inch bore size.

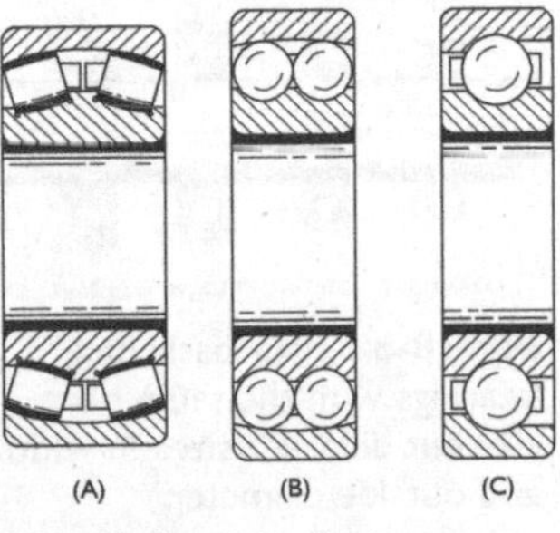

Fig. 18-6 Self-aligning bearings.

Ball Bearing Dimensions

Basic ball bearings, for general use in all industries, are manufactured to standardized dimensions of bore, outside diameter, and width (boundary dimensions). Tolerances for these critical dimensions and for the limited dimensions for corner radii have been established by industry standards. Therefore, all types and sizes of ball bearings made to industry-standardized specifications are satisfactorily interchangeable with other makes of like size and type.

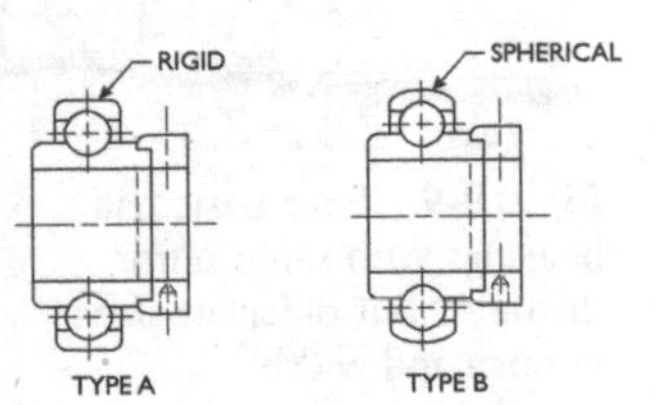

Fig. 18-7 Wide inner ring bearings.

Most basic ball bearings are available in four different series: *extra light, light, medium,* and *heavy.* The names applied to each series are descriptive of the relative proportions and load-carrying capacities of the bearings. This means there are as many as four bearings (one in each series) with the same bore size but with different widths, outside

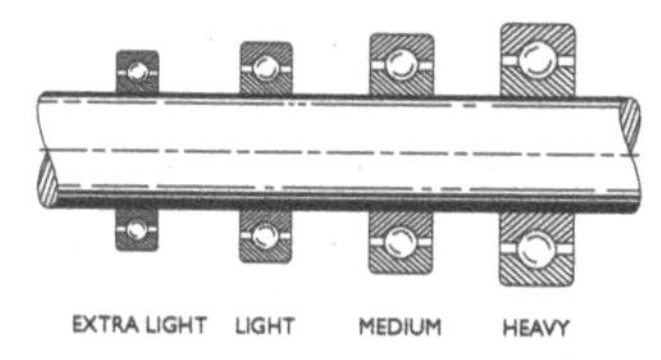

Fig. 18-8 Four basic ball bearings with the same bore size but different sizes in width and outside diameter.

diameter, and load-carrying capacities. The relative proportions of bearings in each series with the same bore are illustrated in Fig. 18-8.

It is also possible to select as many as four bearings with the same outside diameter (one in each series) with four different bore sizes, widths, and load-carrying capacities. Thus, you have a choice of four different shaft sizes without changing the diameter of the housing. This relationship is illustrated in Fig. 18-9.

Bearing manufacturers determine the series using numbers that they incorporate in their basic bearing numbering systems. The *extra light* series is designated as the 100 series, the *light* as 200, the *medium* as 300, and the *heavy* as 400.

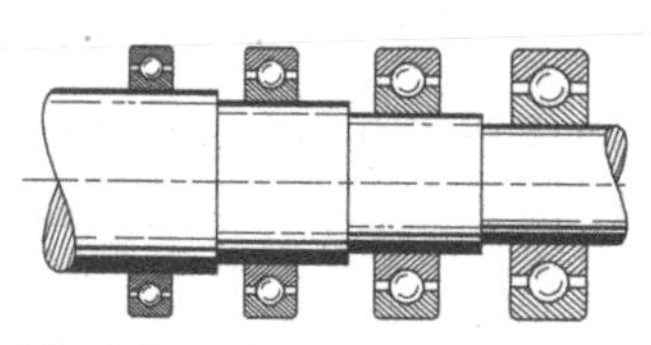

Fig. 18-9 Four basic ball bearings with same outer diameter but different sizes in bore and width.

Extra Light Series (100)

As the name implies, this series is designed to give the smallest widths and outside diameters in proportion to the standard bore sizes. The smaller cross-sectional areas of this series' bearings necessitate the use of smaller balls than the heavier series, and more of them (see Table 18-1).

Light Series (200) and Medium Series (300)

The light and medium series bearings are the most widely used of the four series. They are commonly used in a great variety of machinery and equipment (Tables 18-2 and 18-3).

Heavy Series (400)

The heavy series bearings are used where exceptionally heavy load conditions are encountered. Because their use is rather limited, they are not produced in nearly as wide a selection of types and sizes as the light and medium series (Table 18-4).

The basic ball bearing number is made up of three digits. The first digit indicates the bearing series (i.e., 100, 200, 300, or 400). The second and third digits, from 04 and up, when multiplied by 5, indicate the bore in millimeters. An example in each of the four duty series follows:

108—Extra Light Series—40 mm Bore—1.5748 in.

205—Light Series—25 mm Bore—0.9843 in.

316—Medium Series—80 mm Bore—3.1496 in.

420—Heavy Series—100 mm Bore—3.9370 in.

The bore of bearings with a basic number under 04 is as follows:

00—10 mm—0.3937 in.

01—12 mm—0.4724 in.

02—15 mm—0.5906 in.

03—17 mm—0.6693 in.

Table 18-1 Ball Bearing Dimension Table—100 Series

Basic Bearing Number	*Bore*		*OD*		*Width*	
	mm	*in.*	*mm*	*in.*	*mm*	*in.*
100	10	0.3937	26	1.0236	8	0.3150
101	12	0.4724	28	1.1024	8	0.3150
102	15	0.5906	32	1.2598	9	0.3543
103	17	0.6693	35	1.3780	10	0.3937
104	20	0.7874	42	1.6535	12	0.4724

(continued)

Table 18-1 (continued)

Basic Bearing Number	Bore		OD		Width	
	mm	in.	mm	in.	mm	in.
105	25	0.9843	47	1.8504	12	0.4724
106	30	1.1811	55	2.1654	13	0.5118
107	35	1.3780	62	2.4409	14	0.5512
108	40	1.5748	68	2.6772	15	0.5906
109	45	1.7717	75	2.9528	16	0.6299
110	50	1.9685	80	3.1496	16	0.6299
111	55	2.1654	90	3.5433	18	0.7087
112	60	2.3622	95	3.7402	18	0.7087
113	65	2.5591	100	3.9370	18	0.7087
114	70	2.7559	110	4.3307	20	0.7874
115	75	2.9528	115	4.5276	20	0.7874
116	80	3.1496	125	4.9213	22	0.8661
117	85	3.3465	130	5.1181	22	0.8661
118	90	3.5433	140	5.5118	24	0.9449
119	95	3.7402	145	5.7087	24	0.9449
120	100	3.9370	150	5.9055	24	0.9449
121	105	4.1339	160	6.2992	26	1.0236

Table 18-2 Ball Bearing Dimension Table—200 Series

Basic Bearing Number	Bore		OD		Width	
	mm	in.	mm	in.	mm	in.
200	10	0.3937	30	1.1811	9	0.3543
201	12	0.4724	32	1.2598	10	0.3937
202	15	0.5906	35	1.3780	11	0.4331
203	17	0.6693	40	1.5748	12	0.4724
204	20	0.7874	47	1.8504	14	0.5512
205	25	0.9843	52	2.0472	15	0.5906
206	30	1.1811	62	2.4409	16	0.6299

Basic Bearing Number	Bore		OD		Width	
	mm	in.	mm	in.	mm	in.
207	35	1.3780	72	2.8346	17	0.6653
208	40	1.5748	80	3.1496	18	0.7087
209	45	1.7717	85	3.3465	19	0.7480
210	50	1.9685	90	3.5433	20	0.7874
211	55	2.1654	100	3.9370	21	0.8268
212	60	2.3633	110	4.3307	22	0.8661
213	65	2.5591	120	4.7244	23	0.9055
214	70	2.7559	125	4.9213	24	0.9449
215	75	2.9528	130	5.1181	25	0.9843
216	80	3.1496	140	5.5118	26	1.0236
217	85	3.3465	150	5.9055	28	1.1024
218	90	3.5433	160	6.2992	30	1.1811
219	95	3.7402	170	6.6929	32	1.2598
220	100	3.9370	180	7.0866	34	1.3386
221	105	4.1339	190	7.4803	36	1.4137
222	110	4.3307	200	7.8740	38	1.4961

Table 18-3 Ball Bearing Dimension Table — 300 Series

Basic Bearing Number	Bore		OD		Width	
	mm	in.	mm	in.	mm	in.
300	10	0.0397	35	1.3780	11	0.4331
301	12	0.4724	37	1.4567	12	0.4724
302	15	0.5906	42	1.6535	13	0.5118
303	17	0.6693	47	1.8504	14	0.5512
304	20	0.7874	52	2.0472	15	0.5906
305	25	0.9843	62	2.4409	17	0.6693
306	30	1.1811	72	2.8346	19	0.7480
307	35	1.3780	80	3.1496	21	0.8268

(continued)

Table 18-3 *(continued)*

Basic Bearing Number	*Bore*		*OD*		*Width*	
	mm	*in.*	*mm*	*in.*	*mm*	*in.*
308	40	1.5748	90	3.5433	23	0.9055
309	45	1.7717	100	3.9370	25	0.9843
310	50	1.9685	110	4.3307	27	1.0630
311	55	2.1654	120	4.7244	29	1.1417
312	60	2.3622	130	5.1181	31	1.2205
313	65	2.5591	140	5.5118	33	1.2992
314	70	2.7559	150	5.9055	35	1.3780
315	75	2.9528	160	6.2992	37	1.4567
316	80	3.1469	170	6.6929	39	1.5354
317	85	3.3465	180	7.0866	41	1.6142
318	90	3.5433	190	7.4803	43	1.6929
319	95	3.7402	200	7.8740	45	1.7717
320	100	3.9370	215	8.4646	47	1.8504
321	105	4.1339	225	8.8583	49	1.9291
322	110	4.3307	240	9.4480	50	1.9685
324	120	4.7244	260	10.2362	55	2.1654
326	130	5.1181	280	11.0236	58	2.2835
328	140	5.5118	300	11.8110	62	2.4409
330	150	5.9055	320	12.5984	65	2.5591
332	160	6.2992	340	13.3858	68	2.6772
334	170	6.6929	360	14.1732	72	2.8346
336	180	7.0866	380	14.9606	75	2.9528
338	190	7.4803	400	15.7480	78	3.0709
340	200	7.8740	420	16.5354	80	3.1496
342	210	8.2677	440	17.3228	84	3.3071
344	220	8.6614	460	18.1002	88	3.4646
348	240	9.4488	500	19.6850	95	3.7402
352	260	10.2362	540	21.2598	102	4.0157
356	280	11.0236	580	22.8346	108	4.2520

Table 18-4 Ball Bearing Dimension Table—400 Series

Basic Bearing Number	Bore		OD		Width	
	mm	in.	mm	in.	mm	in.
403	17	0.6693	62	2.4409	17	0.6693
404	20	0.7874	72	2.8345	19	0.7480
405	25	0.9843	80	3.1496	21	0.8268
406	30	1.1811	90	3.5433	23	0.9055
407	35	1.3780	100	3.9370	25	0.9843
408	40	1.5748	110	4.3307	27	1.0630
409	45	1.7717	120	4.7244	29	1.1417
410	50	1.9685	130	5.1181	31	1.2205
411	55	2.1654	140	5.5118	33	1.2992
412	60	2.3622	150	5.9055	35	1.3780
413	65	2.5591	160	6.2992	37	1.4567
414	70	2.7559	180	7.0866	42	1.6535
415	75	2.9528	190	7.4803	45	1.7717
416	80	3.1496	200	7.8740	48	1.8898
417	85	3.3465	210	8.2677	52	2.0472
418	90	3.5433	225	8.8533	54	2.1260
419	95	3.7402	250	9.8425	55	2.1654
420	100	3.9370	265	10.4331	60	2.3622
421	105	4.1339	290	11.4173	65	2.5591
422	110	4.3307	320	12.5984	70	2.5759

Bearing Types

Single-Row Deep-Groove Ball Bearings

Single-row deep-groove ball bearings are the most widely used of all ball bearings and probably of all anti-friction bearings. They can sustain combined radial and thrust loads, or thrust loads alone, in either direction even at extremely high speeds. They are also termed *Conrad* type bearings, for the man who first successfully designed and produced this style of bearing. A cross-section of the single-row deep-groove bearing is illustrated in Fig. 18-10.

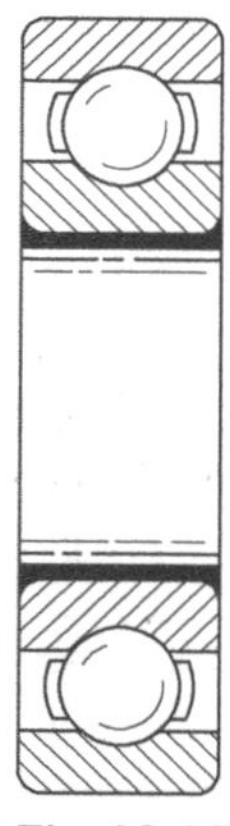

Fig. 18-10 Single-row deep-groove ball bearing.

Double-Row Deep-Groove Ball Bearings

The double-row deep-groove bearing shown in Fig. 18-11 embodies the same principle of design as the single-row bearing. The grooves for the two rows of balls are positioned so that the load through the balls tends to push outward on the outer ring races. This bearing has substantial thrust capacity in either direction and high radial capacity due to two rows of balls.

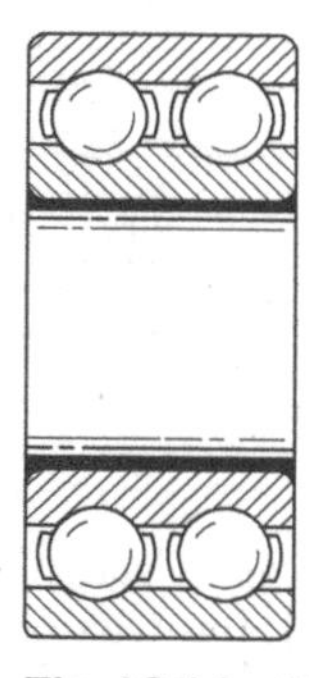

Fig. 18-11 Double-row deep-groove ball bearing.

Angular-Contact Radial-Thrust Ball Bearings

Angular-contact bearings, illustrated in Fig. 18-12, can support radial loads when combined with thrust loads in one direction. The inner and outer rings are made with an extra high shoulder on one side only (thrust side). This type of bearing is designed for combination loads where the thrust component is greater than the capacity of single-row deep-groove bearings. They may be mounted either face to face or back to back and in tandem for constant thrust in one direction. When mounting in pairs, take care that the bearings have been ground for the style of assembly used.

Ball-Thrust Bearings

The ball-thrust bearing can support thrust loads on one direction only—no radial loading. The thrust load is transmitted through the balls parallel to the axes of the shaft, resulting in very high thrust capacity. The rings of these bearings are commonly known as washers (Fig. 18-13). To operate successfully, this type of bearing must be at least moderately thrust-loaded at all times. This style of bearing should not be operated at high speeds because centrifugal force will cause excessive loading of the outer edges of the races.

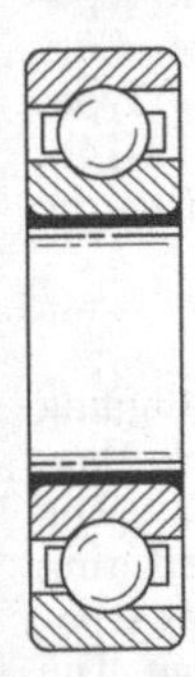

Fig. 18-12 Radial-thrust ball bearing.

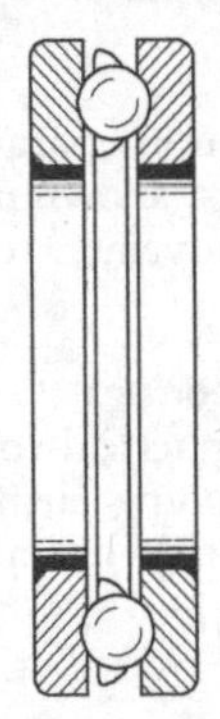

Fig. 18-13 Ball-thrust bearing.

Cylindrical-Roller Bearings

The cylindrical-roller bearing has straight cylindrical-shaped rolling elements. These rolling elements are approximately equal in length and diameter (Fig. 18-14). These equal-dimensioned rollers distinguish cylindrical-roller bearings from other roller bearings that have rollers with a much greater length-to-diameter ratio. They are used primarily for applications where heavy radial loads must be supported—loads beyond the capacities of radial ball bearings of comparable sizes. They are ideally suited for heavy loads in the

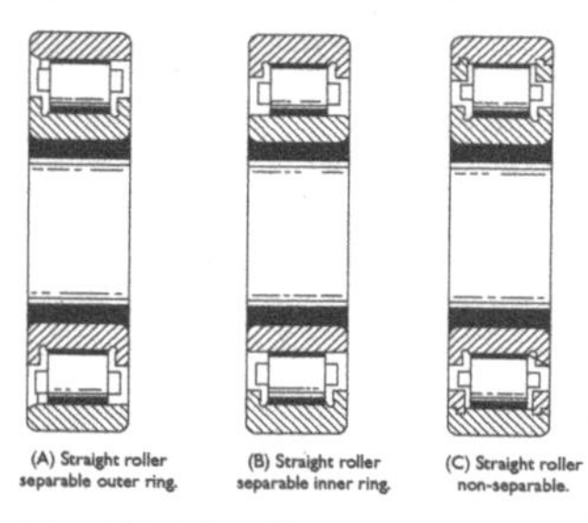

Fig. 18-14 Three types of cylindrical bearings.

moderate speed ranges because there is a line contact between rolling elements and races. Therefore, the rolling elements are deformed less under heavy load conditions than the point contact of ball bearings of comparable size. There are three basic types of straight cylindrical-roller bearings: the separable outer-ring type (Fig. 18-14a), the separable inner-ring type (Fig. 18-14b), and the nonseparable type (Fig. 18-14c). Types with separable rings, shown in Fig. 18-14a and Fig. 18-14b, allow free axial movement of the shaft in relation to the housing.

Spherical-Roller Bearings

The double-row spherical-roller bearing is a self-aligning bearing utilizing rolling elements shaped like barrels (Fig. 18-15). The outer ring has a single spherical raceway. The double-shoulder inner ring has two spherical races separated by a center flange. The rollers are retained and separated by an accurately constructed cage.

Fig. 18-15 Spherical-roller double-row bearing.

This type of bearing is inherently self-aligning because the assembly of the inner unit—that is, the ring, cage, and rollers—is free to swivel within the outer ring. This automatic adjustment allows successful operation under severe misalignment conditions. It will support a heavy radial load and heavy thrust loads from both directions.

Taper-Bore Spherical-Roller Bearings

A specialized style of spherical-roller bearing is the taper-bore style of bearing, designed specifically to ensure maximum inner-ring grip at assembly. Most bearing applications have a rotating inner ring and a stationary outer ring. When correctly assembled, the inner ring is sufficiently tight on the shaft to ensure that both inner ring and shaft turn as a unit and that "creeping" of the ring on the shaft does not occur. Should creeping occur, there will be overheating, excessive wear, and erosion between the shaft and the inner ring. For normal applications, the inner ring is press-fitted to the shaft and/or clamped against a shoulder with a locknut. On applications subjected to severe shock or unbalanced loading, however, the usual press-fit or locknut clamping does not grip tightly enough to prevent creeping. For such applications, a design providing maximum grip of the inner ring on the shaft is required. In providing for the tremendous grip that is possible with the method to be discussed, some very important conditions must be controlled:

1. The stress in the inner ring must remain below the elastic limit of the steel.
2. The internal bearing clearance must not be eliminated.
3. There must be a practical mounting method.

The taper-bore self-aligning spherical-roller bearing incorporates the features required. Size for size, its capacity is greater than that of any other type of bearing. The taper bore provides a simple method of mounting that allows controlled stretching of the inner ring to obtain maximum gripping power. When mounting this bearing, the inner ring is forced up on the taper by tightening a locknut. As the inner ring stretches, and as its grip increases, the internal bearing clearance is reduced. The bearing is manufactured with sufficient internal clearance to allow this stretching of the inner ring.

The grip and the clearance are controlled by checking internal clearance before mounting and by tightening the nut sufficiently to reduce the internal clearance by a specific amount.

The taper-bore bearing also allows the adaptor-sleeve style of mounting. The use of a tapered adapter sleeve lets you mount the taper-bore bearing on straight cylindrical surfaces. It also provides an easy means of locating the bearing. In many cases, this feature is the reason for using taper-bore bearings in applications where loading is relatively light and the tremendous inner-ring grip feature is not required. Fig. 18.16 illustrates the taper-bore spherical-roller bearing with mounting adaptor for assembly to straight cylindrical surfaces.

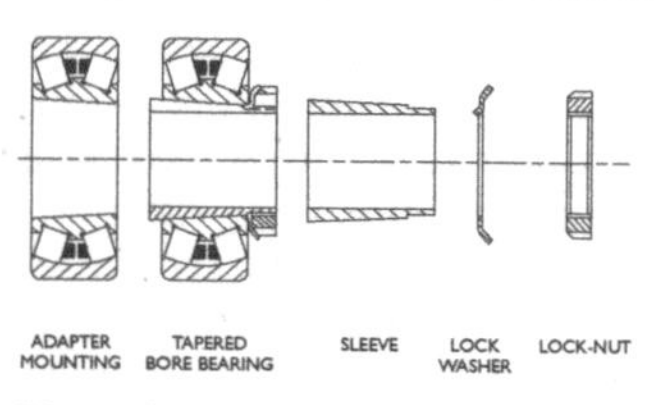

Fig. 18-16 Taper-bore bearing with adaptor.

Taper-Roller Bearings

Taper-roller bearings can be classified as a separate group of roller bearings because of their design and construction. They consist fundamentally of tapered (cone-shaped) rollers operating between tapered raceways. They are so constructed, and the angle of all rolling elements so proportioned, that if straight lines were drawn from the tapered surfaces of each roller and raceway they would meet at a common point on the centerline of the axis of the bearing, as illustrated in Fig. 18-17.

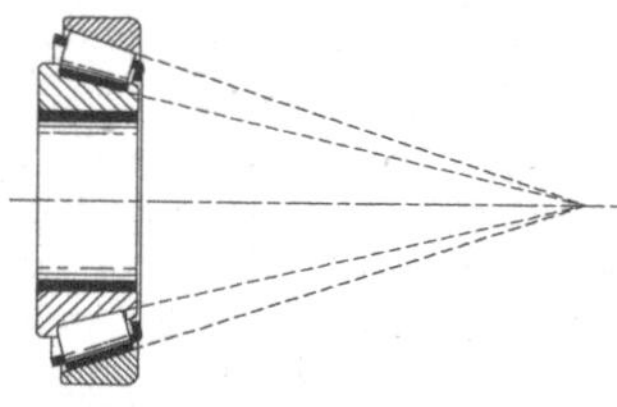

Fig. 18-17 Taper-roller bearing.

One major difference between taper-roller bearings (separable types) and most anti-friction bearings is that taper-rollers are adjustable. In many cases this is a distinct advantage because it permits accurate control of bearing running clearance.

The proper amount of running clearance may be assembled into the bearing at installation to suit the specific application. This feature also permits preloading the bearing for applications where extreme rigidity is required. This adjustable feature also requires that the millwright or the mechanic follows proper procedure at assembly to ensure the correct setting. Another major difference between taper-roller bearings and most other anti-friction bearings is that standard taper-roller bearings are made to inch dimensions rather than metric dimensions.

The basic parts of a taper-roller bearing are the *cone,* or inner race; the *taper rollers;* the *cage* (this is usually called the "retainer" or "separator" in other anti-friction bearings); and the *cup,* or outer race.

The most widely used taper-roller bearing is a single-row type with cone, rollers, and cage factory-assembled into one unit with the cup independent and separable. Fig. 18-18 shows what is commonly called the standard single-row taper-roller bearing. Many additional styles of single- and multiple-row taper-roller bearings are made, including unit assemblies that are factory pre-adjusted.

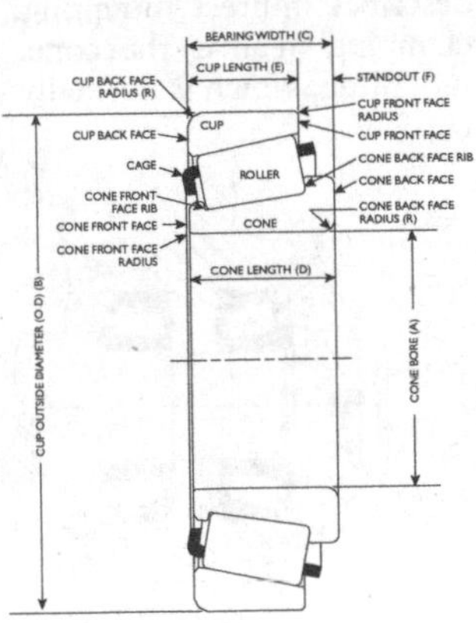

Fig. 18-18 Single-row taper roller bearing.

Taper-Roller Bearing Adjustment

While each single-row taper-roller bearing is an individual unit, their construction is such that they must be mounted in pairs so that thrust may be carried in either direction. Two systems of mounting are employed, indirect and direct (Fig. 18-19 and Fig. 18-20).

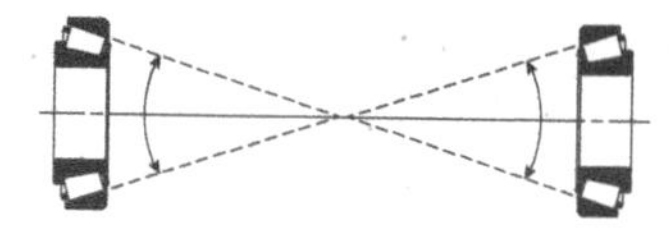

Fig. 18-19 Indirect mounting.

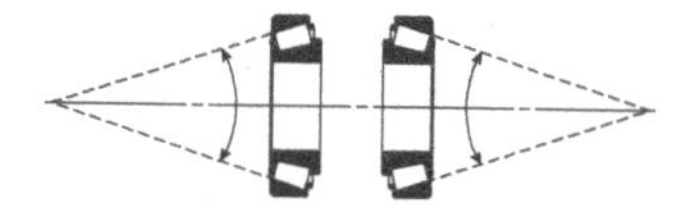

Fig. 18-20 Direct mounting.

Indirect mounting is used for applications where maximum stability must be provided in a minimum width. Direct mounting is used where this assembly system offers mounting advantages and where maximum stability is neither required nor desired.

The terms *face to face* and *back to back* may also be used to describe these systems of mounting. If these terms are used, they must be qualified by stating whether the terms are used in regard to the cup or the cone. Indirect and direct mounting may more appropriately be referred to as *cone-clamped* and *cup-clamped,* respectively. Cone-clamped describes indirect mounting, which is normally secured by clamping against the cone. Cup-clamped describes direct mounting, which is normally secured by clamping against the cup.

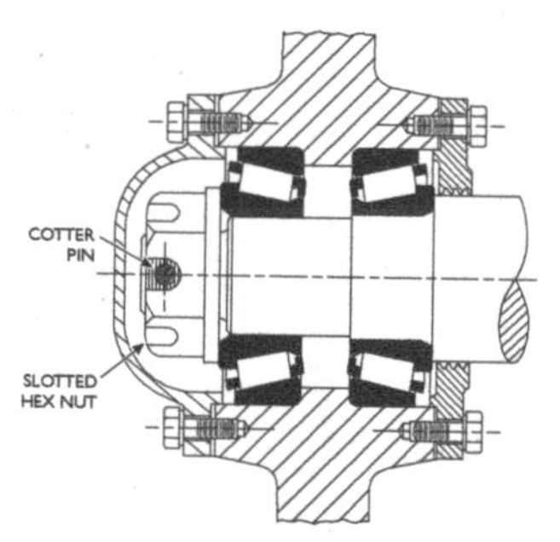

Fig. 18-21 Adjustable mounting.

Many devices are used for adjusting and clamping taper-roller bearing assemblies. Illustrated in Figs. 18-21 through Fig. 18-26 are six of the basic devices used for this purpose. There are many others in general use; however, they are variations of these basic designs.

In Fig. 18-21, the slotted hex nut and the cotter pin are used to adjust bearings. Here the adjustment need not be extremely precise, and there is room for the nut. Simply tighten the nut while rotating the outer assembly until a slight bind is obtained in the bearing. This ensures proper seating of all parts. Then back off the nut one slot and lock with a cotter pin. The result is free-running clearance in the bearing. The bearing is rotated in order to seat rollers against the cone. This should be done when making any bearing adjustment.

In Fig. 18-22, two standard locknuts and a tongued washer are used to adjust the bearing. They provide a much finer adjustment than a slotted hex nut. Pull up the inner nut until there is a slight bind on the bearings, then back off just enough to allow running clearance after the outer, or "jab," nut is tightened. This type of adjustment is used in full-floating rear axles and industrial applications where slotted hex nuts are not practical or desirable.

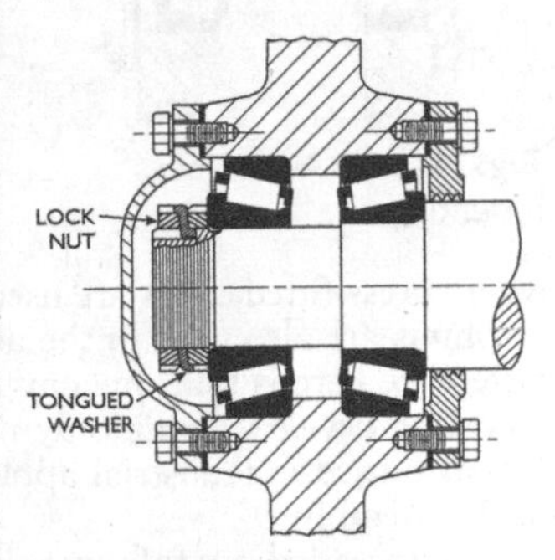

Fig. 18-22 Adjustable mounting.

In Fig. 18-23, shims are used between the end of the shaft and the end plate. The shim pack is selected to give the proper setting of the bearing. (This will vary with the unit in which the bearings are used.) The end plate is held in place by cap screws. The cap screws are wired together for locking. A slot may be provided in the end plate to measure the shim gap.

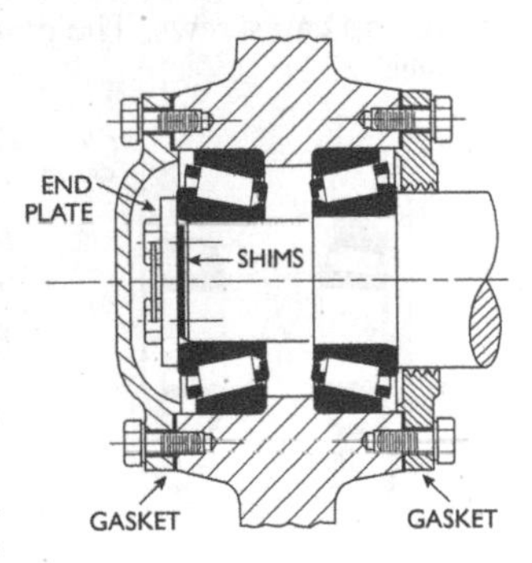

Fig. 18-23 Adjustable mounting.

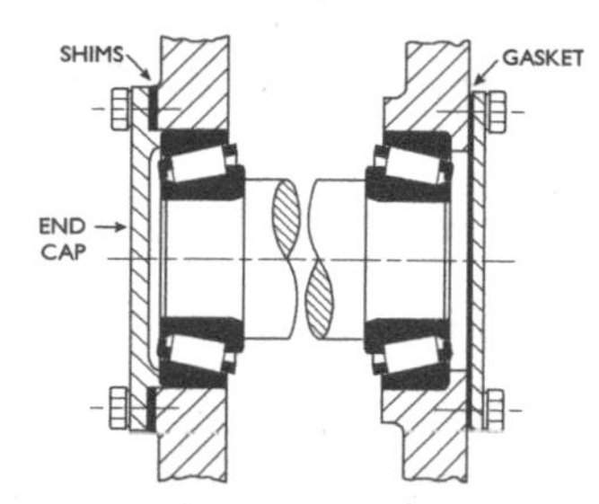

Fig. 18-24 Adjustable mounting.

Fig. 18-24 shows another adjustment using shims. They are located between the end-cap flange and the housing. Select the shim pack that will give the proper bearing running clearance recommended for the particular application. The end cap is held in place by cap screws, which can be locked with a lock washer (as shown) or wired. This easy method of adjustment is used where press-fitted cones are used with loose-fitted cups.

Shims are also used in the adjustment in Fig. 18-25. The difference here is that one cup is mounted in a carrier. This adjusting device is commonly used in gearboxes and drives. It is also used in industrial applications for ease of assembly or disassembly.

The threaded cup follower shown in Fig. 18-26 is another common adjusting device. It is used in gearboxes, drives, and automotive differentials. The follower is locked by means of a plate and cap screws. The plate fits between the lugs of the follower.

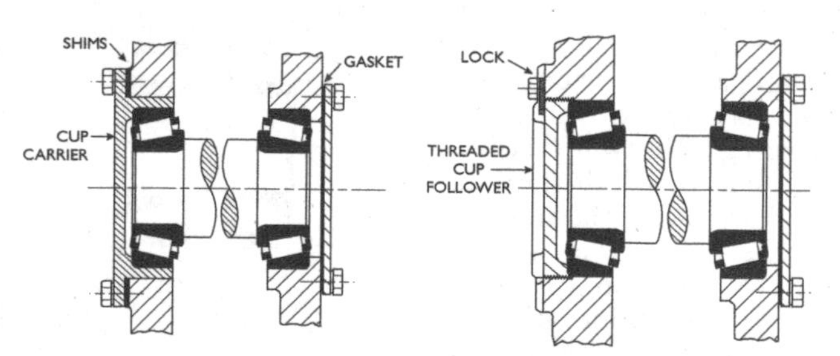

Fig. 18-25 Adjustable mounting.

Fig. 18-26 Adjustable mounting.

19. LIFE OF ANTI-FRICTION BEARINGS

Bearing life is a term used to describe the life expectancy of 90 percent of a group of the same bearings operating under the same load and at the same speed. Because most mechanics are interested in how long a bearing will last, bearing life is usually given in hours of operation. Bearing engineers call the minimum expected life of a bearing the L_{10} life. The term *average life* equals the minimum life, L_{10}, × 5. An average life of 100,000 hours is the same as 20,000 hours L_{10} minimum life.

Any bearing manufacturer's catalog will show the speed limits for each size and style of bearing. These limits are based on specific maximum loads, maximum ambient air temperature, good installation, and proper lubrication. As speed increases for a given size, so does the temperature within a bearing. At some point, either the grease breaks down or the steel begins to destruct.

The life of a bearing depends on many items, but the most important ones are these:

- Speed
- Load
- Proper lubrication
- Proper installation

Life Equation

Each anti-friction bearing has a published basic load rating. This is a standard rating, given by the manufacturer, and can be found in its catalog.

Formulas exist for the life of both a ball bearing and a roller bearing:

For Ball Bearings

$$L_{10} = \frac{1,000,000}{RPM \times 60} \times \left(\frac{C}{P}\right)^3$$

For Roller Bearings

$$L_{10} = \frac{1,000,000}{RPM \times 60} \times \left(\frac{C}{P}\right)^{10/3}$$

Where:

L_{10} is minimum life in hours

C is the published basic load rating in pounds

P is the actual load on the bearing in the machine

For example, consider a ball bearing with a basic load rating of 5,743 lbs, an actual load of 502 lbs, running at 1180 RPM. The L_{10} life is shown in the following calculations:

$$L_{10} = \frac{1,000,000}{1180 \times 60} \times \left(\frac{5743}{502}\right)^3$$

$$L_{10} = 21,148 \text{ Hours}$$

If the same bearing is run under the same load but at half the speed, the life doubles:

$$L_{10} = \frac{1,000,000}{590 \times 60} \times \left(\frac{5743}{502}\right)^3$$

$$L_{10} = 42,296 \text{ Hours}$$

Mathematically, it can be shown that bearing life is inversely proportional to the RPM. If the speed is doubled, the bearing life is halved. If the speed is halved, the bearing life is doubled.

If the actual load on the bearing is changed—for instance, doubled—then the following calculation shows the result. The load used will be 1004 lb—double the 502 that was used originally:

$$L_{10} = \frac{1,000,000}{1180 \times 60} \times \left(\frac{5743}{1004}\right)^3$$

$$L_{10} = 2,644 \text{ Hours}$$

Now the hours are reduced to one-eighth of 21,148, or only 2,644 hours. Load is a very big factor on the life of a bearing. On a ball bearing, if the load is doubled, the life decreases by a factor of 8. On a roller bearing, if the load is doubled, the life decreases by a factor of 10. Controlling the load and the speed is important to extend the life of any machine running on anti-friction bearings, but the load is far more influential than the speed.

Proper Lubrication

To keep wear to a minimum, bearings rely on a thin film of lubricant between balls or rollers, races, and retainers to prevent actual surface-to-surface contact. This lubricant also dissipates heat, prevents corrosion, restricts the entry of contaminants, and flushes out any particles that do result from wear. The quickest and surest way to determine the proper lubricant for a bearing is to check the manufacturer's manual that was supplied with the equipment.

Grease lubrication is suitable for most low to medium speeds but not normally recommended for the highest speeds. Oil is used at higher speeds. Oil, especially when set up in a system—which includes a pump, cooler, and a reservoir—can carry more heat away from a bearing than grease.

Remember that grease is a combination of oil and thickening agent, sometimes called soap. The viscosity and the amount of thickener that is added determine the thickness of the grease. Two greases with the same thickness rating are not necessarily the exact same combination of elements and may not blend together well. Usually it is preferable to choose one brand of grease and stick with it.

Proper Installation

It is absolutely critical to follow the bearing manufacturer's recommendations about properly installing a bearing. Everything must be clean and dry when you install bearings. Take care to ensure that the bearings are not cocked or misaligned in any way. Accidentally dropping a bearing or hitting it with a hammer where the force is transmitted to the rolling elements is a sure way to doom a bearing to an early death.

The seven mistakes to watch for when installing a bearing are these:

1. Allowing dirt or water to enter the bearing.
2. Overheating the bearing to shrink-fit it to the shaft (250° F is usually the maximum).
3. Hitting or pressing the bearing to cause force to be transmitted to the rolling elements.
4. Cocking the bearing on the shaft so that it is not square to the shaft shoulder.
5. Failing to match mark pillow block housings to make a correct reassembly.
6. Failing to use the spanner wrench to tighten the bearing nuts (hammer and punch not acceptable).
7. Forgetting to install oil to the bearing reservoir after a machine is rebuilt.

20. INSTALLATION OF ANTI-FRICTION BEARINGS

Individual Bearing

An anti-friction bearing requires an interference fit on one of the bearing races, usually the inner race. Take care when mounting a bearing onto a shaft; it is critical that the internal clearance (required for the bearing to spin) is not totally removed when the inner race is stretched to fit over the shaft. Dimensional checks are the key to proper mounting. The bearing manufacturer or machine builder provides checking measurements for the shaft. When replacing a bearing, be sure to check that proper fitting between bearing and shaft is obtained. Fig. 20-1 shows an example of the checks to be made on a shaft to ensure proper fitting without loss of internal clearance.

There are two common ways of mounting an individual bearing to a shaft: *arbor (or hydraulic) press mounting* and *thermal mounting.*

In press mounting, a force capable of moving the bearing onto the shaft is applied to the face of the ring having the interference fit—usually the face of the inner race. Always exert force on the ring to be mounted to prevent damaging the bearing. Fig. 20-2 shows the correct and incorrect methods of applying force.

Thermal mounting uses a technique of heating the bearing (causing expansion), cooling the shaft (causing it to shrink), or a combination of the two to mount the bearing onto the shaft. In most applications, the inner ring should be

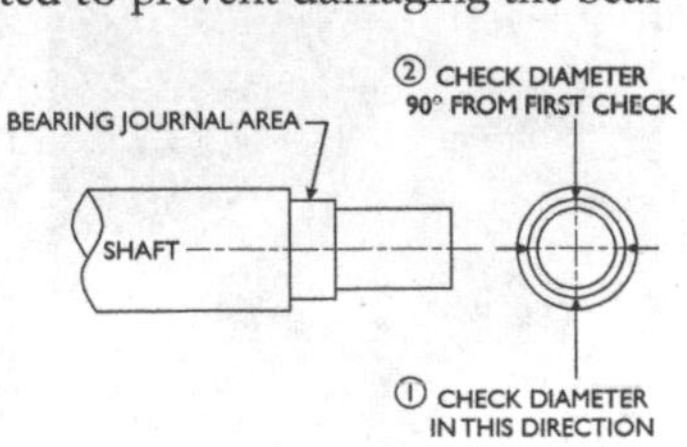

Fig. 20-1

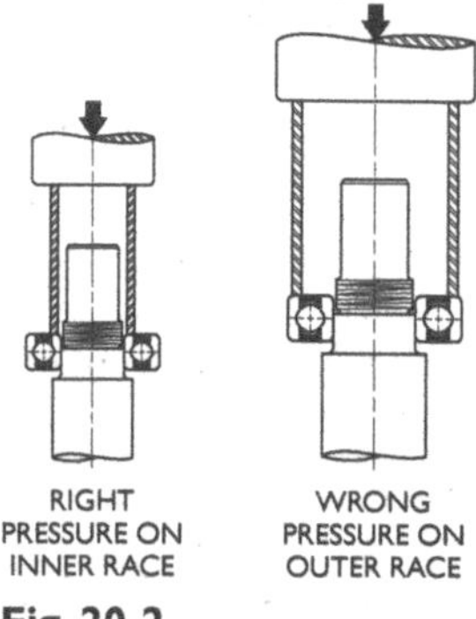

Fig. 20-2

uniformly heated to a temperature not to exceed 250°F. *Never* use a direct flame or hot plate. Heating a bearing to a temperature in excess of 250°F for extended periods of time will anneal the bearing metal and reduce the hardness—a good way to shorten the bearing life drastically.

A bearing can be heated using a cone heater (Fig. 20-3), electric oven, induction heater, hot oil bath, or even an electric light bulb centered in the bearing inner race. Using a cone heater has the advantage that the heater is quick, portable, and easy to operate. It is not uncommon to see the cone heater as the tool of choice in a maintenance shop.

Some cone heaters have a magnetic thermocouple that attaches to the bearing and senses the actual temperature—either alarming when a temperature of 250°F is reached or controlling to keep the temperature from exceeding this limit. A less expensive cone heater makes use of a temperature crayon (temp-stick) that is applied to the inner race. When the crayon just starts to melt and flow, the bearing is at the correct temperature.

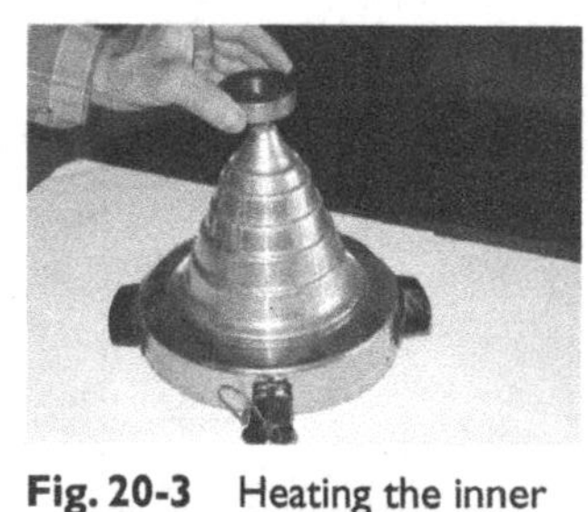

Fig. 20-3 Heating the inner race using a cone heater.
Courtesy of Maintenance Troubleshooting.

When the bearing reaches the proper temperature it should fit easily on the shaft. The example that follows shows the expected clearance by heating a 25-mm bore bearing to install it onto a shaft.

Example

A 50-mm bore bearing must be assembled onto a shaft using heating. The maximum allowable temperature for heating the bearing is 250°F. The measurements of the bearing show the bore to be 1.9680 in., and the shaft is checked with a micrometer and has a diameter of 1.9690 in. in the area of the bearing seat. Obviously, the assembly will give an interference fit. The temperature in the maintenance shop is 70°F.

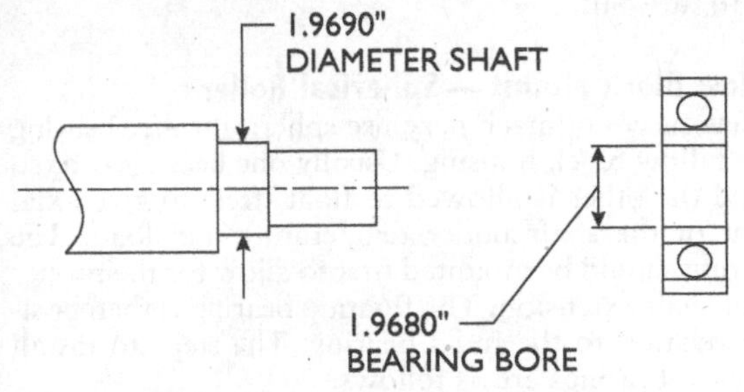

Fig. 20-4 Thermal fitting.

1. The thermal linear coefficient of expansion of steel = .0000063 in./in./°F.
2. The formula for expansion of the bore of a bearing due to heating is:

   Inner Ring Bore Expansion = (Diameter of bore) × (Change in temperature) × (Coefficient of expansion of steel)

   Inner Ring Bore Expansion = (1.9680 in.) × (250°F – 70°F) × (.000063 in./in./°F)

   Inner Ring Bore Expansion = (1.9680 in.) × (180°F) × (.0000063 in./in./°F) = 0.0022 in.

   Heated Inner Ring Bore Diameter = Inner Ring Bore Diameter Cold + Inner Ring Bore Expansion

Heated Inner Ring Bore Diameter = (1.9680 in.) + (0.0022 in.) = 1.9702 in.

Clearance = (Heated Inner Ring Bore Diameter) – (Shaft Diameter)

Clearance = (1.9702) – (1.9690) = 0.0012 in.

3. The inner ring will slide over the shaft with a 0.0012-in. clearance (an easy job).
4. After cooling the inner race will have an interference fit to the shaft.

Split Pillow Block Mount—Spherical Roller

Quite a few pieces of machinery use spherical roller bearing with split pillow block housing. Usually one bearing is fixed (held), and the other is allowed to float (free) to give axial movement of the shaft under temperature and load. The fixed bearing should be mounted first to allow for the proper amount of shaft extension. The floating bearing is then positioned in relation to the fixed bearing. The steps to install these types of bearings are as follows:

1. Slide the inner (inboard) seal ring onto the shaft. It should slide freely into position (Fig. 20-5).
2. Position the adapter sleeve onto the shaft, threads facing outboard, to the approximate location on the shaft to provide the proper center-to-center bearing distance (Fig. 20-6).

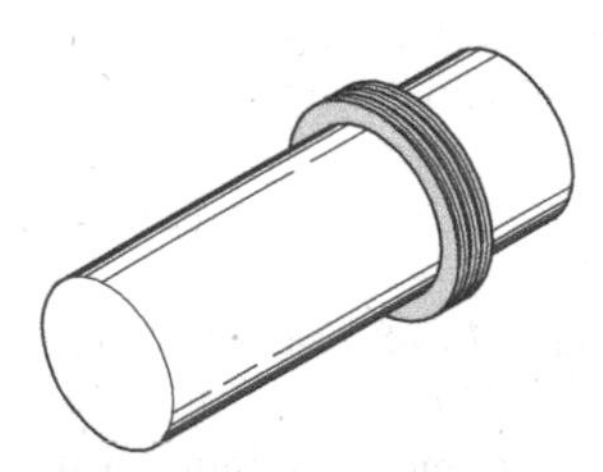

Fig. 20-5 Inner seal.

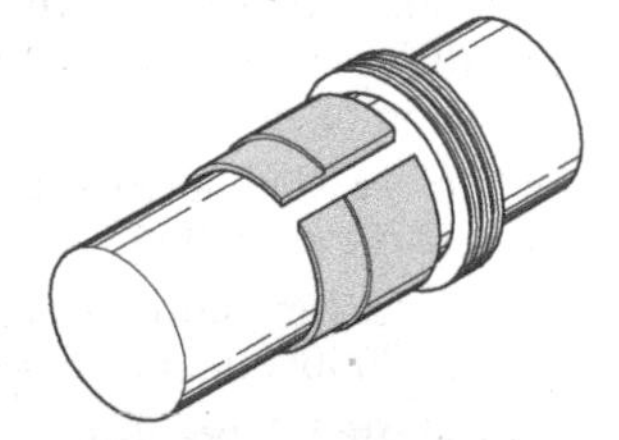

Fig. 20-6 Adapter sleeve.

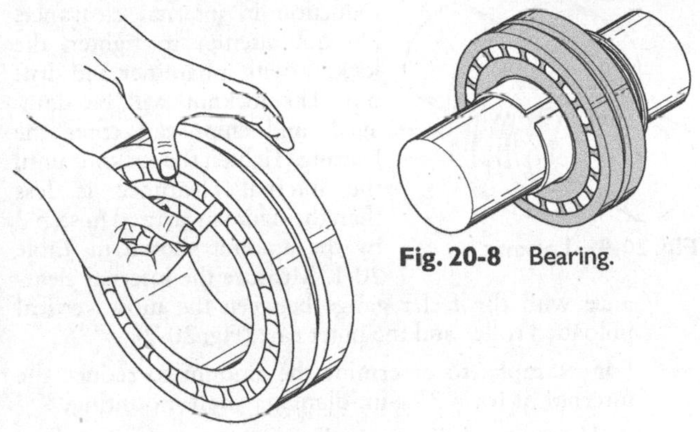

Fig. 20-8 Bearing.

Fig. 20-7 Unmounted clearance.

3. Measure the unmounted internal clearance in the bearing by inserting progressively larger feeler gauge blades the full length of the roller between the most vertical unloaded roller and the outer ring sphere. Don't roll the feeler through the clearance; slide it through. Record the measurement of the largest size feeler that will slide through. This is the unmounted internal clearance (Fig. 20-7).
4. Mount bearing on adapter sleeve, starting with the large bore of the inner ring to match the tape of the adapter. With the bearing hand tight on the adapter, locate the bearing to the proper position on the shaft. *Do not apply lockwasher at this time because the drive-up procedure many damage the locknut.* (Fig. 20-8)
5. Apply the locknut with the chamfered face toward the bearing. Use a light coating of oil on the face of the lock-nut where it contacts the inner-ring face of the bearing. (Use a spanner wrench and a hammer to tighten the locknut.) Large bearings will require a heavy-duty span-ner wrench and sledgehammer to obtain the required

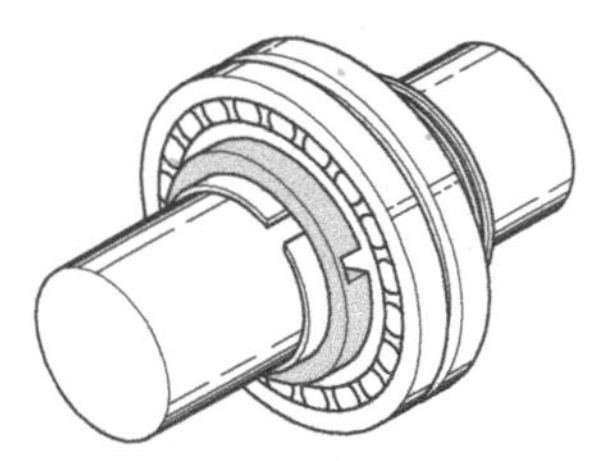

Fig. 20-9 Locknut.

reduction in internal clearance. Do not attempt to tighten the locknut with a hammer and drift pin. The locknut will be damaged, and chips can enter the bearing. Tighten the locknut until the internal clearance is less than the figure measured in step 3 by the amount shown in Table 20-1. Measure the internal clearance with the feeler gauge between the most vertical unloaded roller and the outer ring (Fig. 20-9).

For example, to determine the amount to reduce the internal fit for a 3⁷⁄₁₆-in.-diameter shaft mounting:

a. Unmounted internal clearance was measured as 0.004 in.

b. Reduction in internal clearance from Table 20-1 equals .0015 in.

c. Final mounted internal clearance: (.004 in.) – (.0015 in.) = .0025 in.

Table 20-1 Internal Reduction Clearance for Spherical Roller Bearings

Shaft Diameters, in.		
Over	*Including*	*Clearance Reduction, in.*
1¼	2¼	.001
2¼	3½	.0015
3½	4¼	.002
4¼	5	.0025
5	6½	.003
6½	7¼	.0035
7¼	8¼	.0045
8¼	9¼	.0045
9¼	11	.005
11	12½	.006
12½	14	.007

6. Remove locknut and mount lockwasher on adapter sleeve with the inner prong of the lockwasher toward the face of the bearing and located in the slot of the adapter sleeve. Replace and retighten the locknut with the spanner wrench. Hit the spanner wrench one time with the hammer to seat the locknut firmly. Check to make certain that clearance has not changed (Fig. 20-10).

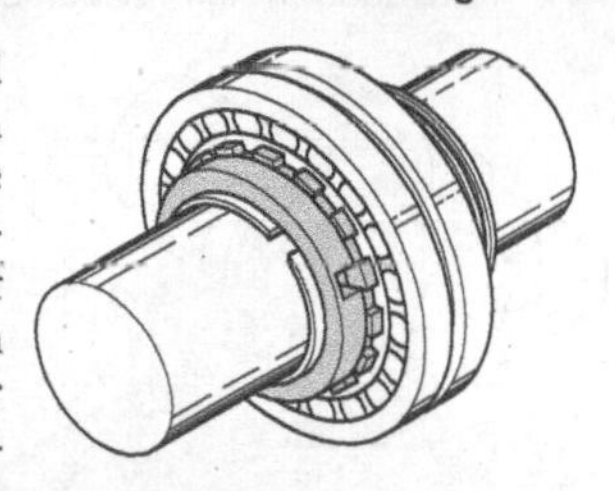

Fig. 20-10 Locknut and lockwasher.

7. Slide the outer seal onto the shaft. Locate both the inner and outer seals to match the labyrinths in the housing (Fig. 20-11).

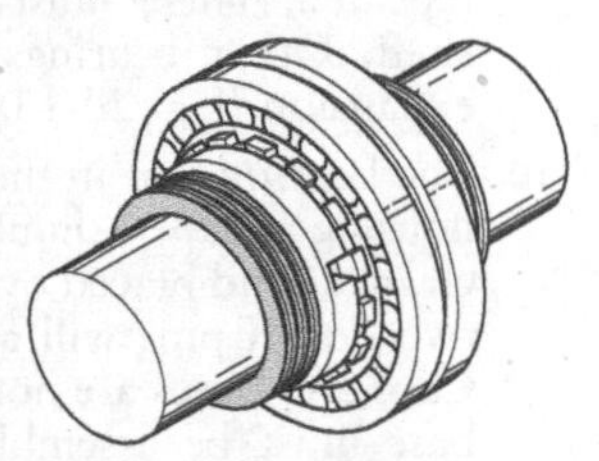

Fig. 20-11 Outer seal.

8. Remove any paint and burrs from the mating surface at the split, and thoroughly clean the pillow block housing. The vertical hole at the bottom of each enclosure groove must be free of foreign matter. Place the shaft with the bearings into the lower half of the housing while carefully guiding the seal rings on the shaft into the seal grooves. Bolt the fixed bearing into place (Fig. 20-12).
9. Move the shaft axially so that the stabilizing ring can be inserted between the fixed bearing outer race and the housing shoulder on the locknut side of the bearing. Center all other bearings on the shaft in their housing seats. If a bearing has to be repositioned after being tightened, it has to be loosened on the adapter sleeve before being moved. Steps 4 through 6 must be

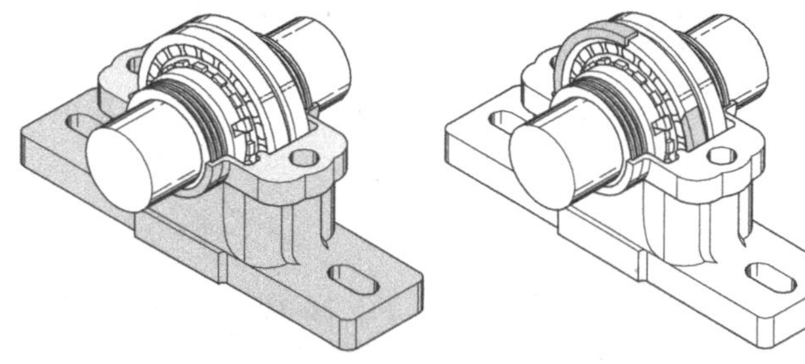

Fig. 20-12 Lower half of housing.

Fig. 20-13 Stabilizing ring (held or fixed bearing).

repeated. There must be only one fixed bearing per shaft. Other bearings must be free to permit shaft expansion (Fig. 20-13).

10. The bearing seat in the upper half of the pillow block housing (cap) should be deburred, thoroughly cleaned, and placed over the bearing (Fig. 20-14). The two dowel pins will align the upper half of the cap. Caps and bases are not interchangeable. Each cap and base must be assembled with its mating half. Then apply lockwashers and cap bolts to housings to complete the assembly (Fig. 20-15).

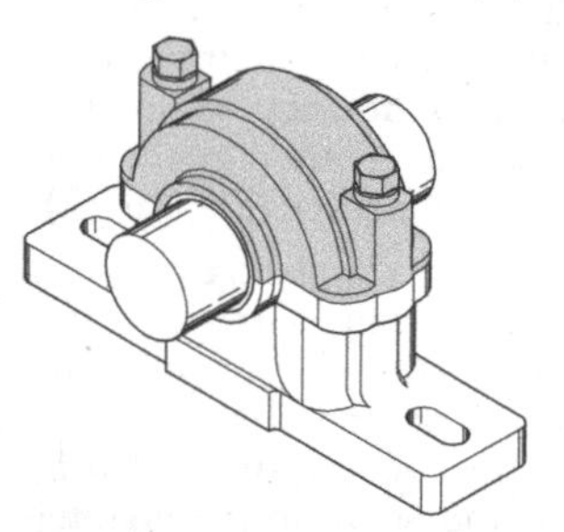

Fig. 20-14 Mounting upper half of pillow block.

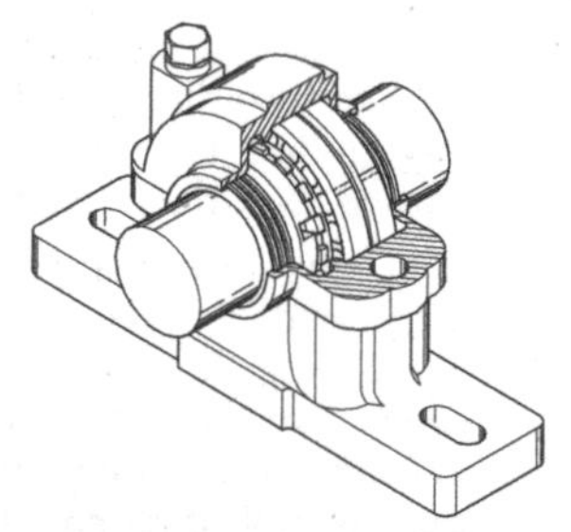

Fig. 20-15 Cutaway illustration.

11. Follow steps 1 through 10 for the floating bearing. In this case, do not install the stabilizer rings. Center the bearings in the housings to allow for expansion from heat or contraction from cold.

One-Piece Pillow Block—Roller Bearing

Roller bearings are either secured to the shaft using set screws (or screws) in the inner ring or by an eccentric collar.

Set Screw Installation

Secure these bearings to the shaft simply by tightening the set screws in the inner-ring extension. In some cases, the shaft can be filed slightly or shallow drilled to give more secure locking of the set screws (Fig. 20-16).

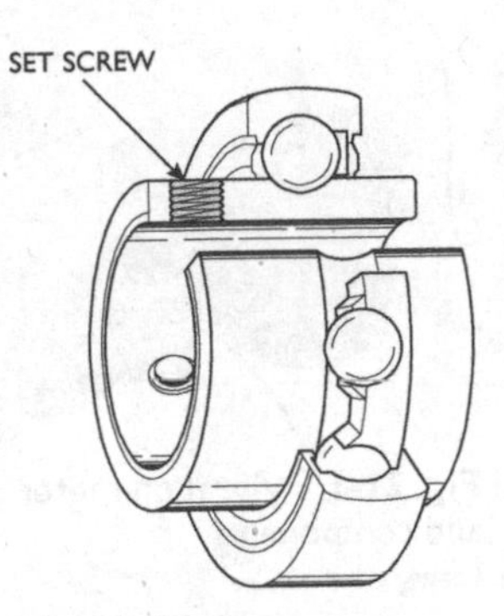

Fig. 20-16

Eccentric Locking Collar

The eccentric locking collar has a recess that is eccentric in relation to the bore. On the inner ring there is a corresponding external eccentric section (Fig. 20-17).

Place the eccentric locking collar in position on the inner ring, and turn it sharply in the same direction that the shaft will rotate until it locks. Tighten the set screw to keep the collar from loosening.

If the direction of rotation is not known, then lock the bearings at either end of the shaft in opposite directions.

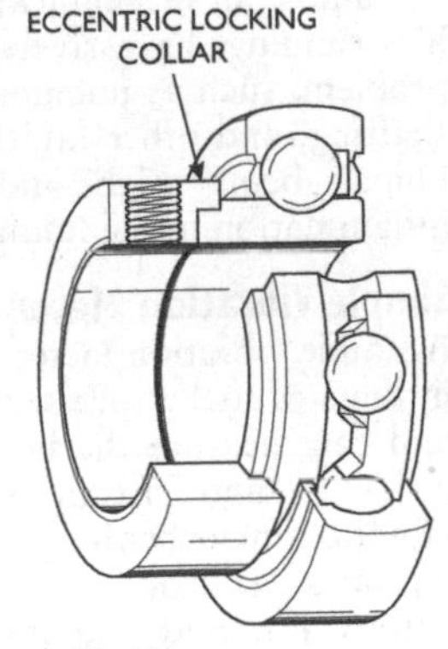

Fig. 20-17

21. VIBRATION MEASUREMENT

Vibration measurement is a simple but effective method of looking at the condition, or "health," of rotating machinery. Periodically checking machine conditions by trending is easy to do with vibration meters. Vibration meters help mechanics spot deteriorating machine conditions before they become critical (Fig. 21-1). By identifying and quantifying a vibration problem, you can take corrective action before the problem becomes significant. Trending with meters allows mechanics or maintenance personnel to plan repairs during normal work hours, rather than scheduling expensive overtime or shutting down production.

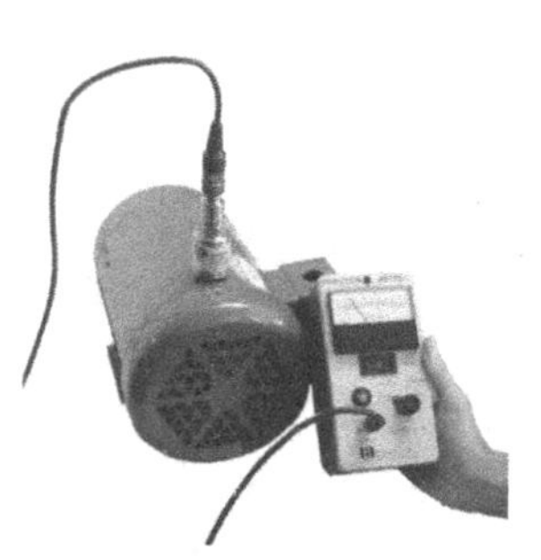

Fig. 21-1 Vibration meter and components.

Copyright Balmac, Inc.

Almost all mechanical rotating equipment vibrates when it is running. Excessively high vibration is a symptom of a problem, such as unbalance, looseness, misalignment, worn shafting, and other faults, just as a high temperature for a human being might indicate infection, organ failure, or inflammation in the human body.

Simple Vibration Meter

A simple vibration meter usually measures vibration velocity in units of inches per second (often shown as in./sec or ips) and can measure the bearing condition to determine flaws inside an anti-friction bearing. Fig. 21-2 shows a simple vibration meter with a magnetic pickup used to take a reading on an electric motor. Most simple vibration meters are battery powered and may use an analog or digital scale.

Many experienced maintenance persons prefer the analog scale using a needle because the swing or variation of the needle can be used to help determine the cause of the vibration.

Fig. 21-2 Simple vibration meter in use. *Copyright Balmac, Inc.*

Types of Readings

The two most useful readings are velocity and bearing Gs. The destructive forces generated in rotating machinery are proportional to vibration velocity. Checking the vibration velocity at key points on a machine is a very good indicator of machinery severity. The bearing G is a measurement of the pulses caused by impacts between bearing parts that have defects and flaws. In most cases, the vibration instrument is capable of obtaining both velocity and bearing Gs readings by flipping a selector switch, thereby changing the circuitry in the instrument to obtain the appropriate reading.

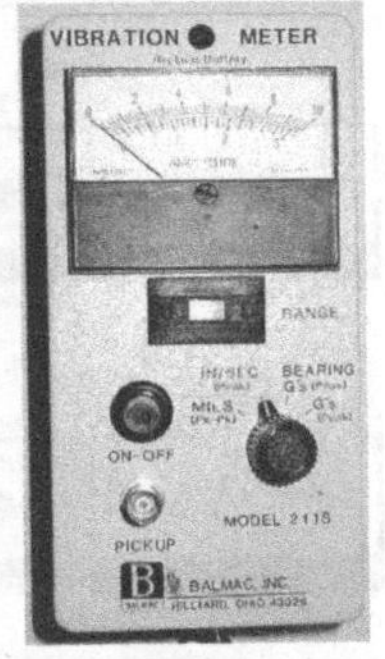

Fig. 21-3 Front panel of vibration instrument. *Copyright Balmac, Inc.*

Fig. 21-3 shows the front panel and face of a typical vibration instrument. Notice the selector switch, which allows readings of velocity and bearing Gs. Other vibration quantities, such as displacement and acceleration, can also be obtained but are less useful for simple machinery "health" determination.

Reading the Instrument

Most simple vibration meters have an amplitude meter with two scales on the meter face. The top scale is marked from 0

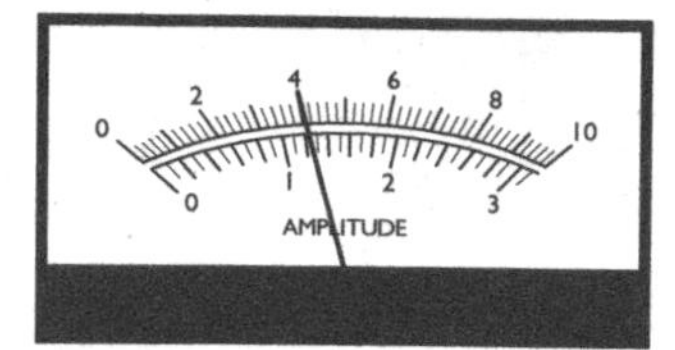

Fig. 21-4

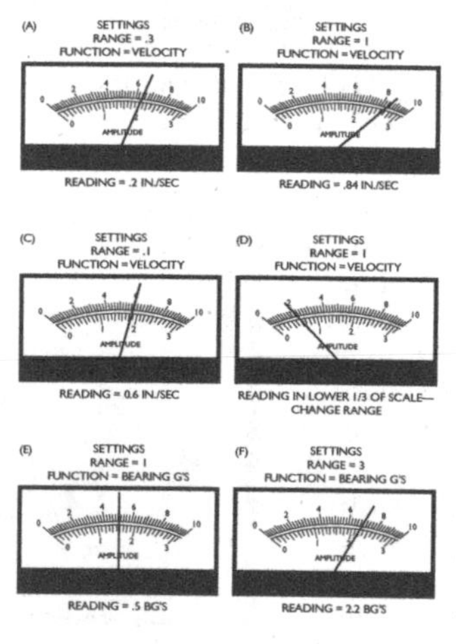

Fig. 21-5

to 10 and the bottom scale from 0 to 3, as shown in Fig. 21-4. When obtaining a reading using the instrument, the scale switch should be set to allow the needle to read in the upper two-thirds of the scale without pegging the meter (going off the scale). Because vibration can be any value on a particular machine, it is all right to allow the meter to peg until the correct scale setting can be obtained.

Setting the range of the instrument determines where to place the decimal point when taking a reading. If the RANGE switch is on any of the ranges that use a "1" (0.1, 1, 10, or 100), then the top portion of the amplitude meter scale should be read. If the RANGE switch is on any of the ranges that use a "3" (.03, 0.3, 3, or 30), then the bottom portion of the amplitude meter scale should be read.

The examples shown in Fig. 21-5 give visual explanations of various measurements and their correct readings.

Obtaining the Readings

Vibration forces may be measured in horizontal, vertical, and axial directions—as shown in Fig. 21-6. The vibration pickup may be hand-held using the extension probe or affixed using the magnetic base. The intent is to detect the direction of the strongest signals. Sometimes the direction of the highest vibration gives a clue as to the cause of the vibration. Looseness will show up as high readings in the vertical direction on most machinery, while misalignment might cause higher readings in the axial directions.

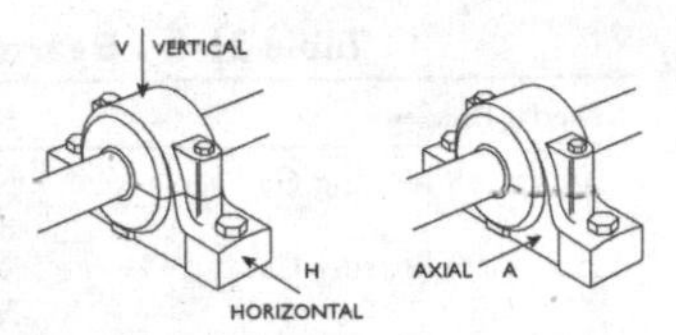

Fig. 21-6

Figure 21-7 provides suggested monitoring points for certain generic classes of machinery.

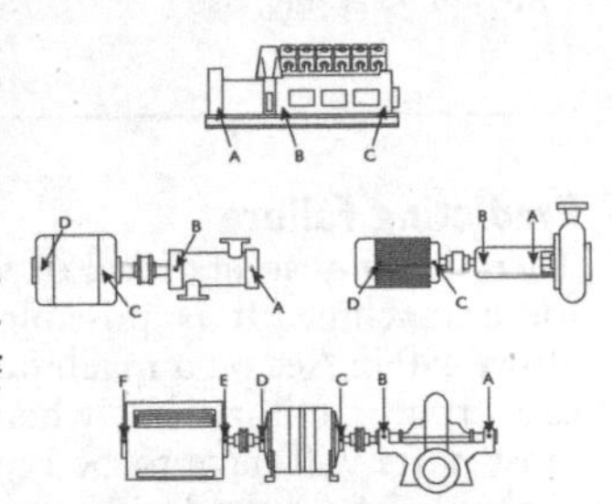

Fig. 21-7 Suggested monitoring points.

Determination of Conditions

Use Table 21-1 to judge the severity of vibration levels for general machinery.

The bearing Gs readings in Table 21-2 may be used to judge the condition of anti-friction bearings for general machinery.

Table 21-1 Machinery Severity Table

Vibration Reading	*Severity Level*	*Action Needed*
.15 in./sec or less	Normal vibration	Continue to monitor.
.15 in./sec to .29 in./sec	Rising vibration	Continue to monitor.
.30 in./sec to .59 in./sec	Alarm	Determine the cause of vibration and make correction.
.60 in./sec to .89 in./sec	Loss of oil film	Shut machine down if possible.
.90 in./sec or more	Hammering of parts	Machinery is undergoing damage.

Table 21-2 Bearing Condition Table

Bearing Reading	*Bearing Condition*
.00 to .49 bearing Gs	No cause for alarm. Bearing is good.
.50 to .99 bearing Gs	Spalling apparent in bearing. Minor damage.
1.0 to 1.49 bearing Gs	Minor to major damage to rolling elements or raceways.
Above 1.50 bearing Gs	Seriously consider bearing replacement before bearing failure occurs.

Predicting Failure

There is no absolute level of vibration that indicates failure for a machine. It is possible to measure vibration levels above .90 in./sec on a machine for weeks and still not have a catastrophic failure, but when the machine is disassembled most parts will have to be replaced and machine work will probably be required to repair worn journals or housing fits. In short, the machine will be taken out of service while it is still running, but the components will be in need of major replacement or repair. Finding the source of the vibration and correcting it when it just starts to be considered excessive will reduce the level of vibration and save a piece of equipment from costly repair.

Determining the cause of vibration with a simple hand-held meter is not always easy. More complex instruments, called vibration analyzers, are often brought to the machine site to assist in determining the specific cause. Vibration analysis is beyond the scope of this book and usually requires advanced training by the instrument manufacturer. Often an outside service may be called on to determine and help correct the problem after the local mechanic has found that a problem exists using the simple hand-held vibration meter.

22. LUBRICATION

Most lubricants used in industry are mineral based and obtained from petroleum by refining processes and further purification and blending.

Functions of a Lubricant

Lubricants have three major functions: limit friction, minimize wear, and dissipate heat.

Limit Friction

Friction is defined as the resistance to motion of contacting surfaces. Even smooth metal surfaces have microscopic rough spots called *asperities*. Friction is increased by the presence of asperities on surfaces. Attempts to overcome the force of friction will increase the localized heat generated between contacting surfaces. This heat can actually create temperatures high enough to weld two surfaces together.

Lubrication prevents peaks of asperities from touching each other through what is called film strength. Molecules of lubricants are naturally bonded together, often in chains. Any attempt to break the chain creates an opposite tension that prevents separation.

Minimize Wear

Wear is the removal of material from one or more moving surfaces in contact with each other. The material removed becomes the source of additional friction and increased wear on the surfaces involved.

A quality lubricant will fill the valleys of the asperities and provide an additional film over the peaks of the asperities. The asperities of the two surfaces are prevented from contacting each other, and wear will be minimized.

Dissipate Heat

Even well-lubricated parts will heat up due to friction and external heat. One advantage of liquid lubricants is their ability to absorb and dissipate point sources of heat.

Types of Industrial Lubricants

Lubricants can be divided into two types: *solid* and *liquid*.

Solid Lubricants

Solid lubricants are materials such as graphite, molybdenum disulfide, and PTFE (polytetrafluoroethylene); they are used in smaller equipment or on surfaces where just a minor amount of movement is expected. Lead, babbitt, silver, gold, and some metallic oxides are solid lubricants that can provide for more movement or pressure between surfaces. Some machine designers use ceramics or inter-metallic alloys to coat the surfaces of moving parts—they can also be considered a lubricant.

Liquid Lubricants

Liquid lubricants, in industry, fall into two categories: greases and oils.

Measuring the Properties of Greases and Oils

Criteria for measuring the properties of grease are these:

> **Hardness.** Greases range from hard to soft. Table 22-1 shows this range. The NLGI is the National Lubricating Grease Institute.

Table 22-1 Hardness of Grease

NGLI Number	*Consistency*	*ASTM Worked Penetration at 25°C (77°F) 10^{-1}mm*
000	Very fluid	445–475
00	Fluid	400–430
0	Semi-fluid	355–385
1	Very soft	310–340
2	Soft	265–295
3	Semi-firm	220–250
4	Firm	175–205
5	Very firm	130–160
6	Hard	85–115

Greases with an NLGI of #000 are like a liquid, while #6 greases are almost solid. The most frequently used greases are #0, #1, and #2.

Dropping point. The temperature at which the grease will change from semi-solid to liquid—basically the melting point.

Water resistance. This determines whether a grease will dissolve in water. This is a very important quality if there is a chance that water will come in contact with the lubricant.

Stability. This property determines the ability of a grease to retain its characteristics with time.

Criteria for measuring the properties of oil are these:

Viscosity. This is the most important characteristic; it refers to the thickness of the fluid and can also be described as the resistance to flow. Viscosity is affected by temperature and decreases as temperature increases. There are many ways of measuring viscosity, but they are all based on the time taken for a fixed volume of oil to pass through a standard orifice under laboratory conditions.

There are three commonly used terms for viscocity: Saybolt Universal Seconds (SUS), centipoise (cP), and centistokes (cSt).

Saybolt Universal Seconds is an indication of the time it takes 60 mm of fluids to flow through a calibrated Saybolt Universal tube (also called a viscosimeter). This is an old method used to describe viscosity, but a lot of American companies still use it.

A *centipoise* is 0.01 poise. A poise is an absolute viscosity unit in the metric system. A *centistoke* is 0.01 stokes. A one-stoke fluid has an absolute viscosity of 1 poise and a density of 1 gram per cubic centimeter. In other words, centistokes differ from centipoise by a density factor.

Some industrial organizations classify oils by the type and viscosity rating:

- International Standards Organization—Viscosity Grade (ISO VG)
- Society of Automotive Engineers (SAE)—Viscosity Number
- SAE—Gear Viscosity Number
- American Gear Manufacturers Association (AGMA)—Lubricant Number

Viscosity Index. The rate of change of viscosity with temperature. A high viscosity index shows that the oil will remain the same over a wide range of temperatures, whereas a low index indicates that the oil will thin out rapidly with an increase in temperature.

Flash Point. The temperature at which the vapor of a lubricant will ignite.

Fire Point. The temperature (higher than the flash point) that is required to form enough vapor from the lubricant to cause it to burn steadily.

Pour Point. The low temperature at which the lubricant becomes so thick it will not flow.

Oxidation Resistance. If oil is exposed to the atmosphere, especially at a higher temperature, oxygen is absorbed into the oil. A chemical change takes place in the oil that drastically reduces its lubricating properties.

Emulsification. The measure of the tendency for oil and water to mix together.

Synthetic oils are being used more frequently today. Current uses include situations of high pressure or vacuum, very high or low temperature, nuclear radiation, and chemical contamination. They have a better film strength than mineral oils and are much better at inhibiting oxidation.

Regreasing Interval for Anti-Friction Bearings

Fig. 22-1 shows lubrication frequency for rolling element bearings.

Example

A 320 series bearing is turning 1800 RPM at 90°C. This bearing is a radial ball bearing with a bore of 100 mm.

1. Find 1800 RPM at the bottom of the chart.
2. Move up the chart to the curve for 100 mm.
3. Move to the left from this point, and read the hours for radial ball bearing = 4000 hours.

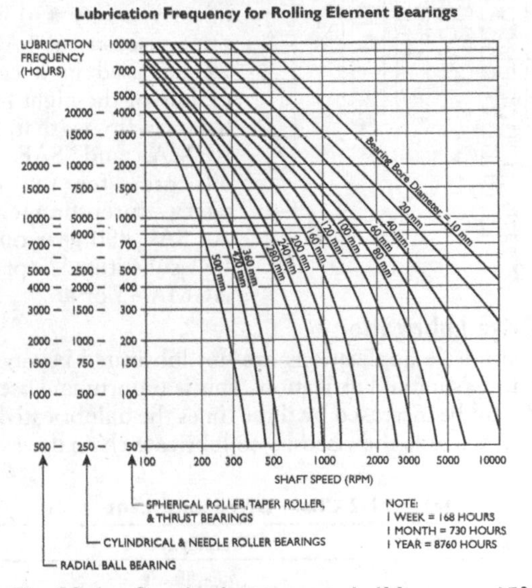

Fig. 22-1 Cut the frequency in half for every 15°C over 70°C. The frequency can be doubled if the temperature is consistently below 55°C, but this is the maximum interval increase recommended.

4. Cut this time in half for the higher temperature.
5. Convert the hours to months or weeks of operation: 2000 hours = 12 weeks or 3 months.

Oil Viscosity Classification Chart

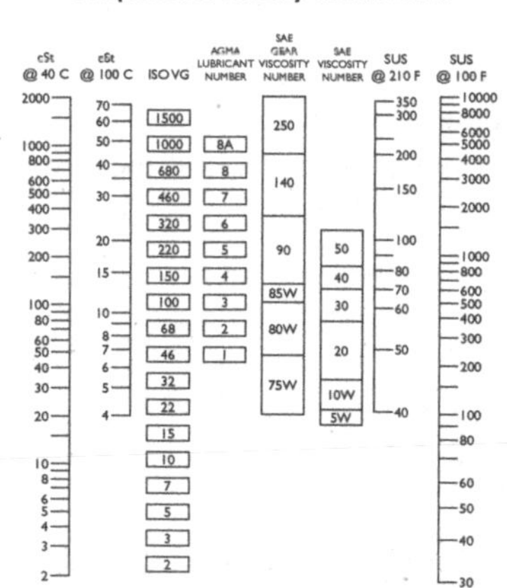

Fig. 22-2

Fig. 22-2 compares oils rated by different standards. Place a straight edge horizontally across the chart to compare one type of reading with another.

For instance, SAE gear oil of viscosity #90 is the same viscosity as SAE 50 weight used in a car. A middle-of-the-night maintenance tip is that ISO, AGMA, and SAE gear oils are—for the most part—interchangeable. An SAE #90 gear oil can be substituted for an AGMA #5 or #6.

Chain Drive Lubrication

Most chain drive applications are not lubricated beyond the manufacturer's initial lubrication. This is unfortunate because chain life can be increased by three times the unlubricated life. Table 22-2 suggests the viscosity to lubricate chain drives.

Table 22-2 Chain Drive Lubricant

Ambient Temperature (in °C)	*Viscosity at 40°C*
–5 – 5	30 – 80 cSt
5 – 40	80 – 120 cSt
40 – 50	120 – 180 cSt
50 – 60	180 – 300 cSt

The most common oil used is SAE 30 because it meets the temperature requirements most often encountered in industry.

Regreasing Schedule for AC Electric Motors

While it is a good idea to check with the manufacturer for regreasing intervals, Table 22-3 can be used for ball bearing motors if no better information exists.

Table 22-3 AC Motor Regreasing Schedule (Ball Bearing Only)

Type of Service	*Typical Examples*	*½ to 7½ Horsepower*	*10 to 40 Horsepower*	*50 to 200 Horsepower*
Easy	Motor operating infrequently—1 hour per day	10 years	7 years	5 years
Standard	Machine tools, fans, pumps, textile machinery	7 years	5 years	3 years
Severe	Motors for continuous operation in key or critical locations	4 years	2 years	1 year
Very Severe	Dirty and vibrating applications where end of shaft is hot from high ambient temperature	9 months	4 months	4 months

Regreasing intervals for motors are usually longer than most people assume they will be. Overgreasing can present a problem, resulting in grease entering the motor. It is particularly important to follow the manufacturer's recommendation for motors.

23. MECHANICAL POWER TRANSMISSION

The three principal systems used for the transmission of rotary mechanical power between adjacent shafts are belts, chains, and gears. Understanding common terms and concepts is a prerequisite to understanding the systems.

Pitch Diameter and Pitch Circle

Pitch is the distance from a point to a corresponding point (Fig. 23-1).

Fig. 23-1

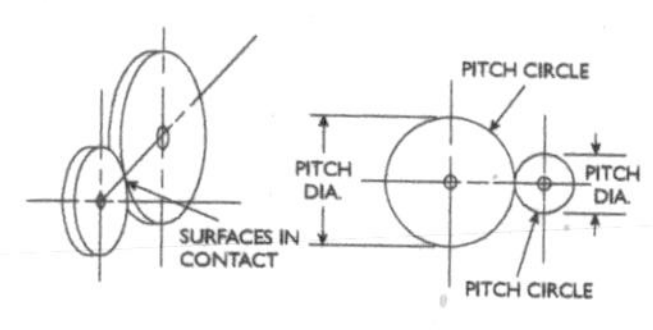

Fig. 23-2

The *pitch diameter* specifies the distance across the center of the *pitch circle*. Pitch diameter dimensions are specific values, even though the *pitch circle* is imaginary. Rotary power transmission calculations are based on the concept of circles or cylinders in contact. These circles are called *pitch circles* (Fig. 23-2).

As shafts rotate, the surfaces of the pitch circles travel equal distances at equal speeds (assuming no slippage). Shafts then rotate at speeds proportional to the circumference of the pitch circles and therefore proportional to the pitch diameters.

In belt and chain drives, the pitch circles are actually separated. This is true because the belt or chain is in effect an extension of the pitch circle surface.

- As rotation occurs the pitch circle surfaces will travel the same distance at the same surface speed.
- Because the circumference of a 4-in. pitch circle is double that of a 2-in. pitch circle, its rotation will be one-half as much.

- The rotation speed of the 4-in. pitch circle will be one-half that of the 2-in. pitch circle (Fig. 23-3).

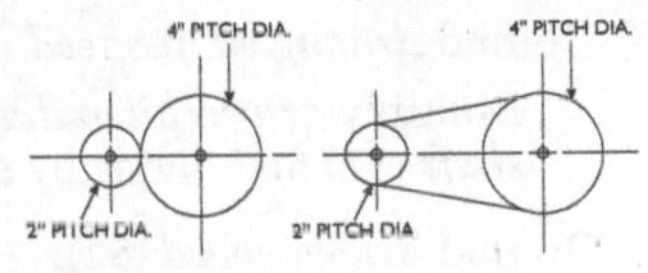

Fig. 23-3

Speed and Ratio Calculations

Rotational speed and pitch diameter calculations for belts, chains, and gears are based on the concept of pitch circles in contact. The relationship that results from this concept may be stated as follows: *Shaft speeds are inversely proportional to pitch diameters.*

In terms of rotational speed and pitch diameters, this relationship may be expressed in equation form as follows:

$$\frac{\text{DRIVER ROTATIONAL SPEED}}{\text{DRIVEN ROTATIONAL SPEED}} = \frac{\text{DRIVEN PITCH DIAMETER}}{\text{DRIVER PITCH DIAMETER}}$$

Where:

S1 for Driver Rotational Speed

S2 for Driven Rotational Speed

P1 for Driver Pitch Diameter

P2 for Driven Pitch Diameter

The basic equation then becomes:

$$\frac{S1}{S2} = \frac{P2}{P1}$$

The equation may be arranged into the following forms, one for each value of the four values. To find an unknown value, substitute the known values in the appropriate equation:

$$S1 = \frac{P2 \times S2}{P1} \quad S2 = \frac{P1 \times S1}{P2} \quad P1 = \frac{S2 \times P2}{S1} \quad P2 = \frac{S1 \times P1}{S2}$$

The preceding equations may also be stated as rules. Following these rules is another convenient way to calculate unknown shaft speeds and pitch diameters.

To find *driving shaft speed* (S1):

Multiply *driven pitch diameter* (P2) by *speed of driven shaft* (S2) and divide by *driver pitch diameter* (P1).

To find *driven speed* (S2):

Multiply *driver pitch diameter* (P1) by *speed of driver shaft* (S1) and divide by *driven pitch diameter* (P2).

To find *driver pitch diameter* (P1):

Multiply *speed of driven shaft* (S2) by *driven pitch diameter* (P2) and divide by *speed of driver shaft* (S1).

To find *driven pitch diameter* (P1):

Multiply *speed of driver shaft* (S1) by *driver pitch diameter* (P2) and divide by *speed of driven shaft* (S2).

In gear and sprocket calculations, the number of teeth is used rather than the pitch diameters. The equations and rules hold true because the number of teeth in a gear or sprocket is directly proportional to its pitch diameter.

Examples

To Find Driver Shaft Speed (S1)
Known values:

P1 Driver Pitch Diameter = 4 in.

P2 Driven Pitch Diameter = 6 in.

S2 Driven Shaft Speed = 750 RPM

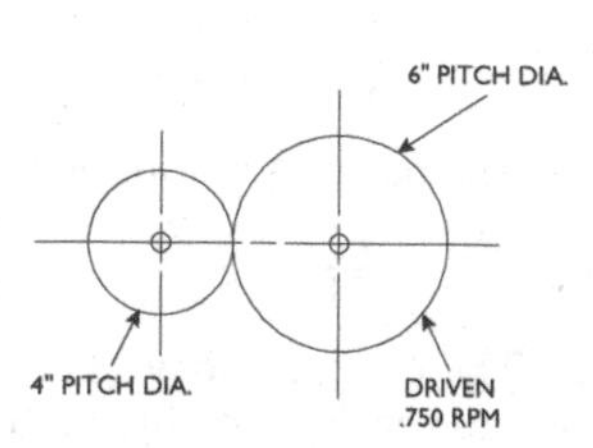

Fig. 23-4

$$S1 = \frac{P2 \times S2}{P1} \text{ or } \frac{6 \times 750}{4} \text{ or } \frac{4500}{4} \text{ or } 1125 \text{ RPM}$$

To Find Driven Shaft Speed (S2)
Known values:

P1 Driver Pitch Diameter = 2 in.

P2 Driven Pitch Diameter = 4 in.

S1 Driver Shaft Speed = 100 RPM

$$S2 = \frac{P1 \times S1}{P2} \text{ or } \frac{2 \times 100}{4} \text{ or } \frac{200}{4} \text{ or } 50 \text{ RPM}$$

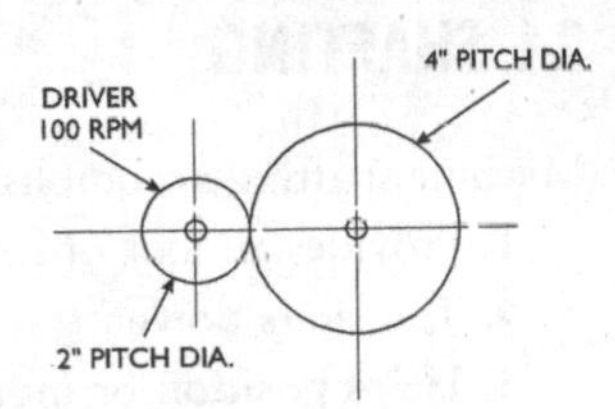

Fig. 23-5

To Find Driver Pitch Diameter (P1)

Known values:

P2 Driven Pitch Diameter = 3 in.

S1 Driver Shaft Speed = 600 RPM

S2 Driven Shaft Speed = 2000 RPM

$$P1 = \frac{S2 \times P2}{S1} \text{ or } \frac{2000 \times 3}{600} \text{ or } \frac{6000}{60} \text{ or } 10\text{" Pitch Circle}$$

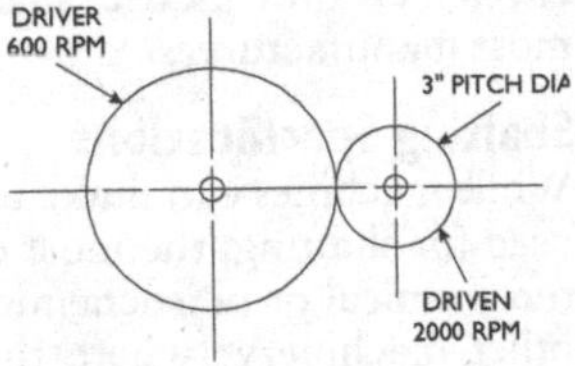

DRIVER
600 RPM
3" PITCH DIA
DRIVEN
2000 RPM

Fig. 23-6

To Find Driven Pitch Diameter (P2)

Known values:

P1 Driver Pitch Diameter = 8 in.

S1 Driver Shaft Speed = 400 RPM

S2 Driven Shaft Speed = 1280 RPM

$$P2 = \frac{S1 \times P1}{S2} \text{ or } \frac{400 \times 8}{1280} \text{ or } \frac{3200}{1280} \text{ or } 2..5\text{" Pitch Circle}$$

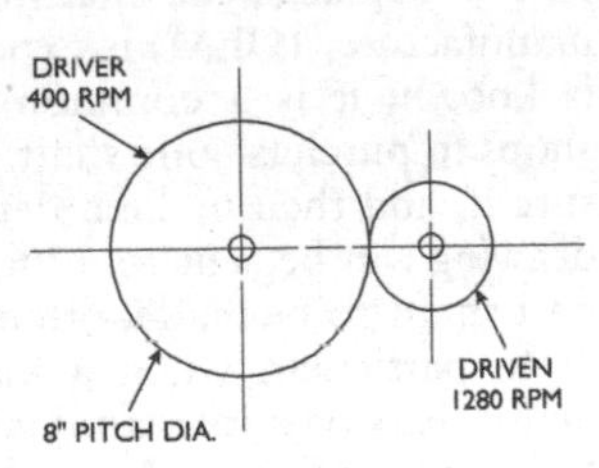

Fig. 23-7

24. SHAFTING

Machine shafting accomplishes three purposes:

1. Provides an axis of rotation.
2. Transmits power.
3. Helps position or mount gears, pulleys, bearings, and other working elements.

Shafting is rated on its ability to accommodate bending and torsional loads. Shaft loads result from external forces caused by misalignment, belt drives, chain drives, flywheels, brakes, or other accessories. Depending on customer preference, inch and metric shaft extensions are available from most manufacturers.

Shafting Specifications

While machines can make use of the many different materials used for shafting, the most common shaft is made of steel. In the chemical or petrochemical industries, shafts for pumps or other machinery (where the wetted surfaces are in contact with the rotating element), stainless steel and other exotic steel alloys for shafting materials might be appropriate.

Purchasing fan shafts, pump shafts, and other replacement parts directly from the manufacturer guarantees that the shaft is exactly what is required for the application. The cost of replacement shafting from the original equipment manufacturer (OEM) is expensive. Where the shaft material is known, it is a common practice in many maintenance shops to purchase one shaft from the OEM, accurately measure it, and then make a detailed mechanical drawing. This drawing can be sent to a local machine shop to allow spare part shafts to be made, often at a greatly reduced cost.

In particular, when a machine has been in service for many years past its manufacturer's warranty period, there is little or no reason why replacement shafts need to be purchase from the OEM.

Emergency Shaft Repair

In an emergency, you might look to used parts to repair a piece of production equipment. If a bearing, seal, or key unexpectedly fails, there may not be adequate time to locate a spare shaft or even to machine one. With today's cost-efficiency concerns, a maintenance department cannot purchase and store a shaft for each piece of equipment in its area of responsibility.

If the used shaft is worn in the area of the bearing seat, where the bearing inside diameter would spin on the seat if the old shaft were used, some emergency techniques may save the day. The shaft can be chucked up in a lathe, and the area of the bearing seat can be knurled. Knurling throws up metal and expands the diameter or the shaft to make up for the undersize area. The knurl is then turned to obtain the correct diameter required for press fit (or shrink fit) of the bearing. Keep in mind that bearing manufacturers do not recommend this procedure at all. For the most part, the knurl will not support the inner ring of a bearing properly for a long time, so the manufacturer is correct. But, knurling might be good enough to allow the machine to operate for a few weeks until a spare shaft is obtained (Fig. 24-1).

Another emergency shaft repair method is spray metallizing. This is a perfectly acceptable repair on the sealing area and other non-load-carrying portions of a shaft, but it is not recommended for use in the area of the bearing journal. Again, the bearing manufacturer advises against this technique as a permanent fix, but spraying can keep a machine in operation while you locate a new spare or build a proper replacement shaft at the machine shop.

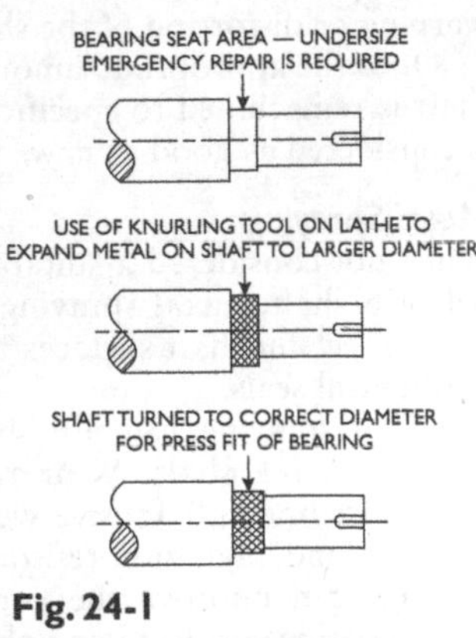

Fig. 24-1

It is important to restate that both of these procedures are *not* recommended by machinery or bearing manufacturers as permanent solutions, but plenty of mechanics have saved the day using these bootstrap methods and following up by replacing the temporary fix with a proper repair as soon as possible.

Permanent Shaft Repair

Obviously, the most permanent shaft repair would be to obtain a replacement part from the OEM or machine a duplicate at a local shop. Often a shaft can be salvaged with a perfectly good quality repair that "makes the shaft as good as new."

Submerged Arc Welding

Shaft repair at the site of a bearing journal can be done with *submerged arc welding*—the highest-quality repair method that eliminates the possible flaking and peeling problems of other, lesser-quality repair techniques.

This special process involves welding new material on worn shaft areas under tightly controlled conditions. As the shaft slowly turns in a lathe, heat generated during welding is evenly distributed over the shaft surface. This heat distribution provides the necessary "stress relief" to prevent warping or distortion of the shaft itself.

Once the appropriate amount of buildup is achieved, the shaft is remachined to specifications, and the repaired shaft is considered as good as new.

Metal Spraying

While not considered a suitable repair for the bearing journal of a shaft, metal spraying is a quick and cost-effective way to reclaim shaft surfaces in the area for seal journals or mechanical seals.

Metal spraying can also be used to improve the wear resistance of a shaft. Wear resistance coating can protect against fretting and abusive wear, and it is impact abrasion-, erosion-, and cavitation-resistant (Fig. 24-2). Other sprayed coatings can improve the corrosion resistance of a shaft. These coatings protect less noble materials—predominantly low-alloy steels—from chemical attack. This type of coating also protects against oxidation, sulfidation, and galvanic corrosion. The chemical and petrochemical processing industries often use this type of thermal spray coating.

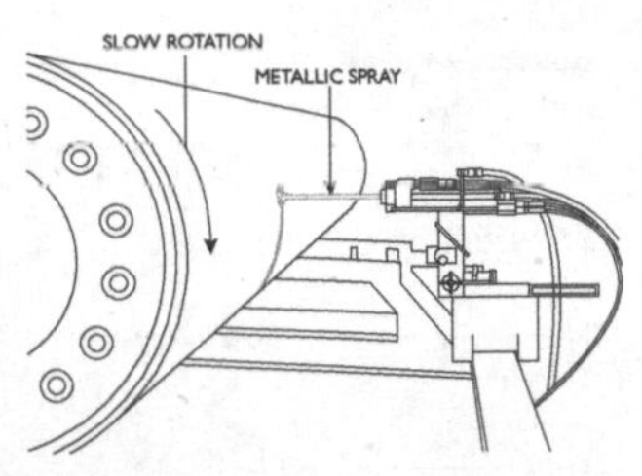

Fig. 24-2

TIG Welding

Another permanent shaft repair technique for the seal and key slot areas is tungsten inert gas (TIG) welding. TIG welding, although slower than MIG or stick welding, offers better control of the heat-affected zone normally responsible for shaft breakage next to a weld.

Hard Chrome Plating

Hard chrome plating is used for hard surfacing specific areas of a shaft to produce a wear-resistant surface. *Hard* chrome is a misnomer. The electroplating industry refers to any deposition of chrome of thickness of 0.005 in. or more as hard chrome; deposits of less than that are called "decorative" plating. The area where wear resistance is required is turned down, chrome plated to an oversize OD, and then ground down to produce the correct dimension and a smooth, almost mirror-like surface for wear resistance. Shaft areas that are prone to packing or seal rubs are good candidates for hard chroming.

Use of a Shaft Sleeve

A shrink-fit metal sleeve is an acceptable shaft repair for a bearing seat (Fig. 24-3). Here, the shaft is turned down in the area where the damage occurred. A suitable sleeve is machined where the ID of the sleeve is bored to be shrunk-fit to the shaft and the OD of the sleeve is made to be larger than the original shaft dimension. The sleeve is heated and shrunk onto the shaft. When the sleeve has cooled, the shaft is again chucked up in the lathe, and the OD of the sleeve is

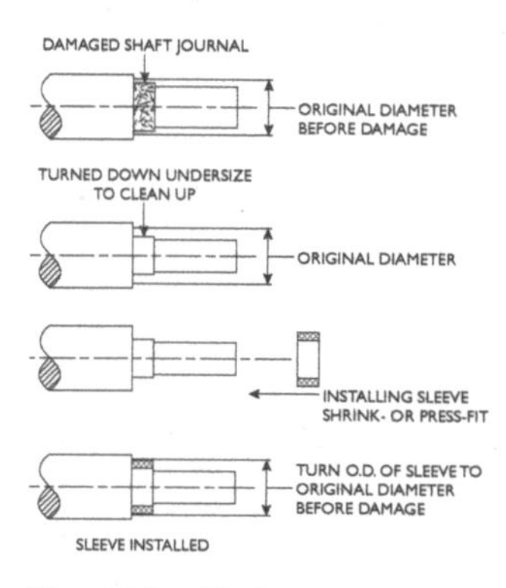

Fig. 24-3 Shaft repair by sleeving.

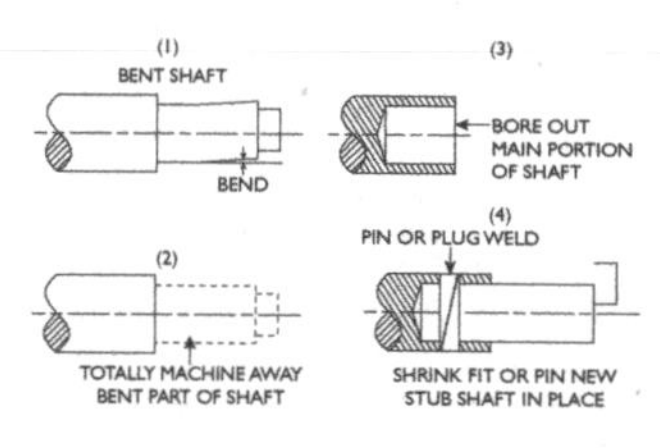

Fig. 24-4

turned down to the proper final dimension.

Use of a Stub Shaft

In many machine failures, the very end of the shaft — where the coupling, sheave, or sprocket used for transmission of power — ends up being bent. The rest of the shaft is in perfect condition and sustains no damage. Machining off the end of the shaft and constructing a stub shaft in the original dimensions is a perfectly good way to save a shaft. Fig. 24-4 shows the steps involved in using the stub shaft approach to repair a shaft that has been bent or broken at the drive end.

Straightening a Bent Shaft

Another common repair problem occurs with long shafting used on deep well pumps or agitation systems. The failure of a bearing or other component causes the shaft to bend when the machine locks up under load. This is different from the sharp bend that might occur at the drive end of a shaft, as mentioned previously. In this case, the shaft may be 4 to 8 ft or more in length and has been bowed. Straightening a shaft is accomplished using a hydraulic press and a dial indicator. The following procedure gives the steps required to straighten and reuse a bowed shaft:

1. Spin the shaft on rollers, which are located near each end of the shaft and, using a dial indicator, check the

total indicator reading (TIR) at several points along the length of the shaft. Use a marker to indicate the high spots along the shaft and determine the area where the bow or bending is at maximum (Fig. 24-5).

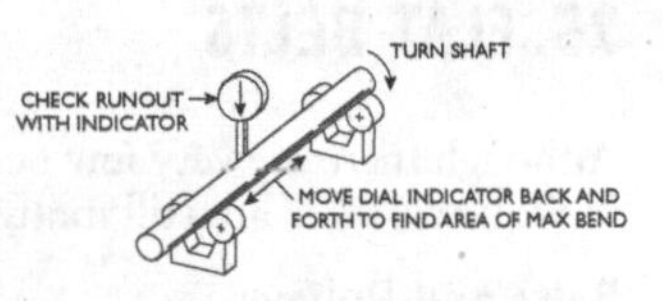

Fig. 24-5

2. Lift the shaft off the rollers and move it onto "V" blocks that are positioned on both sides of a hydraulic press. Apply force at the point on the shaft where the maximum bowing takes place, and use the press to straighten the shaft. Care must be taken to make sure not to "overbend" the shaft.

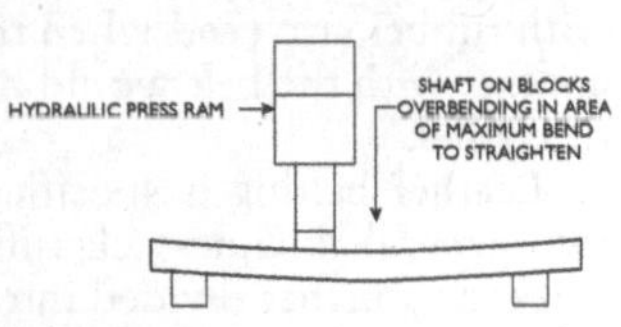

Fig. 24-6

   It may take a few times of bending and checking to develop a feel for the procedure (Fig. 24-6).
3. Remove the shaft, and recheck it with a dial indicator on the rollers. A rule of thumb is that the maximum allowable TIR is about 0.0005 in. per foot of shaft length, but not more than 0.001 in. within and 1 ft of length. For instance, a 6-ft shaft should not have more than $6 \times 0.0005 = 0.003$ in. of runout.
4. If the TIR is still too large, additional marking must be done on the shaft, depicting the results of the new dial indicator readings, and the shaft must be transferred to the press for additional straightening.

Shaft straightening is a straightforward process. Careful work and measurement, combined with slow and steady control of the pressure to the ram on the hydraulic shop press, will result in a shaft that is as good as new.

25. FLAT BELTS

Although there are very few new installations of flat-belt drives in industry, there are still many installations in operation.

Belts and Pulleys

Flat belting may be made from leather, rubber, or canvas. Leather belts are by far the most commonly used. Rubber belting is usually used where there is exposure to weather conditions or moisture, since it does not stretch under these conditions. Canvas or similar fabrics, usually impregnated with rubber, are used when the materials (such as liquids) in contact with the belt would have an adverse effect on leather or rubber.

Leather belting is specified by thickness and width. The two general thickness classifications are single and double. These are further divided into light, medium, and heavy. The thickness specifications for first-quality leather belting are shown in Table 25-1.

Table 25-1 Leather Belting Thickness

Medium	Single	5/32–3/16 in.
Heavy	Single	3/16–7/32 in.
Light	Double	15/64–17/64 in.
Medium	Double	9/32–5/16 in.
Heavy	Double	21/64–23/64 in.

Take precautions to put the flesh side on the outside and the grain, or smooth, side toward the pulley when installing leather belts. Experience shows that when the belt is installed in this manner, it will wear much longer and deliver more power than if it is put on in the reverse way.

Pulleys are normally made from cast iron, fabricated steel, paper, fiber, or wood. Pulleys can be solid or split, and in either case the hub may be split so as to facilitate the assembly of the pulley with the shaft (Fig. 25-1).

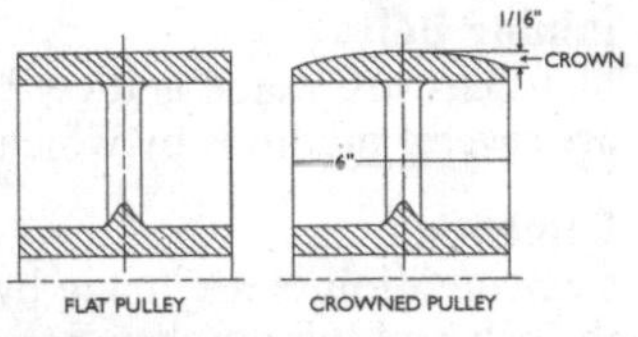

Fig. 25-1

Ordinarily, pulley face widths should slightly exceed the width of the belt they will carry. It is a good practice to use a pulley face approximately *one inch wider* than the belt, for belts under 12-in. wide. For belts 12- to 24-in. wide, the pulley should be *two inches wider*. For belts exceeding 24-in. wide, the pulley should be *three inches wider* than the belt.

All driving pulleys that require belt shifting should have a flat surface; all other pulleys should have a crowned surface.

The main reason for the use of crowned pulleys is that belts *adjust* or *center* themselves better when the pulley is given a slight crown. The height of the pulley crown should be 1⁄96 of the width, or 1⁄8 in. per foot of pulley width.

Slip and Creep

Slip occurs when the load on the belt is increased above a point where the friction between pulley and belt are sufficient to drive the load. Slippage occurs on either or both pulleys and may, if continued, cause the belt to be thrown off.

Creep is a physical characteristic of the belt itself, inherent in power transmission when flat belting is employed. When a flat belt is transmitting power, the pulling part of the belt is under a much greater tension than the slack part and is therefore slightly stretched. As a result of this stretching, a slightly elongated belt is delivered to the driving pulley, which, when it reaches the slack side of the pulley, returns to normal length. The belt is therefore creeping ahead on the pulley. This creeping ahead on the driving pulley causes the slack side and the driven pulley to run at slightly slower surface speed. Because creeping is caused by the elasticity of the belt, it increases with the load; however, in a well-designed drive it is usually in the area of one percent at full load.

Joining Belts

Flat belts are made endless by means of a belt joint. There are several methods by which this is done.

Cementing

Cemented joints are made by carefully tapering the ends of the belt and joining them together with cement and pressure. This is a specialized operation, requiring special tools, clamps, cements, and so on. This process is so seldom required of most mechanics that it is best left to a specialist to perform.

Lacing

Laced joints are used frequently when leather belting is being employed. When properly done, this type of joint is very strong and can be made rapidly. It has no metallic parts, as with metal hook lacing, and therefore it does not present a hand injury hazard.

There are no specific rules in making a laced joint for various sized belts. Two methods are shown in Fig. 25-2 and Fig. 25-3. A general practice is to punch two rows of holes in each end of the belt, the first row one inch from the end of the belt and the second row two inches from the end. The number of holes across the belt should equal the number of inches in the width, with the holes on the side ⅜ in. from the edge.

The lacing starts on one edge of the belt, the ends of the lacing being passed from the outside to the inside. Each end is then passed through the hole in the opposite end of the belt, making two thicknesses of lacing on the inner side, as shown in Fig. 25-2. Each lacing end is then passed through the hole in the back row and again through the same hole as before in the first row, thus

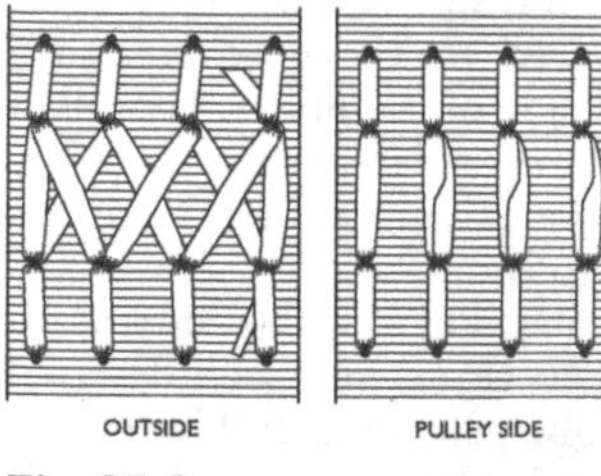

Fig. 25-2

filling the four holes at the side and leaving the lacing straight with the length of the belt.

Both ends of the lacing are now on the outside of the belt. They must be put through the second hole in the first row on the opposite end, causing the lacing to cross on the back or outside of the belt. The ends are now put through these same holes again, opposite end through opposite hole, making two strands of lacing between this set of holes on the inside. The ends again on the back of the belt are now put through the second row of holes and again through the corresponding holes in the first row, again bringing the ends to the back side. The lacings are again crossed, and the same procedure is followed until all four rows of holes have been filled in the same manner.

Use pliers to draw the lacing through the holes and pull it tight. After the pull through the inside hole of the last row to the back side, the ends of the lacing can be cut off (Fig. 25-3).

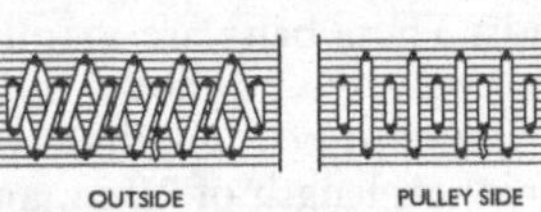

Fig. 25-3

Hooks (Alligator Lacing)

Alligator lacing utilizes a design with teeth that enter parallel to the length of the belt and imbed clear through the belt carcass. The teeth are formed so that when they are driven into the belt, the vital longitudinal carcass fibers are not severed.

Additional strength is provided by the connecting bar and because the fastener strip is one continuous piece. This provides a strong, vise-like hold on the belt and distributes belt tension evenly across the entire belt width. The one-piece fastener strip also ensures that there are no loose pieces to work themselves out of the belt and into the conveyed product. The straight splice obtained with a continuous strip of lacing also helps with hinge pin insertion.

26. V-BELTS

The *"V"-belt* has a tapered cross-sectional shape that causes it to wedge firmly into the sheave groove under load. Its driving action takes place through frictional contact between the sides of the belt and the sheave groove surfaces. While the cross-sectional shape varies slightly with make, type, and size, the included angle of most V-belts is about 42°. There are three general classifications of V-belts: fractional horsepower, standard multiple, and wedge.

Fractional Horsepower

Used principally as single belts on fractional horsepower drives, they are designed for intermittent and relatively light loads. These belts are manufactured in four standard cross-sectional sizes, as shown in Fig. 26-1a.

Standard belt lengths vary by 1-in. increments between a minimum length of 10 in. and a maximum length of 100 in. In addition, some fractional horsepower belts are made to fractional inch lengths.

The numbering system used indicates the cross-sectional size and the nominal outside length. The last digit of the belt number indicates tenths of an inch. Because the belt number indicates length along the outside surface, belts are slightly shorter along the pitch line than the nominal size number indicates.

Standard Multiple

Standard multiple belts are designed for the continuous service usually encountered in industrial applications. As the name indicates, more than one belt provides the required power transmission capacity. Most manufacturers furnish two grades, a standard and a premium quality. The standard belt is suitable for the majority of industrial drives that have normal loads, speeds, center distances, sheave diameters, and operating conditions. The premium quality is made for drives subjected to severe loads, shock, vibration, temperatures, and so on.

The standard multiple V-belt is manufactured in five standard cross-sectional sizes designated "A", "B", "C", "D", and "E", as shown in Fig. 26-1b.

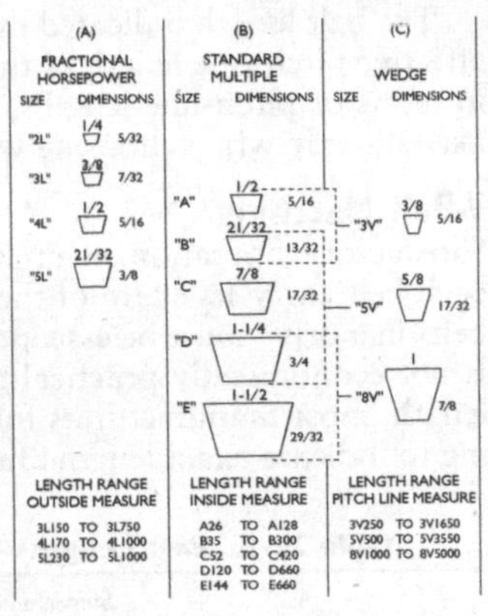

Fig. 26-1 V-Belt types.

The actual pitch length of standard multiple belts may be from one to several inches greater than the nominal length indicated by the belt number because the belt numbers indicate the length of the belt along its inside surface. As belt length calculations are in terms of belt length on the pitch line, a table of pitch line belt lengths is recommended when selecting belts.

Wedge

The wedge belt is an improved design V-belt that makes possible a reduction in size, weight, and cost of V-belt drives. Utilizing improved materials, these multiple belts have a smaller cross-section per horsepower and use smaller diameter sheaves at shorter center distances than is possible with standard multiple belts. Because of the premium-quality, heavy-duty construction, only three cross-sectional belt sizes are used to cover the duty range of the five sizes of standard multiple belts. The dimensions of the three standard "wedge" belt cross-sectional sizes—"3V", "5V", "8V"—are shown in Fig. 26-1c.

The wedge belt number indicates the number of ⅛ in. of top width of the belt. As shown in Fig. 26.1c, the "3V" belt has a top width of ⅜ in., the "5V" a width of ⅝ in., and the "8V" a full 1 in. of top width.

The belt length indicated by the wedge belt number is the effective pitch-line length of the belt. As belt calculations are in terms of pitch-line lengths, nominal belt numbers can be used directly when choosing wedge belts.

V-Belt Matching

Satisfactory operation of multiple belt drives requires that each belt carry its share of the load. To accomplish this, all belts in a drive must be essentially of equal length. Because it is not economically practical to manufacture belts to exact length, most manufacturers follow a practice of code marking to indicate exact length (Tables 26-1 and 26-2).

Table 26-1 Belt Lengths—Standard Multiple Type

	Standard Pitch Lengths				
Belt Number Indicating Nominal Length	*A*	*B*	*C*	*D*	*E*
26	27.3				
31	32.3				
33	34.3				
35	36.3	36.8			
38	39.3	39.8			
42	43.3	43.8			
46	47.3	47.8			
48	49.3	49.8			
51	52.3	52.8	53.9		
53	54.3	54.8			
55	56.3	56.8			
60	61.3	61.8	62.9		
62	63.3	63.8			
64	65.3	65.8			
66	67.3	67.8			
68	69.3	69.8	70.9		
71	72.3	72.8			
75	76.3	76.8	77.9		
78	79.3	79.8			
80	81.3				
81		82.8	83.9		
83		84.8			
85	86.3	86.8	87.9		

Standard Pitch Lengths

Belt Number Indicating Nominal Length	A	B	C	D	E
90	91.3	91.8	92.9		
96	97.3		98.9		
97		98.8			
105	106.3	106.8	107.9		
112	113.3	113.8	114.9		
120	121.3	121.8	122.9	123.3	
128	129.3	129.8	130.9	131.3	
136		137.8	138.9		
144		145.8	146.9	147.3	
158		159.8	160.9	161.3	
162			164.9	165.3	
173		174.8	175.9	176.3	
180		181.8	182.9	183.3	184.5
195		196.8	197.9	198.3	199.5
210		211.8	212.0	213.3	214.5
240		240.3	240.9	240.8	241.0
270		270.3	270.9	270.8	271.0
300		300.3	300.9	300.8	301.0
330			330.9	330.8	331.0
360			360.9	360.8	361.0
390			390.9	390.8	391.0
420			420.9	420.8	421.0
480				480.8	481.0
540				540.8	541.0
600				600.8	601.0
660				660.8	661.0

Each belt is measured under specific tension and marked with a code number to indicate its variation from nominal length. The number 50 is commonly used as the code number to indicate a belt within tolerance of its nominal length. For each ⅒ of an inch over nominal length, the number 50 is increased by 1. For each ⅒ of an inch under nominal length, 1 is subtracted from the number 50. Most manufacturers code mark as shown in Fig. 26-2.

For example, if the 60-in. "B" section belt shown is manufactured 3⁄10 of an inch longer, it will be code marked 53 rather than the 50 shown. If it is made 3⁄10 of an inch shorter, it will

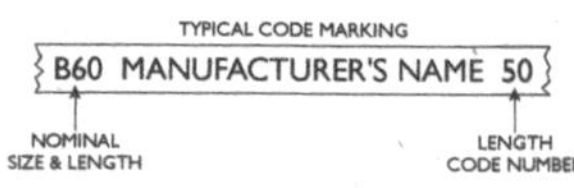

Fig. 26-2 Reading the belt code number.

be code marked 47. Although both of these belts have the belt number B60, they cannot be used satisfactorily in a set because of the difference in their actual lengths.

It is possible for the length of belts to change slightly during storage. Under satisfactory conditions, however, changes will not exceed measuring tolerances. Therefore, belts may be combined by matching code numbers. Ideally, sets should be made up of belts having the same code numbers; however, the resiliency of the belts allows some length variation. Table 26-3 lists the maximum recommended variations for standard multiple belts when making up matched belt sets.

Matchless Belts

In recent years certain top-shelf manufacturers have improved their process of making "V" belts. Tighter tolerances on the rubber, fiber, carbon, and other raw ingredients, plus much closer process control in making and curing the belts, have led to a whole new class of "matchless" belts. This quality of belt does not require length matching because each belt is exactly the same as any other for a given size. Obviously, purchasing belts from different manufacturers for use on the same machine is not a good idea because the standards of each manufacturer for "matchless" belts will vary.

Sheave-Groove Wear

All "V" belts and sheaves will wear to some degree with use. As wear occurs, the belts will ride lower in the grooves. Generally, a new belt should not seat more than 1/16 in. below the top of the groove. While belt wear is usually noticed, sheave-groove wear is often overlooked.

As wear occurs at the contact surfaces on the sides of the grooves, a dished condition develops. This results in reduced wedging action, loss of gripping, loss of power, and accelerated wear as slippage occurs. Installing new belts in worn grooves will give temporary improvement in operation, but belt wear

Table 26-2 Belt Lengths — Wedge Type

Effective Pitch Length	Belt Number 3V	Belt Number 5V	Belt Number 8V	Effective Pitch Length	Belt Number 3V	Belt Number 5V	Belt Number 8V
25	3V250			112	3V1120	5V1120	8V1120
26½	3V265			118	3V1180	5V1180	8V1180
28	3V280			125	3V1250	5V1250	8V1250
30	3V300			132	3V1320	5V1320	8V1320
30½	3V305			140	3V1400	5V1400	8V1400
33½	3V335			150		5V1500	8V1500
37½	3V375			160		5V1600	8V1600
40	3V400			170		5V1700	8V1700
42½	3V425			180		5V1800	8V1800
45	3V450			190		5V1900	8V1900
47½	3V475			200		5V2000	8V2000
50	3V500	5V500		212		5V2120	8V2120
53	3V530	5V530		224		5V2240	8V2240
56	3V560	5V560		236		5V2360	8V2360
60	3V600	5V600		250		5V2500	8V2500
63	3V630	5V630		265		5V2650	8V2650
67	3V670	5V670		280		5V2800	8V2800
71	3V710	5V710		300		5V3000	8V3000
75	3V750	5V750		315		5V3150	8V3150
80	3V800	5V800		335		5V3350	8V3350
85	3V850	5V850		355		5V3550	8V3350
90	3V900	5V900		375			8V3750
95	3V950	5V950		400			8V4000
100	3V1000	5V1000	8V1000	425			8V4250
106	3V1060	5V1060	8V1060	450			8V4500

will be rapid. When changing belts, therefore, sheave-groove wear should be checked with gauges or templates.

Care must be taken when checking grooves that the correct gauge or template with respect to type, size, and pitch diameter is used. As sheave grooves are designed to conform to the belt cross-section change as it bends, small diameter sheaves have less angle than larger diameter sheaves. The variation in sheave-groove includes angles ranging from 34° for small diameter sheaves up to 42° for the largest diameter sheaves.

Table 26-3 Variation

	Belt Lengths				
	A	*B*	*C*	*D*	*E*
2	26-180	35-180	51-180		
3		195-315	195-255	120-255	144-240
4			270-360	270-360	270-360
6			390-420	390-660	390-660

Sheave Grooves

When sheave-groove wear becomes excessive, shoulders will develop on the groove sidewalls. If the sheave is not repaired or replaced, these shoulders will quickly chew the bottom corners off new belts and ruin them.

The more heavily loaded a drive, the greater the effect of groove wear on its operation. Light to moderately loaded drives may tolerate as much as 1/32-in. wear, whereas heavily loaded drives will be adversely affected by .010 to .015 in. of wear. Wear should be checked with the appropriate gauge at the point illustrated in Fig. 26-3.

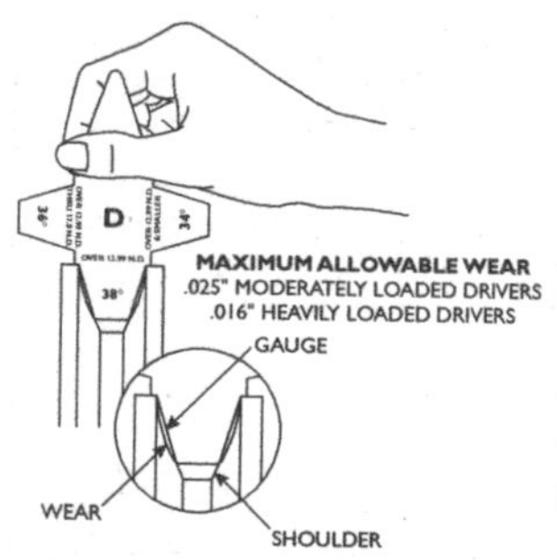

Fig. 26-3 Checking for wear.

Checks for Runout

All installed sheaves should be checked for eccentricity (outside diameter runout) or wobble (face runout), as shown in Fig. 26-4.

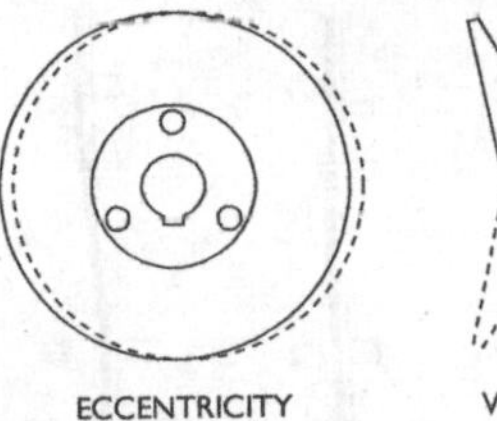

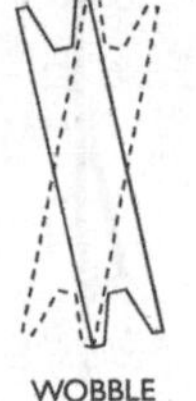

Fig. 26-4 Check for runout.

Improperly bored hubs, bent shafts, or improperly installed bushings are the common causes for these conditions. The following standards should be used when checking for wobble or eccentricity using a dial indicator.

1. Radial runout (eccentricity) limits for sheaves
   Up through 10-in. diameter—0.010 in. maximum
   For each additional inch of diameter—Add 0.005 in.
2. Axial runout (wobble) limits for sheaves
   Up through 5-in. diameter—0.005 in. maximum
   For each additional inch of diameter—Add 0.001 in.

V-Belt Drive Alignment

The life of a V-belt is dependent on, first, the quality of materials and manufacture and, second, installation and maintenance. One of the most important installation factors in influencing operating life is belt alignment. In fact, excessive misalignment is probably the most frequent cause of shortened belt life.

While V-belts, because of their inherent flexibility, can accommodate themselves to a degree of misalignment not tolerated by other types of power transmission, they still must be held within reasonable limits. Maximum life can be attained only with true alignment, and as misalignment increases, belt life is proportionally reduced. If misalignment is greater than 1/16 in. for each 12 in. of center distance, very rapid wear will result.

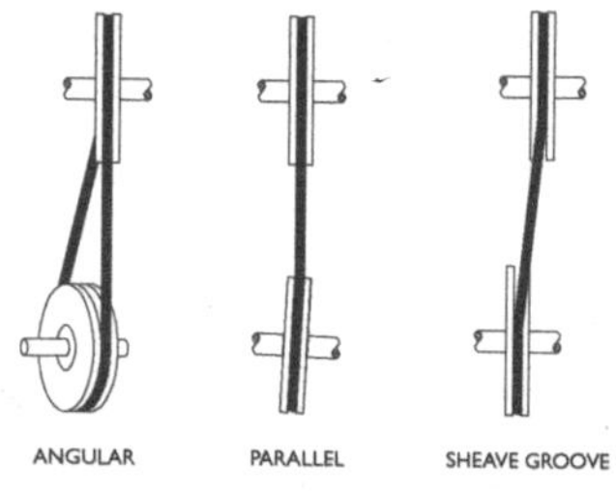

Fig. 26-5 V-belt misalignment.

Misalignment of belt drives results from shafts being out of angular or parallel alignment, or from the sheave grooves being out of axial alignment. These three types of misalignment are illustrated in Fig. 26-5.

Because the shafts of most V-belt drives are in a horizontal plane, angular shaft alignment is easily obtained by leveling the shafts. In those cases where shafts are not horizontal, a careful check must be made to ensure that the angle of inclination of both shafts is the same.

V-Belt Alignment

The most satisfactory method of checking parallel-shaft and axial-groove alignment is with a straightedge. It may also be done with a taut line; however, when using this method, exercise care because the line is easily distorted. Both the straightedge and taut-line checking methods are illustrated in Fig. 26-6, with arrows indicating the four check points. When sheaves are properly aligned, no light should be visible at these four points.

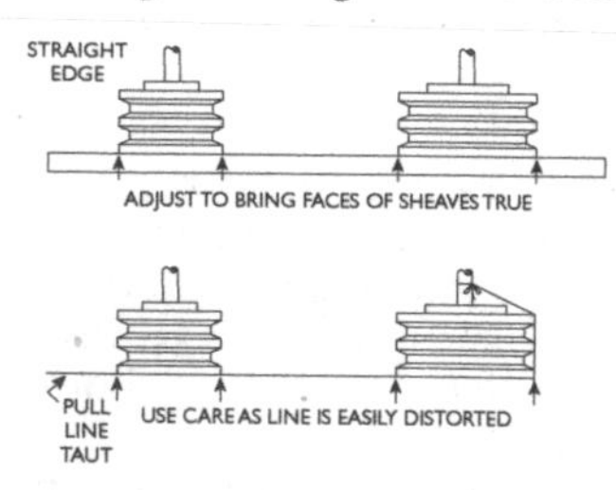

Fig. 26-6 Checking alignment.

V-Belt Installation

V-belts should never be "run on" to sheaves (Fig. 26-7). Doing so places excessive stress on the cords, usually straining or breaking some of them. A belt damaged in this manner will flop under the load and turn over in the sheave groove. The proper installation method is to loosen the adjustable mount, reduce the center distance, and slip the belts loosely into the sheave grooves.

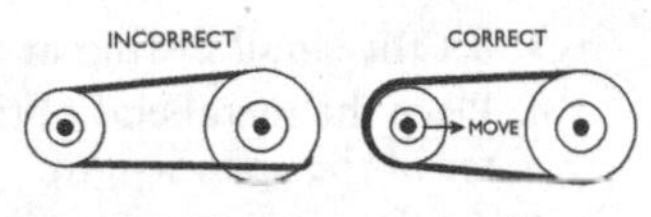

Fig. 26-7 Never stretch belts over sheaves.

Follow these six general rules when installing V-belts:

1. Reduce centers so belts can be slipped on sheaves.
2. Have all belts slack on the same side (top of drive).
3. Tighten belts to the approximately correct tension.
4. Start unit and allow belts to seat in grooves.
5. Stop and retighten belts to correct tension.
6. Recheck belt tension after 24 to 48 hours of operation.

Proper Belt Tensioning

Belt tension is a vital factor in operating efficiency and service life. Too low a tension results in slippage and rapid wear of both belts and sheave grooves. Too high a tension stresses the belts excessively and unnecessarily increases bearing loads. The following instructions provided by Maintenance Troubleshooting show the force-deflection method of tensioning and give an example of proper use of a tension tester for correctly setting belt tension (Fig. 26-8).

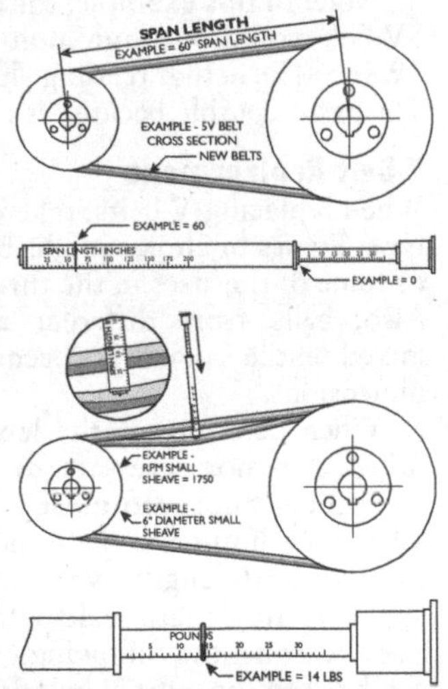

Fig. 26-8

Following are the nine steps for setting the tension of a V-belt:

1. Measure the belt span (tangent to tangent) as shown.
2. Set the large O-ring on the span scale.

3. Set the small O-ring at zero on the force scale.
4. Place the metal end of the tester on one belt at the center of the span length.
5. Apply force to the plunger until the bottom of the large O-ring is even with the top of any adjacent belt.
6. Read the force scale under the small O-ring to determine the force used to give the deflection.
7. Compare the force scale reading with the correct value for either a new belt or an old belt using Table 26-4.
8. If the force shown on the tester is less than the minimum, tighten the belts. If the force shown on the tester is greater than the maximum, loosen the belts.
9. If the chart calls for more tension than one tension tester can deliver, use two or more testers side-by-side and add the forces together.

Note: In this example, the actual force required for a *new* 5V belt would be a minimum of 13.2 lb and a maximum of 19.8 lb. The actual reading shown in the example was 14 lb. This is acceptable because it is within the range as shown.

V-Belt Replacement

When replacing V-belts, take care that you select the correct type. Errors in choice might be made because the top width of some of the sizes in the three types is essentially the same. Also, belts from different manufacturers should not be mixed on the same drive because of variations from nominal dimensions.

When determining the length of belt required for most drives, it is not necessary to be exact: first, because of the adjustment built into most drives and second, because belt selection is limited to the standard lengths available. Because the standard lengths vary in steps of several inches, an approximate length calculation is usually adequate. For these reasons, the following easy method of belt calculation can be used for most V-belt drives:

Table 26-4

Belt X-Section	Smaller Sheave Range		Maximum/Minimum Recommended Deflection Force			
			New Belt		Old Belt	
	Diameter, in.	RPM	Maximum	Minimum	Maximum	Minimum
A	3 to 3⅝	1000–2500	8.3	5.5	5.5	3.7
		2501–4000	6.3	4.2	4.2	2.8
	3⅞ to 4⅞	1000–2500	10.2	6.8	6.8	4.5
		2501–4000	8.6	5.7	5.7	3.8
	5 to 7	1000–2500	12.0	8.0	8.0	5.4
		2501–4000	10.5	7.0	7.0	4.7
B	3⅜ to 4¼	850–2500	10.8	7.2	7.2	4.9
		2501–4000	9.3	6.2	6.2	4.2
	4⅜ to 5⅝	850–2500	11.9	7.9	7.9	5.3
		2501–4000	10.0	6.7	6.7	4.5
	5⅞ to 8⅝	850–2500	14.1	9.4	9.4	6.3
		2501–4000	13.4	8.9	8.9	6.0
C	7 to 9	500–1740	25.5	17.0	17.0	11.5
		1741–3000	20.7	13.8	13.8	9.4
	9½ to 16	500–1740	31.5	21.0	21.0	14.1
		1741–3000	27.8	18.5	18.5	12.5
D	12 to 16	200–850	55.5	37.0	37.0	24.9
		851–1500	47.0	31.3	31.3	21.2
	18 to 20	200–850	67.8	45.2	45.2	30.4
		851–1500	57.0	38.0	38.0	25.6

(continued)

Table 26-4 (continued)

	Smaller Sheave Range		Maximum/Minimum Recommended Deflection Force: New Belt		Old Belt	
Belt X-Section	**Diameter, in.**	**RPM**	**Maximum**	**Minimum**	**Maximum**	**Minimum**
E	$21^5/_8$ to 24	100–450	71.0	47.0	47.0	31.3
		451–900	48.0	32.0	32.0	21.3
3V	$2^1/_4$ to $2^3/_8$	1000–2500	7.4	4.9	4.9	3.3
		2501–4000	6.5	4.3	4.3	2.9
	$2^5/_8$ to $3^5/_8$	1000–2500	7.7	5.1	5.1	3.6
		2501–4000	6.6	4.4	4.4	3.0
	$4^1/_8$ to $6^7/_8$	1000–2500	11.0	7.3	7.3	4.9
		2501–4000	9.9	6.6	6.6	4.4
5V	$4^3/_8$ to $6^5/_8$	500–1740	22.8	15.2	15.2	10.2
		1741–3000	19.8	13.2	13.2	8.8
	$7^1/_8$ to $10^7/_8$	500–1740	28.4	18.9	18.9	12.7
		1741–3000	25.0	16.7	16.7	11.2
	$11^7/_8$ to 16	500–1740	35.1	23.4	23.4	15.5
		1741–3000	32.7	21.8	21.8	14.6
8V	$12^1/_2$ to 17	200–850	74.0	49.3	49.3	33.0
		851–1500	59.8	39.9	39.9	26.8
	18 to $22^3/_8$	200–850	88.8	59.2	59.2	39.6
		851–1500	79.0	52.7	52.7	35.3

1. Add the pitch diameters of the sheaves and multiply by 1½.
2. To this, add twice the distance between centers.
3. Select the nearest *longer* standard belt.

This method should not be used when centers are fixed or if there are extreme pitch-diameter differences on short centers.

Micro-V and Common Backed Belts

In the past few years the micro-V belt has gained popularity. Often used as a drive on an automobile engine to power the alternator, air conditioner, and other accessories from the crankshaft pulley, the micro-V belt appears as a tiny row of "V" shapes molded to a flat common backing.

Wedge-type and conventional V-belts are also made in a banded configuration (Fig. 26-9). These belts provide more uniform power transmission on multiple belt drives. Two or more belts are connected at the top by a layer of rubber and fabric. The connecting layer helps to stabilize the belt.

Troubleshooting V-Belt Problems

A properly designed and maintained V-belt drive should give belt life of 3 to 5 years and sheave service of 10 to 20 years, depending on the application. Service life that falls far short of these applications warrants investigation. Table 26-5 will help prevent a recurrence of many V-belt problems.

Fig. 26-9 Banded belts.

Table 26-5 Troubleshooting V-Belt Problems

Symptom	*Cause*	*Correction*
Broken belt	Shock loads or heavy starting load	Check tension. It may be necessary to redesign the drive for increased belt capacity.

(continued)

Table 26-5 (continued)

Symptom	*Cause*	*Correction*
Excessive heat	Belt slip	Check tension.
	Misalignment	Check alignment of sheaves and shafts.
	Objects falling on belt or getting wedged into sheave groove	Protect the drive with a fine-wire mesh guard.
	Broken tension cords	Probably caused by prying belt onto sheave. Follow correct procedure for belt installation.
Belt squeal	Not enough tension	Increase tension. Check with tension tester.
	Insufficient arc of contact (belt does not have enough wraparound sheave to transmit the load)	Investigate the possibility of redesigning the drive for a decreased sheave ratio.
Premature sheave wear	Dirty environment	Use a closed-at-the-top belt guard.
Irregular belt vibration	Insufficient tension	Increase tension using tension tester.
	Unbalanced shaft or rotors	Use vibration analysis equipment to test for unbalance. Send out for balancing or perform field balance in place.
	Sheaves not balanced	Purchase dynamically balanced sheaves.
	Framework or shafting too light	Investigate possibility of replacing with heavier components or beefing up existing ones.
Belt turnover	Broken tension cords	Probably because of poor installation where prying was done. Follow correct installation procedures.
	Overloaded drive	Consider redesign.
	Misalignment of sheave or shaft	Perform corrective alignment with straightedge or string.
	Worn sheave grooves	Replace sheaves.

27. POSITIVE-DRIVE BELTS

Timing Belts

Positive-drive belts, also called timing belts and gear belts, combine the flexibility of belt drives with the advantages of chain and gear drives. Power is transmitted by positive engagement of belt teeth with pulley grooves, as in chain drives, rather than by friction, as in belt drives. This positive engagement of belt teeth with pulley grooves eliminates slippage and speed variations. There is no metal-to-metal contact, and no lubrication is required.

Belt Construction

The positive-drive belt is constructed with gear-like teeth that engage mating grooves in the pulley. Unlike most other belts, they do not derive their tensile strength from their thickness. Instead, these belts are built thin, their transmission capacity resulting from the steel-cable tension members and tough molded teeth.

Components of Positive-Drive Belts

High-strength synthetic materials are used in positive-drive construction. The three basic components are shown in Fig. 27-1: neoprene backing, steel tension members, and neoprene teeth with nylon facing. The nylon covering on the teeth is highly wear resistant, and after a short "run-in" period, it becomes highly polished with resultant low friction. Belts are so constructed that tooth strength exceeds the tensile strength of the belt when six or more teeth are meshed with a mating pulley.

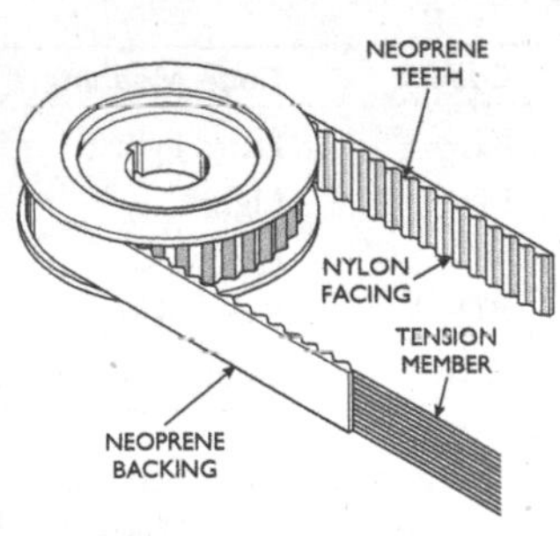

Fig. 27-1

Drive Specifications

With positive-drive belts, as with gear and chain drives, pitch is a fundamental consideration. In this case, *circular pitch* is the distance between tooth centers (measured on the pitch line of the belt) or the distance between groove centers (measured on the pitch circle of the pulley), as indicated in Fig. 27-2.

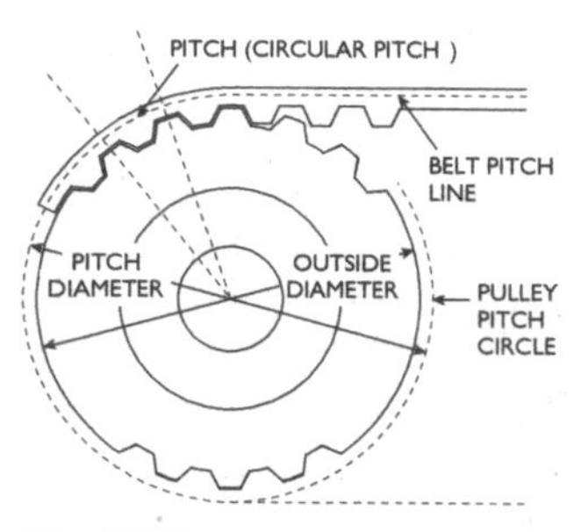

Fig. 27-2

The *pitch line* of a positive-drive belt is located within the cable tension member. The *pitch circle* of a positive-drive pulley coincides with the *pitch line* of the belt mating with it. The pulley *pitch diameter* is *always greater* than its face diameter. All positive-drive belts must run with pulleys of the same pitch. A belt of one pitch *cannot* be used successfully with pulleys of a different pitch.

Positive drive belts are made in five stock pitches. The following code system is used to indicate the pitch of positive-drive systems:

Code	Code Meaning	Pitch (in Inches)
XL	Extra light	1/5
L	Light	3/8
H	Heavy	1/2
XH	Extra heavy	7/8
XXH	Double extra heavy	1 1/4

The standard positive-drive belt numbering system is made up of three parts: first, the pitch length of the belt, which is the actual pitch length multiplied by 10; second, the code for the pitch of the drive; and third, the belt width multiplied by 100.

Examples

390L100—39.0 Pitch Length—Light Size ⅜-in. pitch—1-in. wide

480H075—48.0 Pitch Length—Heavy Size ½-in. pitch—¾-in. wide

560XH200—56.0 Pitch Length—Extra Heavy Size ⅞-in. pitch—2-in. wide

The following list is the stock sizes of positive-drive belts carried by most manufacturers. In addition, much wider and longer drives can be furnished on special order:

Code	Pitch (in Inches)	Stock Widths (in Inches)	Length Range (in Inches)
XL	⅕	¼–5⁄16–⅜	6–26
L	⅜	½–¾–1	12–60
H	½	¾–1–1½–2–3	24–170
XH	⅞	2–3–4	50–175
XXH	1¼	2–3–4–5	70–180

Because of a slight side thrust of positive-drive belts in motion, at least one pulley in a drive must be flanged. When center distance between shafts is eight or more times the diameter of the small pulley, or when the drive is operating on vertical shafts, both pulleys should be flanged.

When installing positive-drive belts, the center distance should be reduced and the belt placed loosely on the pulleys. Belts should not be forced in any way over the pulley or flange, as damage to the belt will result. The belt should be tightened to a snug fit. Because the positive-drive belt does not rely on friction, there is no need for high initial stress. If torque is unusually high, however, a loose belt may "jump grooves." In such cases, the tension should be increased gradually until satisfactory operation is attained. Take care to avoid disturbing shaft parallelism while doing this. On heavily loaded drives, with wide belts, it may be necessary to use a tension measuring tool to tighten the belt accurately. Belt manufacturers should be consulted for their recommendations of equipment and procedures to follow for special situations of this type.

High Torque Drive (HTD) Belts

A new device used in the positive-drive arena is the high torque drive or curvilinear drive. Similar to a timing belt, an HTD drive uses a circular tooth design to allow teeth to enter and leave the mating sprocket grooves in a smooth rolling manner, functioning in much the same manner as the teeth in a gear. The circular tooth design improves stress distribution and allows higher overall loading (Fig. 27-3). These belts are available in 8-, 14-, and 20-mm widths.

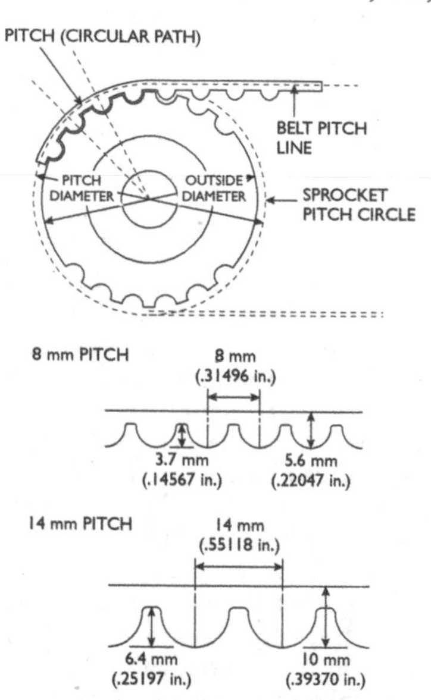

Fig. 27-3 Typical HTD drive

28. CHAIN DRIVES

A chain drive consists of an endless chain whose links mesh with toothed wheels called sprockets. Chain drives maintain a positive ratio between the driving and driven shafts, as they transmit power without slip or creep. The *roller chain* is the most widely used of the various styles of power transmission chain.

Roller Chain

Roller chain is composed of an alternating series of *roller links* and *pin links*. The roller links consist of two pin-link plates, two bushings, and two rollers. The rollers turn freely on the bushings, which are press-fitted into the link plates. The pin links consist of two link plates into which two pins are press-fitted.

In operation, the pins move freely inside the bushings while the rollers turn on the outside of the bushings. The relationship of the pins, bushings, rollers, and link plates is illustrated in Fig. 28-1.

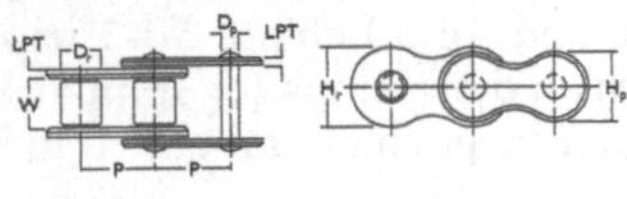

Fig. 28-1

Roller Chain Dimensions

The principal roller chain dimensions are *pitch, chain width,* and *roller diameter.* These dimensions are standardized, and, although there are slight differences between manufacturers' products, because of this standardization chains and sprockets from different manufacturers are interchangeable.

Pitch is the distance from a point on one link to a corresponding point on an adjacent link. *Chain width* is the minimum distance between link plates of a roller link. *Roller diameter* is the outside diameter of a roller, and it is approximately 5/8 of the pitch.

Standard series chains range in pitch from 1/4 to 3 in. There is also a heavy series chain ranging from 3/4- to 3-in.

pitch. The heavy series dimensions are the same as the standard, except that the heavy series has thicker link plates.

Standard Roller Chain Numbers

The standard roller chain numbering system provides complete identification of a chain by number. The right-hand digit in the chain number is 0 for chain of the usual proportions, 1 for lightweight chain, and 5 for rollerless bushing chain. The number to the left of the right-hand figure denotes the number of 1/8 in. in the pitch. The letter H following the chain number denotes the heavy series. The hyphenated 2 suffixed to the chain number denotes a double strand chain, 3 a triple strand, and so on.

For example, the number 60 indicates a chain with six 1/8s, or 3/4-in. pitch. The number 41 indicates a narrow lightweight 1/2-in. pitch chain, the number 25 indicates a 1/4-in. pitch rollerless chain, and the number 120 chain has 12 xth's, or 1 1/2 in., pitch. In multiple-strand chains 50-2 designates two strands of 50 chain, 50-3 triple strand, and so on. General chain dimensions for standard roller chain from 1/4-in. pitch to 3-in. pitch are tabulated in Table 28-1.

Roller Chain Connections

A length of roller chain before it is made endless will normally be made up of an even number of pitches. At either end will be an unconnected roller link with an open bushing. A special type of pin link called a *connecting link* is used to connect the two ends. The partially assembled connecting link consists of two pins press-fitted and riveted in one link plate. The pinholes in the free link plate are sized for either a slip fit or a light press-fit on the exposed pins. The plate is secured in place either by a cotter pin, as shown in Fig. 28-2a, or by a spring clip, as shown in Fig. 28-2b.

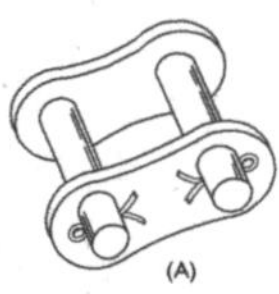

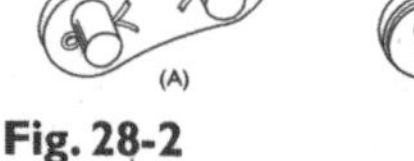

Fig. 28-2

Table 28-1 American Standard Roller Chain Dimensions

U.S. Standard Chain Number					Pin	Thickness Link Plate (LPT)	
Standard	Heavy	Pitch (P)	Max Roller Diameter (D_R)	Width (W)	Diameter (D_P)	Standard	Heavy
25*	—	1/4	0.130*	1/8	0.0905	0.030	—
35*	—	3/8	0.200*	3/16	0.141	0.050	—
41[A]	—	1/2	0.306	1/4	0.141	0.050	—
40	—	1/2	5/16	5/16	0.156	0.060	—
50	—	5/8	0.400	3/8	0.200	0.080	—
60	60H	3/4	15/32	1/2	0.234	0.094	.125
80	80H	1	5/8	5/8	0.312	0.125	.156
100	100H	1 1/4	3/4	3/3	0.375	0.156	.187
120	120H	1 1/2	7/8	1	0.437	0.187	.219
140	140H	1 3/4	1	1	0.500	0.219	.250
160	160H	2	1 1/8	1 1/4	0.562	0.250	.281
180	180H	2 1/4	1 13/32	1 13/32	0.687	0.281	.312
200	200H	2 1/2	1 9/16	1 1/2	0.781	0.312	.375
240	240H	3	1 7/8	1 7/8	0.937	0.375	.500

*Without rollers.
[A]Lightweight chain.

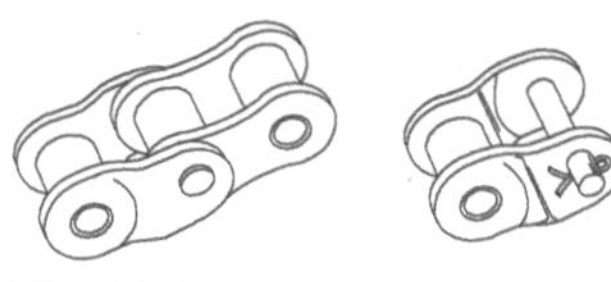

Fig. 28-3

If an odd number of pitches is required, an *offset link* may be substituted for an end roller link. A more stable method of providing an odd number of pitches is using the *offset section.* A standard connecting link is used with the offset section (Fig. 28-3).

Roller Chain Sprockets

Roller chain sprockets are made to standard dimensions, tolerances, and tooth form. The standard includes four types (Fig. 28-4): Type A has a plain sprocket without hubs; type B has a hub on one side; type C has a hub on both sides; type D has a detachable hub.

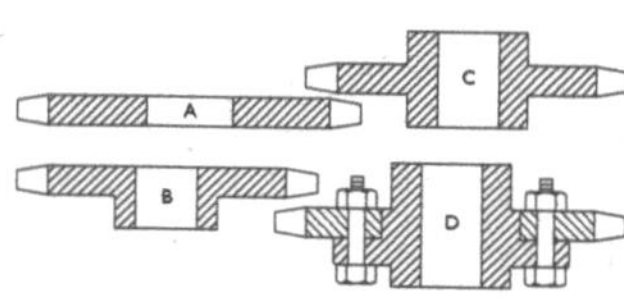

Fig. 28-4

Roller Chain Installation

Correct installation of a roller chain drive requires that the shafts and the sprockets be accurately aligned. Shafts must be set level, or if inclined from a level position, both shafts must be at exactly the same angle. The shafts must also be positioned parallel within very close limits. The sprockets must be in true axial alignment for correct sprocket tooth and chain alignment.

Horizontal shafts may be aligned with the aid of a spirit level. The bubble in the level will tell when they are both in the exact horizontal position. Shafts may be adjusted for parallel alignment, as shown in Fig. 28-5. Any suitable measuring device such as calipers or feeler bars may be used. The distance between shafts on both sides of the sprockets should be equal. For an adjustable shaft drive, make the distance less than the final operating distance for easier chain installation. For drives with fixed shafts, the center distance must be set at the exact dimension specified.

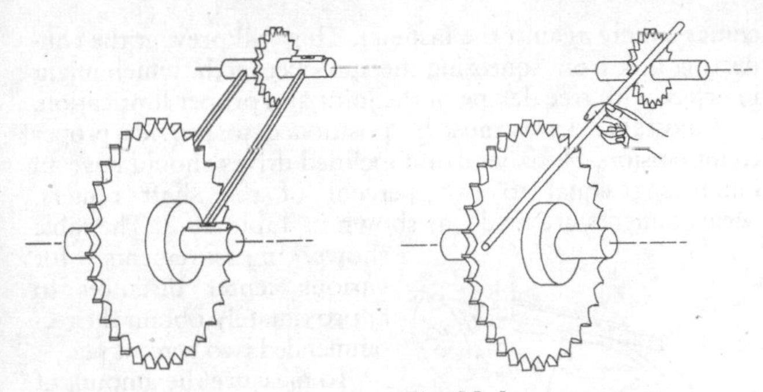

Fig. 28-5

Fig. 28-6

To set axial alignment of the sprockets, apply a straightedge to the machined side surfaces, as shown in Fig. 28-6. Tighten the set screws in the hubs to hold the sprockets and keys in position. If one of the sprockets is subject to end float, locate the sprocket so that it will be aligned when the shaft is in its normal running position. If the center distance is too great for the available straightedge, a taut piano wire may be used.

To install roller chain, fit it on both sprockets, bringing the free chain ends together on one sprocket. Insert the pins of the connecting link in the two end links of the chain, as shown in Fig. 28-7; then install the free plate of the connecting link. Fasten the plate with the cotters or spring clip, depending on the type used. When fastened, tap back the ends of the connecting link pins so that the outside of the free plate

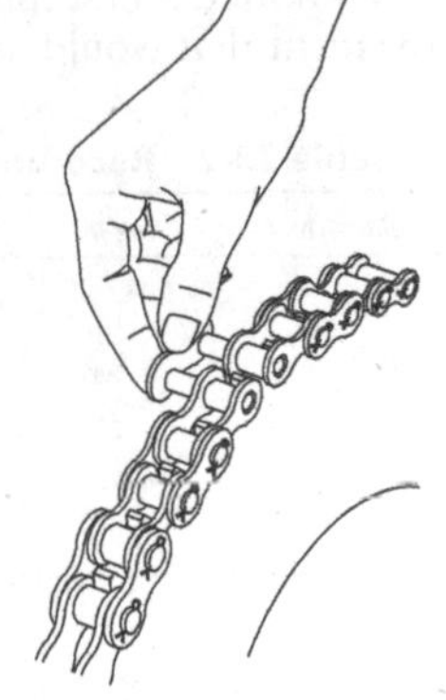

Fig. 28-7

comes snugly against the fastener. This will prevent the connecting link from squeezing the sprocket teeth, which might interfere with free flexing of the joint and proper lubrication.

Adjustable drives must be positioned to provide proper chain tension. Horizontal and inclined drives should have an initial sag equal to two percent of the shaft centers. Measurements are made, as shown in Table 28-2. The table shows measurements for various center distances to approximately obtain the recommended two percent sag.

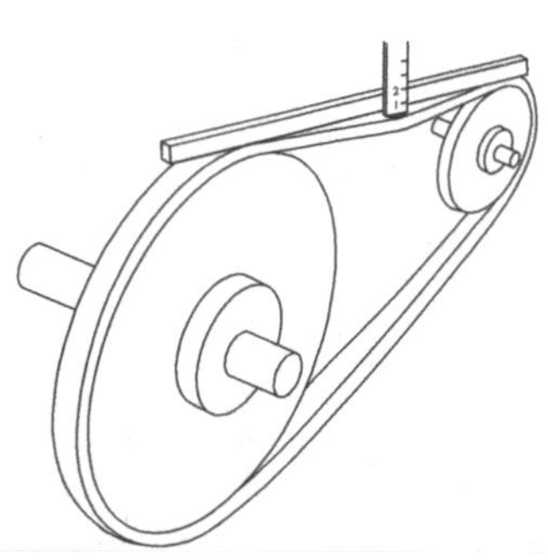

Fig. 28-8

To measure the amount of sag, pull the bottom side of the chain taut so that all of the excess chain will be in the top span. Pull the top side of the chain down at its center, and measure the sag, as shown in Fig. 28-8; then adjust the centers until the proper amount is obtained. Make sure the shafts are rigidly supported and securely anchored to prevent deflection or movement that would destroy alignment.

Table 28-2 Recommended Allowable Chain Sag

Shaft Centers, in.	*Sag, in.*	*Shaft Centers, in.*	*Sag, in.*
18	3/8	54	1 1/8
24	1/2	60	1 1/4
30	5/8	70	1 1/2
36	3/4	80	1 5/8
42	7/8	90	1 7/8
48	1	100	2
54	1 1/8	125	2 1/2

Silent Chain

Silent chain, also called *inverted-tooth* chain, is constructed of leaf links having inverted teeth designed so that they engage cut tooth wheels in a manner similar to the way a rack engages a gear. The chain links are alternately assembled, either with pins or a combination of joint components.

Silent chain and sprockets are manufactured to a standard that is intended primarily to provide for interchangeability between chains and sprockets from different manufacturers. It does not provide for a standardization of joint components and link plate contours, which differ in each manufacturer's design; however, all manufacturers' links are contoured to engage the standard sprocket tooth, so joint centers lie on the pitch diameter of the sprocket. The general proportions and designations of a typical silent chain link are shown in Fig. 28-9.

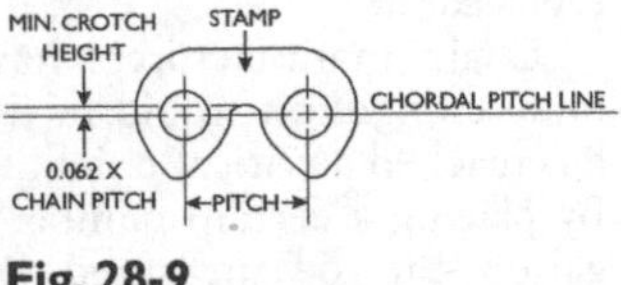

MIN. CROTCH HEIGHT
STAMP
CHORDAL PITCH LINE
0.062 X CHAIN PITCH
PITCH

Fig. 28-9

Silent chain is manufactured in a wide range of pitches and widths in various styles. Chain under ¾-in. pitch has outside guide links that engage the sides of the sprocket. The most widely used style is the *middle-guide* design with one or more rows of guide links that fit guide grooves in the sprockets. Some manufacturers use the term "wheel" rather than sprocket.

Silent chains are designated by a combined letter and number symbol as follows:

- A two-letter symbol: SC
- One or two numerical digits indicating the pitch in eighths of inches (usually stamped on each chain link)
- Two or three numerical digits indicating the chain width in quarter inches

For example, the number "SC302" designates a silent chain of 3/8-in. pitch and 1/2-in. width. Or the number "SC1012" designates a silent chain of 1¼-in. pitch and 3-in. width.

Chain Replacement

During operation, chain pins and bushings slide against each other as the chain engages, wraps, and disengages from the sprockets. Even when parts are well lubricated, some metal-to-metal contact does occur, and these parts eventually wear. This progressive joint wear elongates chain pitch, causing the chain to lengthen and ride higher on the sprocket teeth. The number of teeth in the large sprocket determines the amount of joint wear that can be tolerated before the chain jumps or rides over the ends of the sprocket teeth. When this critical degree of elongation is reached, the chain must be replaced.

Chain manufacturers have established tables of maximum elongation to aid in the determination of when wear has reached a critical point and replacement should be made. By placing a certain number of pitches under tension, elongation can be measured (Fig. 28-10). When elongation reaches the limits recommended in the table, the chain should be replaced.

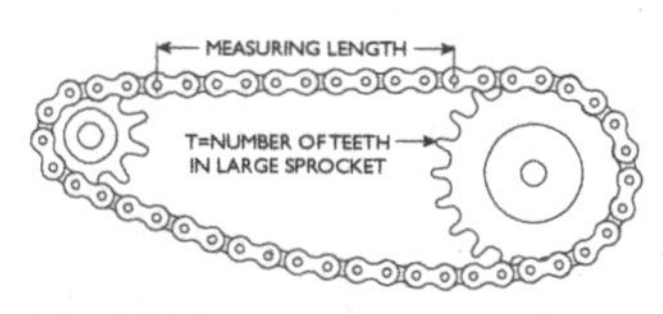

Fig. 28-10

The recommended measuring procedure is to remove the chain and suspend it vertically with a weight attached to the bottom. When the chain must be measured while on sprockets, remove all slack and apply sufficient tension to keep the chain section that is being measured taut.

Troubleshooting Chain Drive Problems

Table 28-4 will help you troubleshoot many common chain drive problems.

Table 28-3 Chain Elongation Lengths

Chain Number	Pitch, in.	Measuring Length: Number of pitches	Measuring Length: Nominal length, in.	Length of chain when replacement is required* — Number of teeth in largest sprocket (T): Up to 67	68–73	74–81	82–90	91–103	104–118	119–140	141–173
35	3/8	32	12	12 3/8	12 11/32	12 5/16	12 9/32	12 1/4	12 7/32	12 3/16	12 5/32
40	1/2	24	12	12 3/8	12 11/32	12 5/16	12 9/32	12 1/4	12 7/32	12 3/16	12 5/32
50	5/8	20	12 1/2	12 7/8	12 19/32	12 13/16	12 25/32	12 3/4	12 25/32	12 11/16	12 21/32
60	3/4	16	12	12 3/8	12 11/32	12 5/16	12 9/32	12 1/4	12 7/32	12 3/16	12 5/32
80	1	24	24	24 3/4	24 11/16	24 5/8	24 9/16	24 1/2	24 7/16	24 3/8	24 5/16
100	1 1/4	20	25	25 3/4	25 11/16	25 5/8	25 9/16	25 1/2	25 7/16	25 3/8	25 5/16
120	1 1/2	16	24	24 3/4	24 11/16	24 5/8	24 9/16	24 1/2	24 7/16	24 3/8	24 5/16
140	1 3/4	14	24 1/2	25 1/4	25 3/16	25 1/8	25 1/16	25	24 15/16	24 7/8	24 13/16
160	2	12	24	24 3/4	24 11/16	24 5/8	24 9/16	24 1/2	24 7/16	24 3/8	24 5/16
180	2 1/4	11	24 3/4	25 1/2	25 7/16	25 3/8	25 5/16	25 1/4	25 3/16	25 1/8	25 1/16
200	2 1/2	10	25	25 3/4	25 11/16	25 3/8	25 9/16	25 1/2	25 7/16	25 3/8	25 5/16
240	3	8	24	24 3/4	24 11/16	24 5/8	24 9/16	24 1/2	24 7/16	24 3/8	24 5/16

*Valid for drives with adjustable centers or drives employing adjustable idler sprockets.

Table 28-4 Chain Troubleshooting Guide

Problem	*Cause*	*Solution*
Excessive noise	Sprocket misalignment	Realign
	Loose bearings	Tighten set-screws
	Chain or sprocket wear	Replace components
	Chain pitch size too large	Recalculate drive selection
	Too much or too little slack	Adjust take-up
Vibrating chain	High load fluctuation	Make use of fluid coupling or torque converter
	Resonance	Change speed of machine
Chain climbs sprockets	Heavy overload	Reduce load or install stronger chain
	Excessive slack	Adjust take-up
	Worn chain	Replace chain
Excessive wear of sprockets link plate and one side of sprocket teeth	Misalignment	Realign shafts and
Broken pins, roller, install, or bushings	Speed too high for sprocket size and pitch	Use shorter-pitch chain or larger-diameter sprockets
	Shock loading	Reduce source of shock load or use stronger chain
	Inadequate lubrication	Perform proper PM lubrication at correct intervals
	Sprocket or chain corrosion	Install stainless or other corrosion-resistant chain and sprockets
	Material build up in sprocket teeth	Install material relief sprockets that allow material to spill out

Problem	*Cause*	*Solution*
Chain clings to sprocket	Center distance too long or high load fluctuation	Adjust center distance or install idler take-up
	Excessive chain wear	Replace chain
	Excessive chain slack	Adjust center distance or install idler take-up
Breakage of link plates	Vibration	Install vibration dampener
	Shock load	Install a shock absorber
Stiff chain	Inadequate lubrication	Perform proper PM lubrication at correct intervals
	Misalignment	Realign sprockets and shafts
	Corrosion	Replace with non-corrosive components

29. GEAR DRIVES

The spur gear might be called the basic gear, as all other gears were developed from it. Its teeth are straight and parallel to the bore centerline. Spur gears may run together with other spur gears on parallel shafts, with internal gears on parallel shafts, and with a rack. A rack is in effect a straight-line gear. The smaller of a pair of gears is often called a *pinion.* On large, heavy-duty drives, the larger gear in a pair is often called a *bullgear.*

Involute Systems

The involute profile or form is commonly used for gear teeth. It is a curve that is traced by a point on the end of a taut line unwinding from a circle (Fig. 29-1). The larger the circle, the straighter the curvature, and for a rack—essentially, an infinitely large gear—the form is straight or flat.

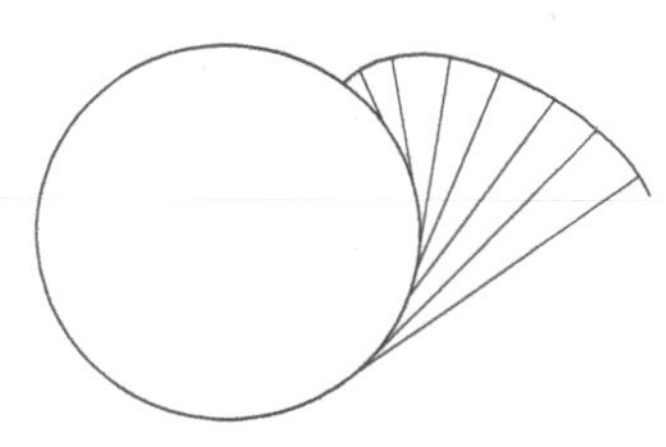

Fig. 29-1 Involute profile.

The involute gearing system is based on a rack having straight or flat-sided teeth. The sides of each tooth incline toward the center top at an angle called the *pressure angle* (Fig. 29-2). The 14½° pressure angle was standard for many years; however, the use of the 20° pressure angle has grown. Today, 14½° gearing is generally limited to replacement work. The advantages of 20°

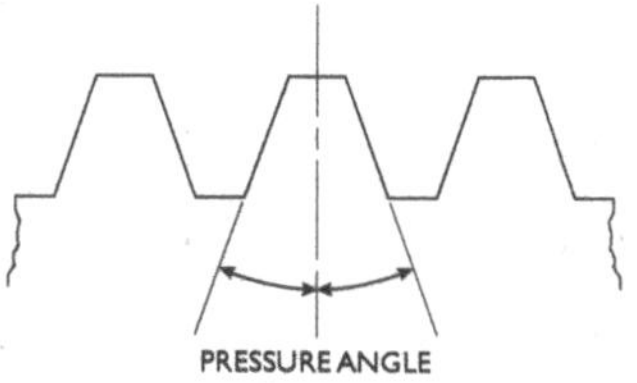

Fig. 29-2

gearing are greater strength and wear resistance, plus its use of pinions with fewer teeth.

It is extremely important that the pressure angle be known when gears are mated, as all gears that run together *must have the same pressure angle.*

Many types and designs of gears developed from the spur gear. Even though they are commonly used in industry, many are complex in design and manufacture. Some of the types in wide use are bevel gears, helical gears, herringbone gears, and worm gears. Each type, in turn, has many specialized design variations.

Pitch Diameters and Center Distances

Pitch circles are the imaginary circles that are in contact when two standard gears are in correct mesh (Fig. 29-3). The diameters of these circles are the pitch diameters of the gears. The center distance of two gears in correct mesh is equal to one-half the sum of the two pitch diameters (Fig. 29-4).

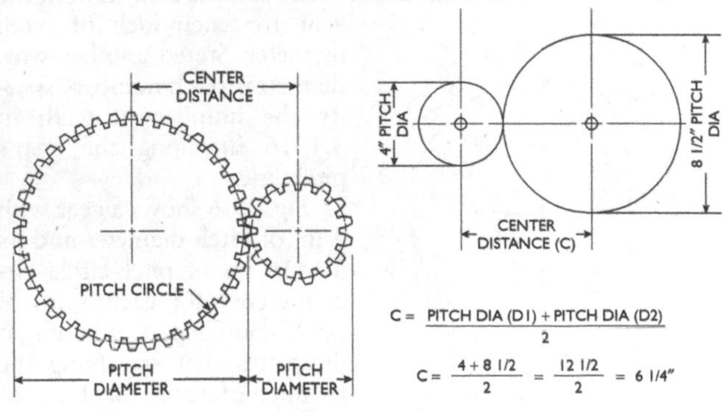

Fig. 29-3 **Fig. 29-4**

Circular Pitch

The size and proportions of gear teeth are designated by a specific type of pitch. In gearing terms, there are specific types of pitch. They are *circular pitch* and *diametral pitch.* Circular pitch is simply the distance from a point on one tooth to a corresponding point on the next tooth, measured along the pitch line, as illustrated in Fig. 29-5. Large gears are frequently made to circular pitch dimensions.

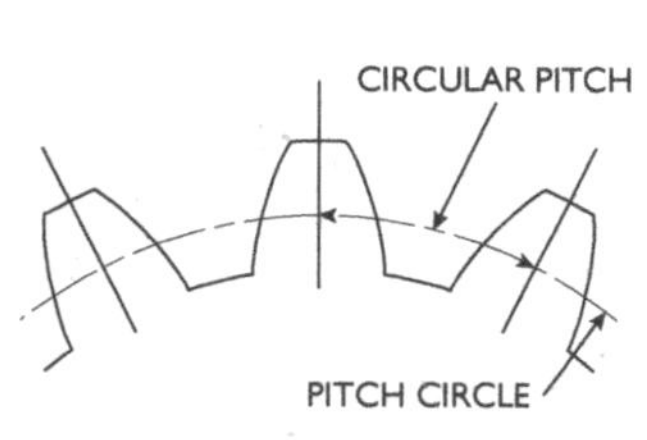

Fig. 29-5

Diametral Pitch

The *diametral-pitch* system is the most widely used gearing system, with practically all common size gears being made to diametral-pitch dimensions. Diametral-pitch numbers designate the size and proportions of gear teeth by specifying the number of teeth per inch of pitch diameter. For instance, a 12 diametral-pitch number indicates that there are 12 teeth in the gear for each inch of pitch diameter. Stated another way, diametral-pitch numbers specify the number of teeth in 3.1416 in along the gear's pitch line.

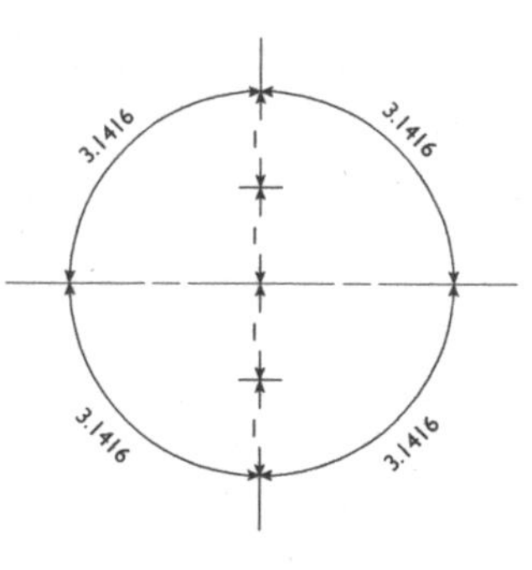

Fig. 29-6

Fig. 29-6 shows a gear with 4 in. of pitch diameter and its 3.1416 in. of pitch-circle circumference for each 1 in. of pitch diameter. In addition it illustrates that specifying the number of teeth for 1 in. of pitch diameter is, in fact, specifying the number of teeth in 3.1416 in. along the pitch line. For each 1 in. of pitch diameter there are pi ( ) in. or 3.1416 in. of pitch-circle circumference.

The fact that the diametral-pitch number specifies the number of teeth in 3.1416 in. along the pitch line may be more easily visualized when applied to the rack. In Fig. 29-7, the pitch line of a rack is a straight line, and a measurement may be made along it easily.

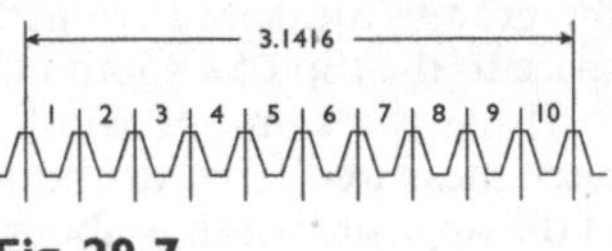

Fig. 29-7

Approximating Diametral Pitch

Determining the diametral pitch of a gear can be accomplished without precision measuring tools, templates, or gauges. Measurements need not be exact because diametral-pitch numbers are usually whole numbers. Therefore, if an approximate calculation results in a value close to a whole number, that whole number is the diametral-pitch number of the gear. There are three easy methods of determining the approximate diametral pitch of a gear. A common steel rule, preferably flexible, is adequate to make the required measurements.

Method 1

Count the number of teeth in the gear, add 2 to this number, and divide by the outside diameter of the gear. For example, the gear shown in Fig. 29-8 has 40 teeth and its outside diameter is about 4⁷⁄₃₂ in. Adding 2 to 40 gives 42; dividing 42 by 4⁷⁄₃₂ in gives an answer of 9³¹⁄₃₂ in. As this is approximately 10, it can be safely stated that the gear is a 10 diametral-pitch gear.

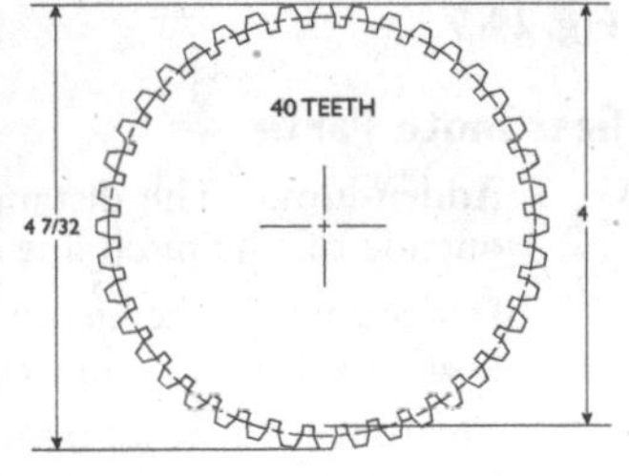

Fig. 29-8

Method 2

Count the number of teeth in the gear, and divide this number by the measured pitch diameter. The pitch diameter of the gear is measured from the root or bottom of a tooth space to the top of a tooth on the opposite side of the gear.

Using the same 40-tooth gear shown in Fig. 29-8, the pitch measured from the bottom of a tooth space to the top of the opposite tooth is about 4 in. Dividing 40 by 4 gives an answer of 10. In this case, the approximate whole-number pitch-diameter measurement results in a whole-number answer. This method also indicates that the gear is a 10 diametral-pitch gear.

Method 3

Using a flexible scale, measure approximately 3⅛ in. along the pitch line of the gear (Fig. 29-9). To do this, bend the scale to match the curvature of the gear and hold it about midway between the base and the top of the teeth. This will place the scale approximately on the pitch line of the gear. If the gear can be rotated, draw a pencil line on the gear to indicate the pitch line; this will aid in positioning the scale. Count the teeth in 3⅛ in. to determine the diametral pitch number of the gear.

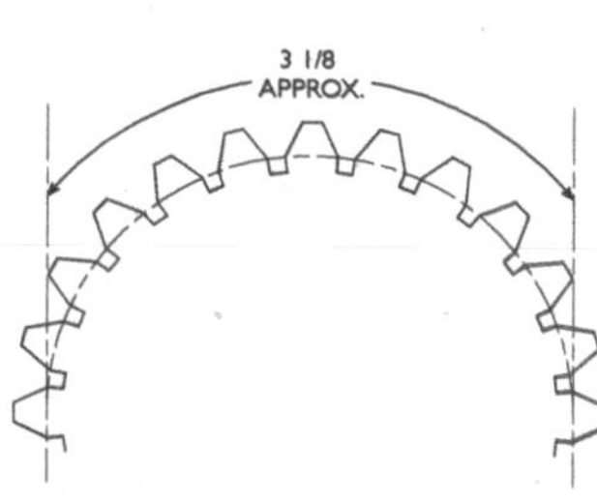

Fig. 29-9

Gear Tooth Parts

Addendum. The distance a tooth projects above, or outside of, the pitch line or circle.

Dedendum. The depth of a tooth space below, or inside of, the pitch line or circle.

Clearance. The amount by which the addendum of a gear tooth exceeds the addendum of a mating gear tooth.

Whole depth. The total height of a tooth or the total depth of a tooth space.

Working depth. The depth of tooth engagement of two mating gears. It is the sum of their addendums.

Tooth thickness. The distance along the pitch line or circle from one side of a gear tooth to the other.

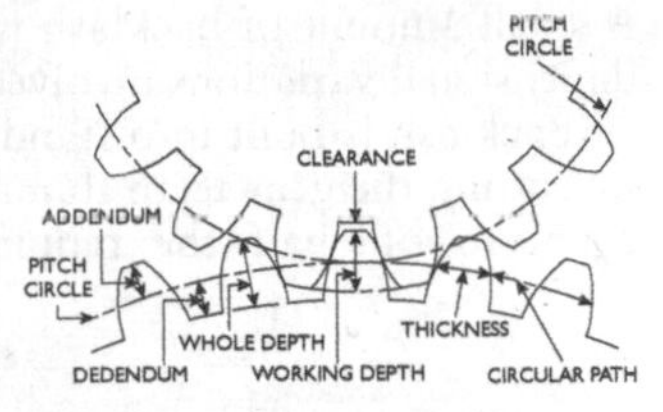

Fig. 29-10 Names of gear parts.

Gear Dimensions

The *outside diameter* of a spur gear is the pitch diameter plus two addendums.

The *bottom,* or *root diameter,* of a spur gear is the outside diameter minus two whole depths.

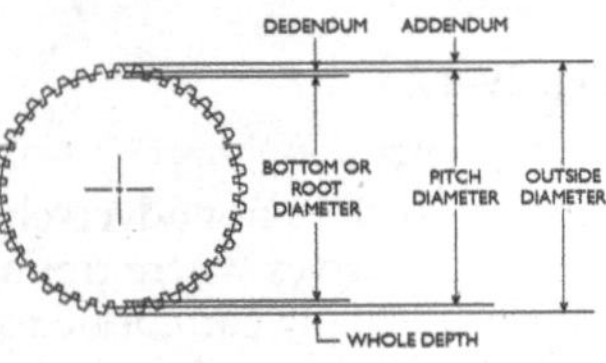

Number of Teeth = N

Diametral Pitch = $P = \frac{N}{D}$

Pitch Diameter = $D = \frac{N}{P}$

Addendum = $A = \frac{1}{P}$

Whole Depth = $WD = \frac{2.2}{P} + .002''$

Outside Diameter = $OD = D + 2A$

Root Diameter = $RD = OD - 2WD$

Fig. 29-11

Backlash

Backlash in gears is the play between teeth that prevents binding. In terms of tooth dimensions, it is the amount by which the width of tooth spaces exceeds the thickness of the mating gear teeth. Backlash may also be described as the distance, measured along the pitch line, that a gear will move when engaged with another gear that is fixed or unmovable.

Normally there must be some backlash present in gear drives to provide running clearance. This is necessary as the binding of mating gears can result in heat generation, noise, abnormal wear, possible overload, and/or failure of the drive.

A small amount of backlash is also desirable because of the dimensional variations involved in manufacturing tolerances.

Backlash is built into standard gears during manufacture by cutting the gear teeth thinner than normal by an amount equal to one-half the required figure. When two gears made in this manner are run together, at standard center distance, their allowances combine to provide the full amount of backlash required (Fig. 29-12).

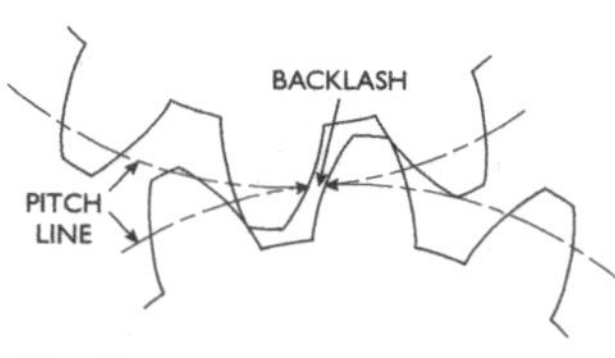

Fig. 29-12

On non-reversing drives, or drives with continuous load in one direction, the increase in backlash that results from tooth wear does not adversely affect operation. On reversing drives and drives where timing is critical, however, excessive backlash usually cannot be tolerated.

Table 29-1 lists the suggested backlash for a pair of gears operating at the standard center distance.

Table 29-1 Backlash Suggestions

Pitch	*Backlash*	*Pitch*	*Backlash*
3 P	.013	8-9 P	.005
4 P	.010	10-13 P	.004
5 P	.008	14-32 P	.003
6 P	.007	33-64 P	.0025
7 P	.006		

Gearboxes

In most industrial locations gears are supplied—rather than exposed—within a gearbox. Boxes can be purchased that are single-, double-, or triple-reduction, depending on the number of gears and shafts. In most cases, the gearbox is filled to a certain level with a high-quality gear oil, and the teeth run into the oil reservoir at the bottom of the box and carry the lubricant up to the top. Most boxes include a sight glass on the side that can be inspected to see the level of oil, and in some cases a mechanic can get an idea whether the oil looks dirty.

Most large gearboxes have an inspection panel near the top of the box that can be removed for partial visual inspection of the gears. Most boxes are built so that the top half can be removed and a complete inspection can be done including backlash checking.

Troubleshooting Gear Problems

Gear problems such as excessive wear are easy to spot on exposed gearing, but they are not so obvious when the gears operate with a closed gearbox. Of course, the box inspection plate may be removed, or the box may be disassembled for inspection, but these procedures take time and may result in excessive machinery downtime for the production department.

Spectrographic oil analysis is an excellent tool for checking potential internal gear or bearing problems on a gearbox. A small amount of oil is properly sampled and sent to a vendor laboratory for analysis. Results may show high ferrous content (indicating gear wear), or high silicon levels (indicating dirty oil), or other high levels of chrome or nickel (indicating potential bearing wear). Pulling an oil sample every six months and checking for non-problems until a problem occurs that is worthy of shutting down the box and opening it for a full investigation can save time in the long run.

30. SHAFT COUPLINGS

Couplings

Power transmission couplings are the usual means of connecting coaxial shafts so that one can drive the other. For example, they are used to connect an electric motor to a pump shaft or to the input shaft of a gear reducer, or to connect two pieces of shafting together to obtain a long length, as with line shafting. Power transmission couplings for such shaft connections are manufactured in a great variety of types, styles, and sizes. They may, however, be divided into two general groups or classifications—*rigid* couplings, also called *solid* couplings, and a second group called *flexible* couplings.

Solid Couplings

Rigid or *solid* couplings, as the names indicate, connect shaft ends together rigidly, making the shafts so connected into a single continuous unit. They provide a fixed union that is equivalent to a shaft extension. They should be used only when *true* alignment and a solid or rigid coupling are required, as with line shafting, or where provision must be made to allow parting of a rigid shaft. A feature of rigid couplings is that they are self-supporting and automatically align the shafts to which they are attached when the coupling halves on the shaft ends are connected. Follow two basic rules to obtain satisfactory service from rigid couplings:

1. Use force fit in assembly of the coupling halves to the shaft ends.
2. After assembly, make a runout check of all surfaces of the coupling, and any surface found to be running out must be machined true.

Checking surfaces is especially necessary if the coupling halves were assembled by driving or bumping, rather than pressing. Rigid couplings should *not* be used to connect shafts of independent machine units that must be aligned at assembly. Fig. 30-1a shows a typical rigid-type coupling. Fig. 30-1b shows the mating surfaces that hold the coupling in rigid alignment when assembled.

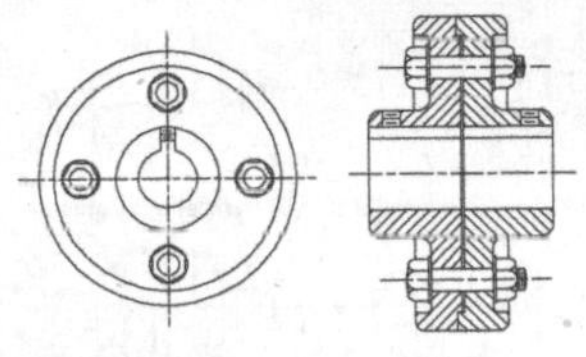

Fig. 30-1

Flexible Couplings

The transmission of mechanical power often requires the connection of two independently supported coaxial shafts so one can drive the other. Prime movers, such as internal combustion engines and electric motors, connected to reducers, pumps, variable speed drives, and so on, are typical examples. For these applications a flexible coupling is used because perfect alignment of independently supported coaxial shaft ends is practically impossible. In addition to the impracticality of perfect alignment, there is always wear and damage to the connected components and their shaft bearings, as well as the possibility of movement from temperature changes and external forces.

In addition to enabling coaxial shafts to operate satisfactorily with slight misalignment, flexible couplings allow some axial movement, or end float, and may, depending on the type, allow torsional movement as well. Another benefit that may result when flexible couplings have nonmetallic connecting elements is electrical insulation of connected shafts. In summary, four conditions may exist when coaxial shaft ends are connected. Flexible couplings are designed to compensate for some or all of these conditions:

- Angular misalignment
- Parallel misalignment
- Axial movement (end float)
- Torsional movement

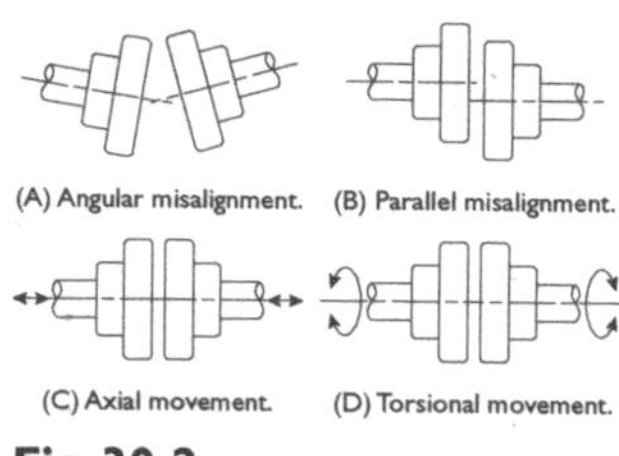

Fig. 30-2

Figure 30-2 shows the four types of flexibility provided by flexible couplings.

Flexible couplings are intended to compensate only for the slight unavoidable minor misalignment that is inherent in the design of machine components and the practices followed when aligning coaxial shafts of connected units. If the application is one where misalignment must exist, universal joints, some style of flexible shafting, or special couplings designed for offset operation may be necessary. Flexible couplings should not be used in attempting to compensate for deliberate misalignment of connected shaft ends.

Most flexible couplings are made up of three basic parts—two hubs that attach to the shaft ends to be connected and a flexing member or element that transmits power from one hub to the other. There are a variety of flexible coupling designs, all having characteristics and features to meet specific needs. Most flexible couplings fall into one or more of the following groupings.

Chain and Gear Couplings

Angular and parallel misalignment is allowed by couplings in this group, as well as end float and limited torsional flexibility. Lubrication is usually required. A chain coupling is essentially two sprockets with hubs for attachment to the shafts. The two sprockets are connected by assembling around them a length of double chain having a number of links that corresponds to the number of sprocket teeth. Gear couplings are made up of meshing internal and external gears or splines. Single-engagement types use one of each. Double-engagement types—the most widely used—use two. Flexibility in these results from both the fit of the teeth and the special shaping of the teeth that permits them to pivot relatively. In general, these couplings permit more angular deflection, parallel misalignment, and end float than chain types. With few exceptions, both chain and gear couplings must be lubricated, and they are designed with a complete housing, seals, and provisions for relubrication.

Jaw and Slider Couplings

Couplings in this category have tightly fitted parts that are constructed to allow sliding action. These couplings have two hubs that attach to the shaft ends and are constructed with surfaces designed to receive a sliding member. This sliding element, which provides the coupling's flexibility, is referred to as the *slider.* The slider is driven by jaws, keys, or openings, and it, in turn, drives the other hub by the same method. Usually slider couplings have to be lubricated.

Resilient Element Couplings

This style of coupling is probably made by more manufacturers and in more design variations than the two other styles combined. The flexible elements used in these couplings may be metal, rubber, leather, or plastic. Most widely used are metal and rubber. Couplings in this category will allow angular and parallel misalignment and end float. Torsional flexibility covers a wide range. Some have virtually none, while certain of those using rubber flexible elements have more torsional flexibility than other types of couplings. Most of the couplings in this grouping do not need lubrication, excepting those using flexible metallic elements.

Coupling Alignment Overview

Coupling manufacturers are proud to proclaim that their coupling can accept large amounts of misalignment, but the bearings and mechanical seals in the coupled equipment cannot take this amount of misalignment without failing prematurely. In fact, the large cause of premature bearing failure is misalignment. Misalignment forces, measured using a vibration meter, can be drastically reduced if coupled machines are properly aligned.

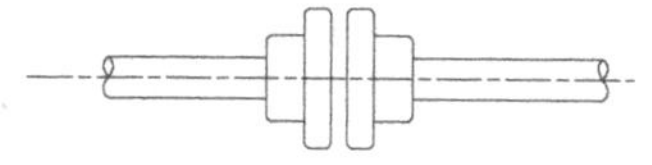

Fig. 30-3

Proper shaft alignment is achieved when the two *centerlines* of coupled machinery are the same or *collinear* (Fig. 30-3). Shaft alignment is often referred to as coupling alignment because the required measurements are commonly taken on the coupling rim or face.

There are two types of coupling or shaft misalignment: angular misalignment and parallel misalignment (Fig. 30-4). Angular misalignment is a condition where two shaft centerlines are at an angle to each other. Parallel misalignment is a condition where no angular misalignment exists and the shafts are parallel, but *offset* from each other. In most cases, the misalignment is a combination of both.

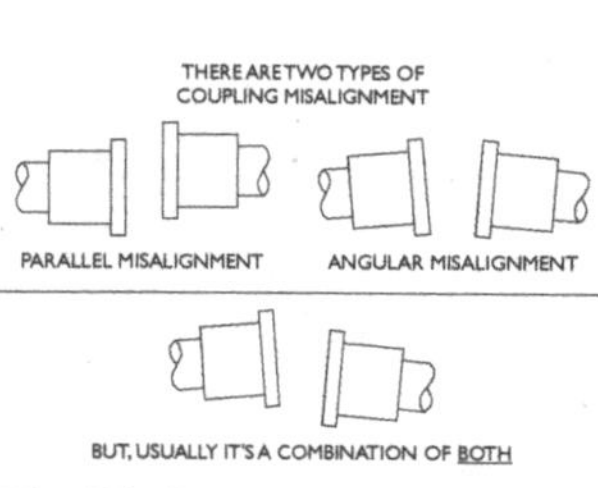

Fig. 30-4

Four basic steps are required to correct misalignment:

1. Correct for angular misalignment in the side view plane (Fig. 30-5).
2. Correct for parallel misalignment in the side view plane (Fig. 30-6).
3. Correct for angular misalignment in the top view plane (Fig. 30-7).
4. Correct for parallel misalignment in the top view plane (Fig. 30-8).

When performing a shaft alignment, one machine will be moved and shimmed into place. Usually the driven equipment should not be moved. The driver, usually the motor, starts out lower than the driven equipment (a pump, for

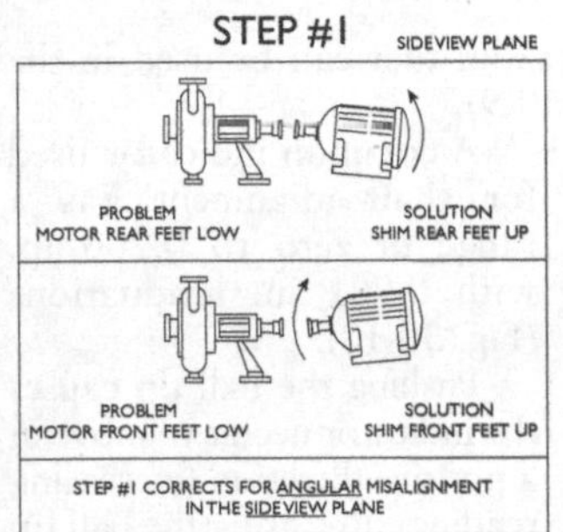

Fig. 30-5

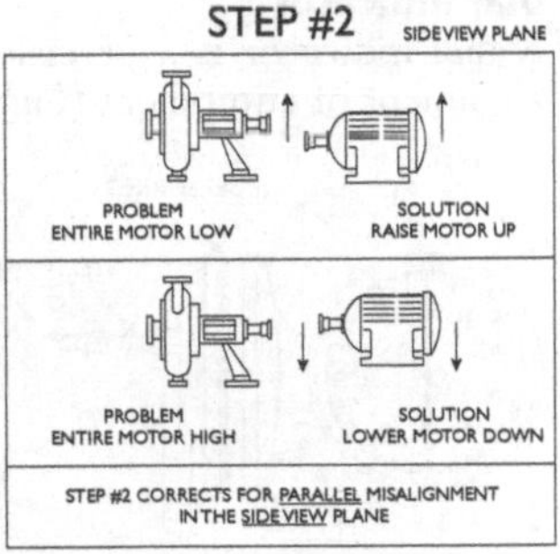

STEP #2
SIDEVIEW PLANE
PROBLEM
ENTIRE MOTOR LOW
SOLUTION
RAISE MOTOR UP
PROBLEM
ENTIRE MOTOR HIGH
SOLUTION
LOWER MOTOR DOWN
STEP #2 CORRECTS FOR PARALLEL MISALIGNMENT
IN THE SIDEVIEW PLANE

Fig. 30-6

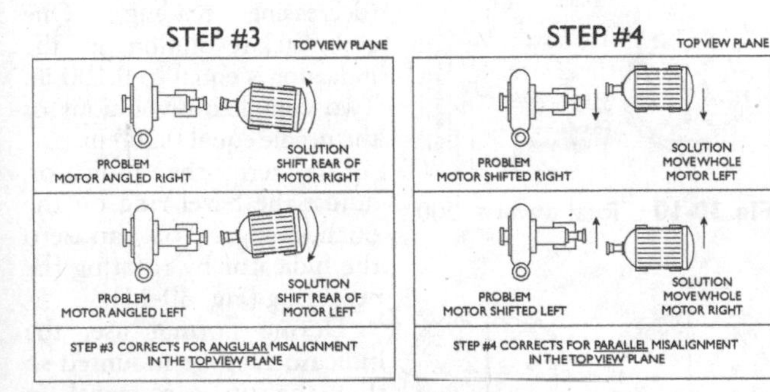

Fig. 30-7

Fig. 30-8

instance) in most conventional machinery combinations and is usually the machine that will be moved during the alignment procedure. The boltholes in a standard motor are about $\frac{1}{16}$ to $\frac{3}{16}$ in. larger in diameter than the bolts used to hold the motor down. The extra space usually leaves enough room for lateral movement of the motor during the alignment procedure.

Dial Indicator

A dial indicator is a precision tool that can be used in the alignment of equipment (Fig. 30-9).

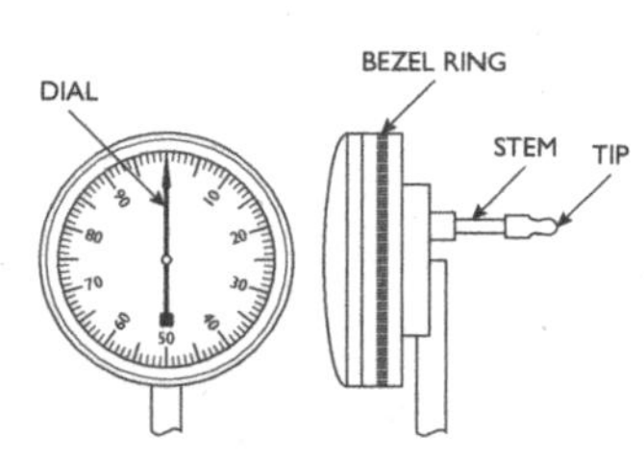

Fig. 30-9 Dial indicator.

A common indicator used for shaft alignment has a range of zero to 0.200 in. with 0.001 in. graduations (Fig. 30-10).

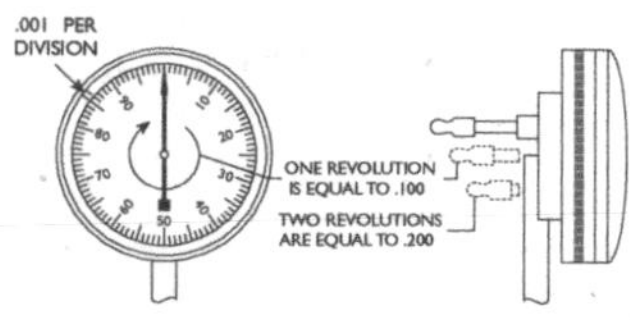

Fig. 30-10 Total travel = .200".

Pushing the ball tip causes the indicator needle to move in a positive direction (increasing reading). Releasing the ball tip causes the indicator needle to move in the negative direction (decreasing reading). One complete revolution in the indicator is equal to 0.100 in. Two complete revolutions of the needle equal 0.200 in.

To zero the indicator, adjust the bezel ring on the outside edge. You can zero the indicator by rotating the bezel ring (Fig. 30-11).

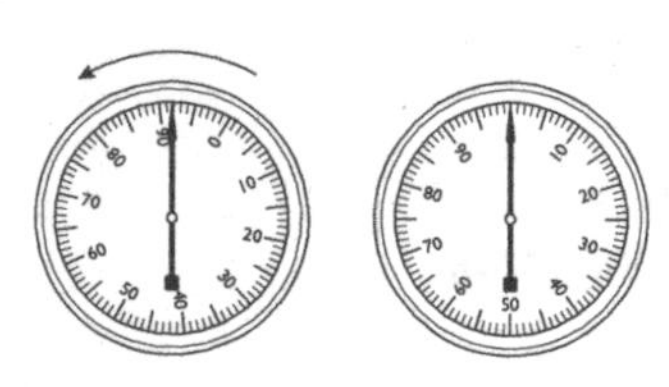

Fig. 30-11 Move bezel to zero indicator.

During normal use, the indicator will be mounted so that the tip can travel in either direction. Usually, the indicator mounting assembly is tightened so that the tip can travel 0.100 in. in either direction. Pushing in the tip will cause a positive movement, and releasing the tip will move the needle in the negative direction.

Coupling Alignment Procedure

Working through an example of a pump and a motor coupled with a flexible coupling will provide an understanding of proper alignment. To perform an alignment, follow eight important steps.

Conventions

Alignment readings are usually represented on a circle. The position of the indicator is shown in the center of the circle. This is usually represented as M-P, indicating that the readings are being taken from the motor to the pump, or P-M, indicating that the readings are being taken from the pump to the motor. Rim readings are recorded on the outside of the circle, and face readings are recorded on the inside of the circle (Fig. 30-12).

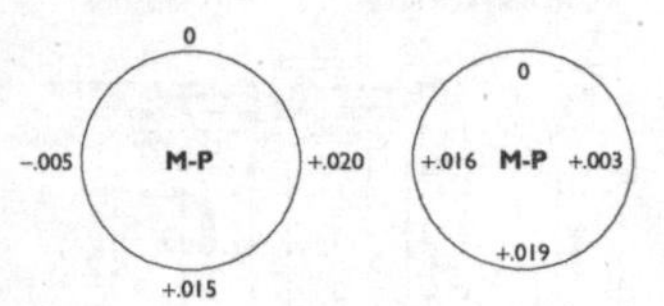

Fig. 30-12

Using the degrees of a circle as a reference, the data positions on the circle are often referred to as the 0°, 90°, 180°, and 270° readings. Another convention in use refers to the hour positions on a clock face. The corresponding data positions are then referred to as the 12 o'clock, 3 o'clock, 6 o'clock, and 9 o'clock readings (Fig. 30-13).

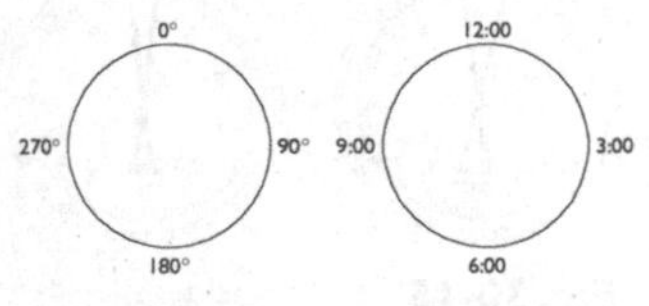

Fig. 30-13

1. Lock out and tag the machinery for safety.

Make sure the power to the equipment is shut off and locked out with a padlock. Press the start button on the machinery before beginning work to ensure that the power is turned off.

2. Check and eliminate runout due to a bad coupling or a bent shaft.

To check for coupling runout, a dial indicator can be mounted on an immovable surface, such as the machinery base. Next, depress the tip of the indicator against the coupling rim. A magnetic base is best for this purpose. If this is not available, an arrangement using the driver and the driven machinery shafts may be employed (Fig. 30-14).

The clamp is mounted on the pump, and the indicator tip is depressed on a machined surface on the motor coupling rim. The whole assembly is then tightened. With the indicator needle zeroed, the pump shaft is held steady and the motor shaft turned slowly. Runout is measured by turning the motor shaft 360° while noting the total indicator travel, in both the positive and negative directions. The total positive and negative movement is called the total indicator reading, or TIR. Fig. 30-15 shows an example.

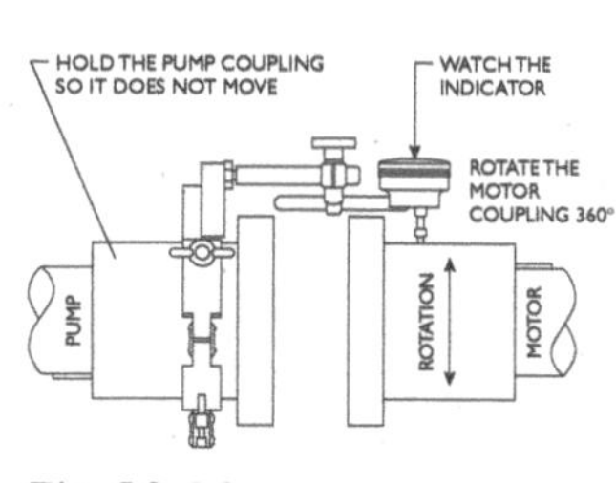

Fig. 30-14

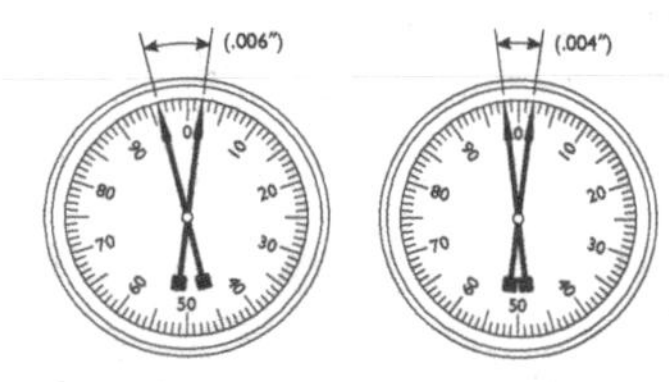

Fig. 30-15

A problem exists if the TIR is greater than 0.005 in. The cause of the problem is either a bent shaft or a bad coupling. To isolate the problem, the coupling half should be removed and the indicator tip depressed on the motor shaft (Fig. 30-16). The motor shaft is turned 360° while noting the indicator needle travel in both the positive and negative directions. If the TIR is greater than 0.005 in., then the shaft is bent and must be repaired or replaced prior to attempting an alignment

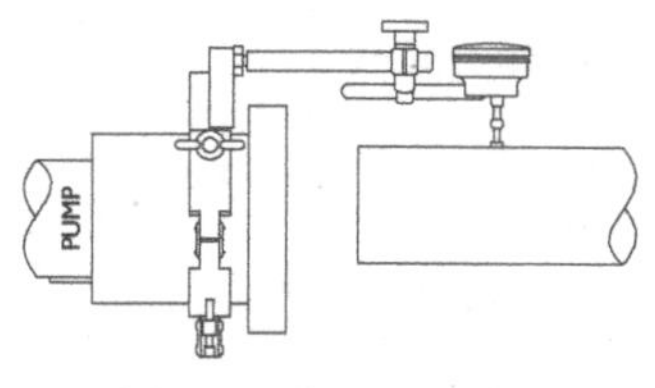

Fig. 30-16 Checking motor shaft for runout.

procedure. If the TIR is less than or equal to 0.005 in., the coupling half may be bored off center and should be replaced.

Next, check the pump shaft by mounting the clamp on the motor shaft and repeating the steps that were described to check the motor shaft and motor coupling. The same standards used to judge the motor shaft apply to replacement of the pump shaft (or repairing) and/or replacing the pump coupling.

3. Find and correct the soft foot.

A machine with four legs tends to distribute its weight more on three feet than on the fourth. This phenomenon is called a soft foot. One foot may actually lift off the base. This condition must be corrected before two shafts can be aligned.

To find and correct the soft foot, clean the area under all the motor feet and the base plate mating surfaces. Starting with a 0.002-in. feeler gauge, attempt to slide the gauge under each foot of the motor (Fig. 30-17). If the gauge slides freely under any foot, a soft foot exists. Increase the size of the feeler gauge, and determine the shim size required to correct the soft foot.

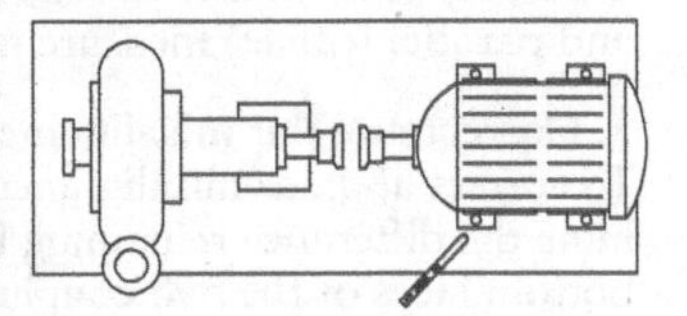

Fig. 30-17 Check for soft foot.

4. Perform a rough alignment.

Rough alignment helps eliminate errors and speeds up the use of the dial indicator. A very close rough alignment will result in relatively small corrections measured with the dial indicators.

First measure the angular misalignment in the side-view plane. Do so by measuring the difference between the gap at the top and the gap at the bottom of the coupling. Use shims under the two front motor feet or the two back motor feet to reduce the difference between both measurements as little as possible.

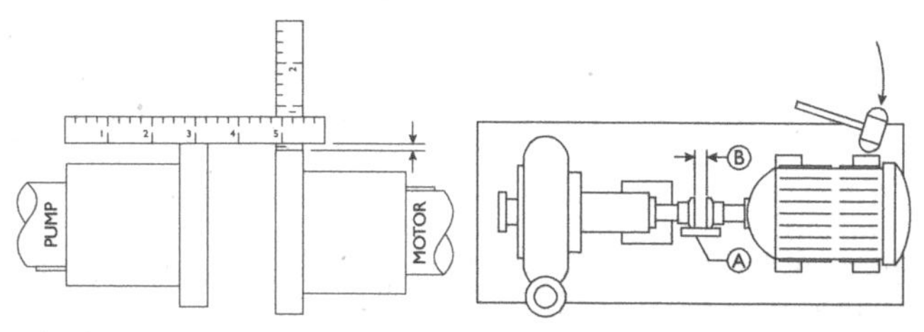

Fig. 30-18 Check parallel misalignment.

Fig. 30-19 Move motor with soft blow mallet.

Next, measure the parallel misalignment (offset) between the two coupling halves (Fig. 30-18).

Insert shims under all four motor feet to reduce this offset.

Repeat the procedure in the top-view plane. In this case, use a soft blow mallet to reduce the difference in the angular and parallel (offset) measurements (Fig. 30-19).

5. Correct angular misalignment in the side-view plane.
To correct angular misalignment in the side-view plane, determine the difference remaining in the gap between the top and bottom faces of the two coupling halves. The gap between the two faces is either wider at the top or wider at the bottom. Mount the dial indicator to measure face readings on the coupling halves. This is often easier to accomplish with the indicator mounted on a "wiggler" (Fig. 30-20).

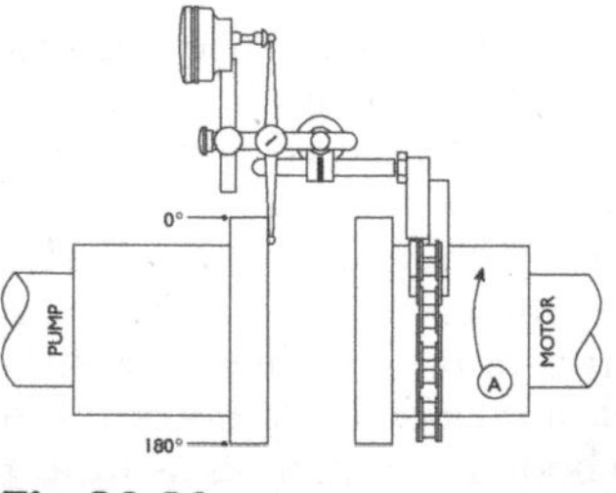

Fig. 30-20

The most common convention is to take readings from the motor to the pump. You can obtain the face readings on the inside or outside face of the coupling half.

Set the indicator or wiggle tip on the top of the face. Mark this point, called 0°, so that you can locate it easily again. Zero the indicator, and slowly turn the motor coupling, noting the direction that the indicator moves. Stop turning when the ball of the wiggler reaches the 180° mark on the face (Fig. 30-21).

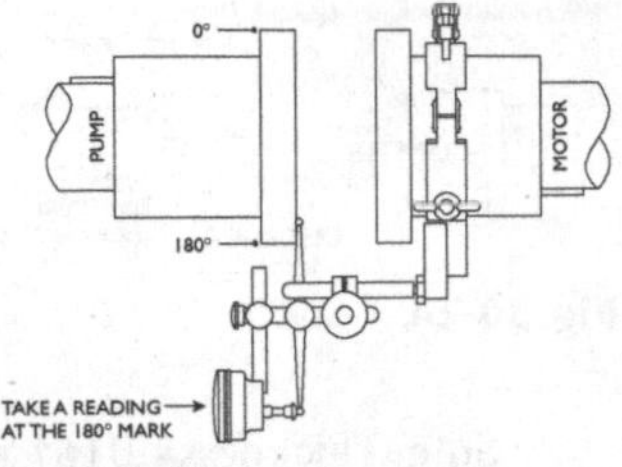

0°
PUMP
MOTOR
180°
TAKE A READING
AT THE 180° MARK
Fig. 30-21

Read the dial indicator, noting the direction of travel of the needle. The indicator reading is not a measurement of the actual gap, but rather the difference in the gap.

If the indicator moves in the positive direction, the ball on the wiggler moved *toward* the motor. The coupling halves are closer at the bottom than at the top, and the back of the motor needs to be shimmed up to reduce the difference in the gap (Fig. 30-22).

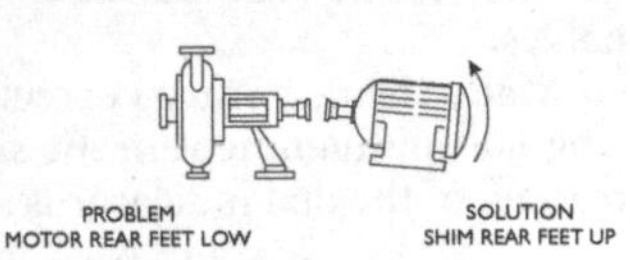

PROBLEM
MOTOR REAR FEET LOW
SOLUTION
SHIM REAR FEET UP
Fig. 30-22

If the indicator needle moves in a negative direction, the ball on the wiggle moved *away* from the motor. The coupling halves are further away at the bottom than at the top, and the front of the motor needs to be shimmed up to reduce the difference in the gap (Fig. 30-23).

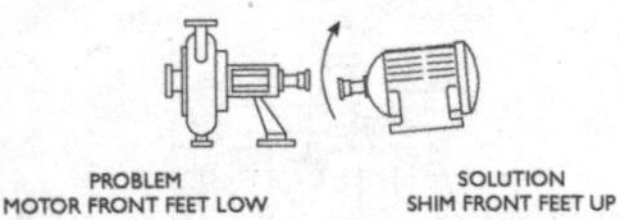

PROBLEM
MOTOR FRONT FEET LOW
SOLUTION
SHIM FRONT FEET UP
Fig. 30-23

The shims required to correct the angular misalignment can be calculated from two dimensions, the *diameter of the coupling* (or, more accurately, the diameter of the circle drawn by the indicator tip) and the *distance from the front bolthole to the back bolthole of the motor feet*. Fig. 30-24 shows the use of the proper formula and where the data must be collected.

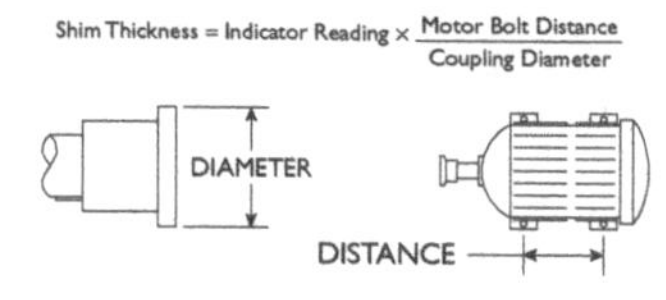

Fig. 30-24

As an example, if the indicator reading was 0.016 in. in the positive direction, the coupling diameter was 4 in. and the motor bolt distance was 12 in., then:

$$\text{Shim Thickness} = .016" \times \frac{12"}{4"} = .048$$

The back end of the motor must be raised 0.048 in. to correct for the 0.16 in. difference between the two coupling halves.

Make the necessary correction, and recheck the readings. Angular misalignment in the side plane is corrected when the reading of the dial indicator is as close to 0.000 in. as possible.

6. Correct parallel misalignment (offset) in the side-view plane.

When angular misalignment has been corrected, the two shaft centerlines are parallel but are not necessarily the same in the side-view plane. Make an accurate measurement to determine the offset between the two centerlines.

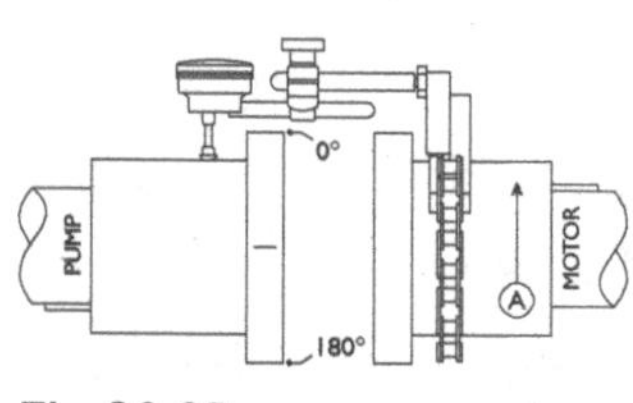

Fig. 30-25

Mount the dial indicator so that the tip is depressed against the rim of the pump coupling (Fig. 30-25).

Zero the needle, and slowly rotate the motor shaft from 0 to 180 in., noting the direction that the needle moves. Record the reading at the 180-in. mark.

To correct for the offset between the two centerlines, an equal number of shims must be added to or removed from all four motor feet.

If the indicator reading is positive, then the indicator tip was pushed in, so shims must be removed to lower the motor (Fig. 30-26).

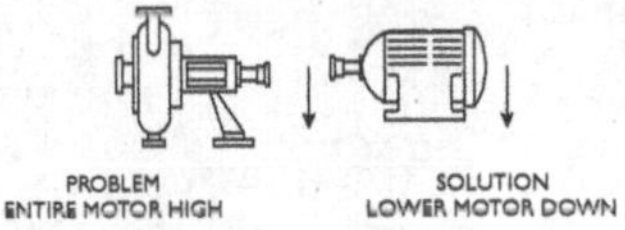

PROBLEM
ENTIRE MOTOR HIGH

SOLUTION
LOWER MOTOR DOWN

Fig. 30-26

If the indicator reading is negative, the indicator tip was let out, so shims must be added to raise the motor (Fig. 30-27).

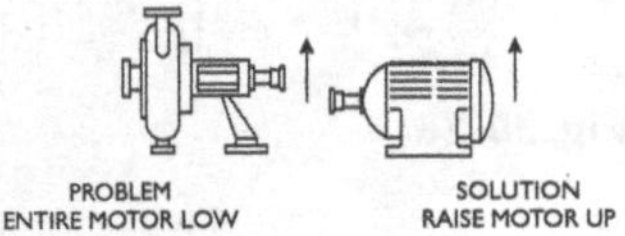

PROBLEM
ENTIRE MOTOR LOW

SOLUTION
RAISE MOTOR UP

Fig. 30-27

Calculate the shims that must be added or removed from under the motor using the following formula:

Shim Thickness = ½ × Dial Indicator Reading

As an example, assume the indicator had a reading of (negative) –0.064 in. at 180°:

Shim thickness = ½ × 0.064 in. = 0.032 in.

So, 0.032 in. of shims must be added under each motor foot to bring the motor up and correct for the offset between the two shafts.

After making the required correction, check the readings again. The idea is to make the indicator reading as close to 0.000 in. as possible.

7. Correct angular misalignment in the top-view plane. Once angular and parallel corrections are made in the side-view plane, corrections in the top-view plane should be made. The motor needs to be moved from side to side.

To correct for angular misalignment in the top-view plane, determine the difference remaining in the *gap between either side of the coupling halves*. Mount the dial indicator to measure face readings on the coupling halves using the wiggler (Fig. 30-28).

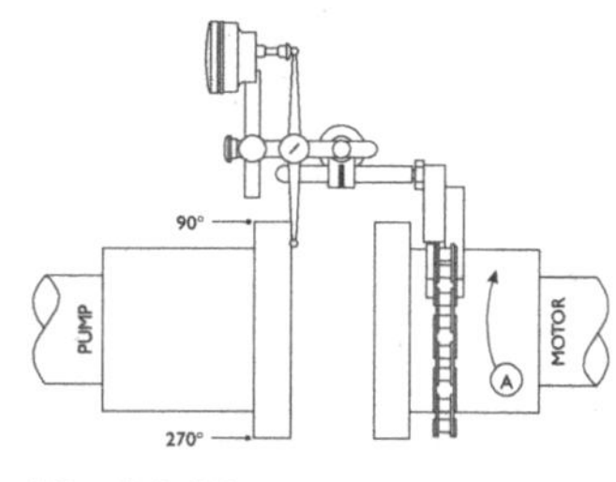

Fig. 30-28

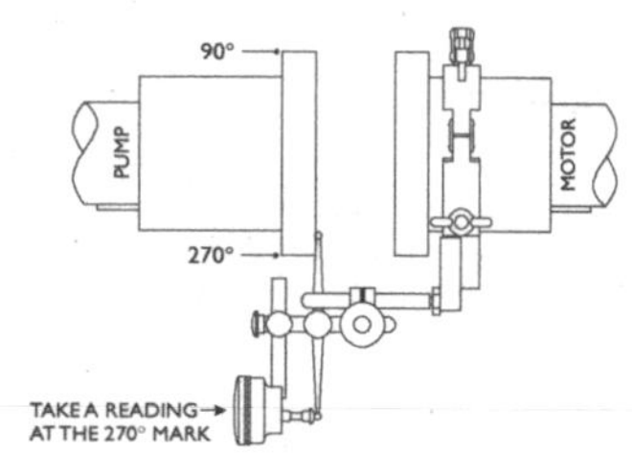

Fig. 30-29

Set the indicator or wiggle tip at a point on the face of one side of the coupling. This point should be 90° from the 0° point used in the side-view plane correction that was accomplished before. Zero the indicator, and slowly turn the motor coupling, noting the direction of movement of the indicator. Stop turning when the wiggler ball reaches 270° on the face (Fig. 30-29).

If the indicator needle moved in the positive direction, the ball on the wiggler has moved toward the motor. The coupling halves are closer at 270° than at 90°, and the back of the motor needs to be moved over to the right to reduce the difference in the gap (Fig. 30-30).

If the indicator needle moved in a negative direction, the ball on the wiggle has moved away from the motor. The coupling halves are further apart at 270° than at 90°, and the back of the motor needs to be moved over to the left to reduce the difference in the gap (Fig. 30-31).

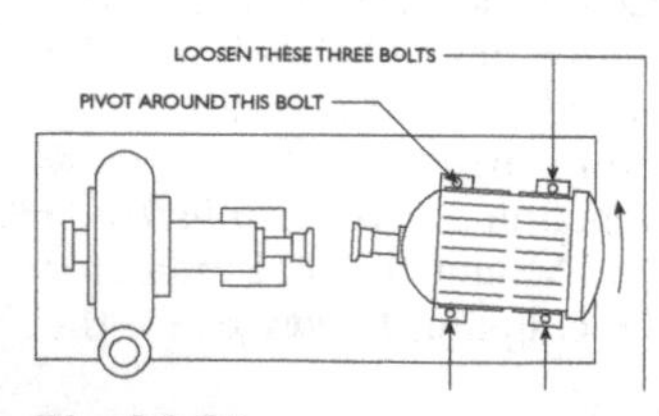

Fig. 30-30

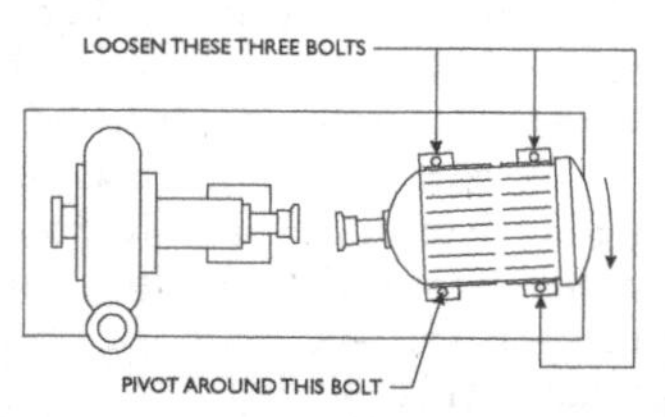

Fig. 30-31

Use a soft-blow mallet to move the back of the motor. Make the necessary corrections, and recheck the readings. Angular misalignment in the top-view plane is corrected when the dial indicator reading is as close to 0.000 in. as possible.

8. Correct parallel misalignment in the top-view plane. When angular misalignment has been corrected in the top-view plane, the two shaft centerlines are parallel but not necessarily the same. Make accurate measurements to determine parallel misalignment (offset) between the two centerlines.

Mount the dial indicator so that the tip is depressed against the rim of the pump coupling (Fig. 30-32).

Zero the needle, and slowly rotate the motor shaft from 90° to 270°, noting the direction that the needle moves. Record the reading at 270°.

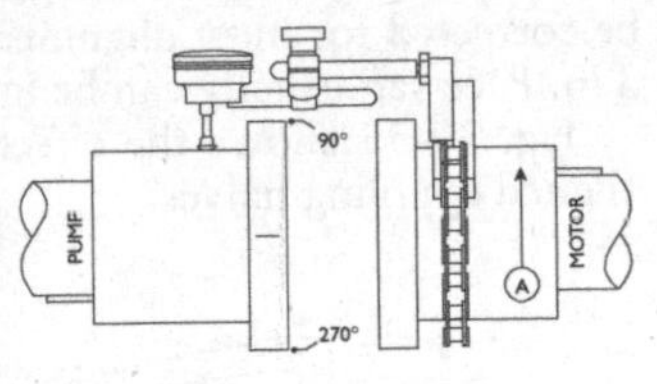

Fig. 30-32

To correct for the parallel misalignment (offset) between the two centerlines, the whole motor must be moved to one side or the other. Take care to move the motor in a straight line perpendicular to its axis of rotation.

If the indicator reading is positive, the indicator tip was pushed in, so the motor must be moved to the left (Fig. 30-33).

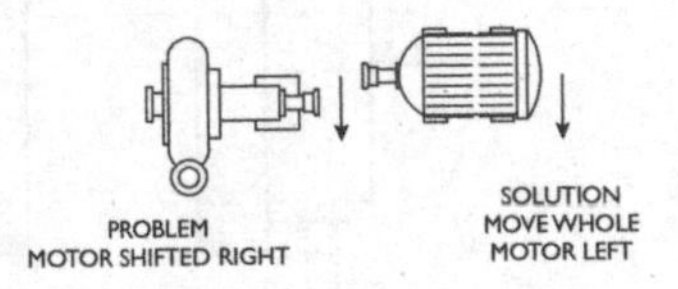

Fig. 30-33

If the indicator reading is negative, the indicator tip was let out, so the motor must be moved to the right (Fig. 30-34).

Use a soft-blow mallet to move the motor over carefully.

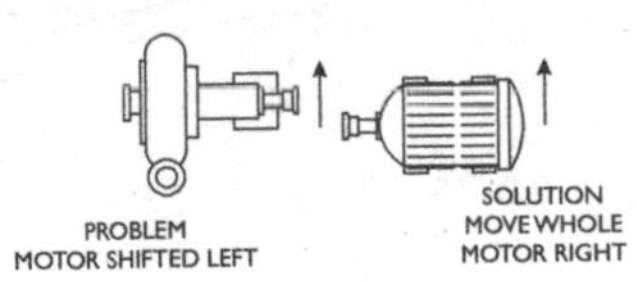

Fig. 30-34

When the necessary corrections are made, recheck the dial indicator to produce a reading as close to 0.000 in. as possible. Good craftsmanship dictates checking all the readings—side view and top view—to make sure that nothing has shifted before calling the job complete.

Bar Sag Corrections

When an alignment assembly spanning across two coupling halves is too long, the assembly tends to bend down due to weight. This is called bar sag. This bend can introduce errors in the final alignment readings. The rim reading error should be corrected for most alignment assembly spans longer than 3 in. Face sag usually can be ignored.

Fig. 30-35 shows the effect of bar sag on two perfectly aligned coupling halves.

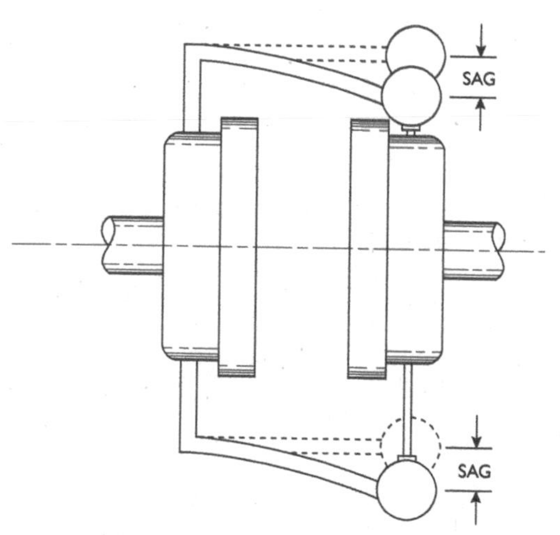

Fig. 30-35

The indicator assembly sags when on the top. With the indicator zeroed at the 0° point, the assembly is turned to the 180° point. The indicator assembly now sags in the other direction. The total indicator reading will be the sum of the two sags.

Bar sag is easy to determine. After taking the parallel readings in the side-view plane, remove the assembly and remount it on a shaft or piece of pipe. With the indicator zeroed, turn the rig upside down and read the sag indicated. As an example, if the indicator reads –0.008 in., the sag value is 0.008 in. To correct for the readings for bar sag, simply add the sag value to the 180° reading, or:

Corrected 180° Rim Reading – 180° Rim Reading (TIR) + Sag Value

Temperature Change Compensation

To compensate for temperature differences between installation conditions and operating conditions, it may be necessary to set one unit high or low when aligning. For example, centrifugal pumps handling cold water, and directly connected to electric motors, require a low motor setting to compensate for expansion of the motor housing as its temperature rises. If the same units were handling liquids hotter than the motor operating temperature, it might be necessary to set the motor high. Follow manufacturers' recommendations for the initial setting when compensation for temperature change is made at cold installation.

Make the final alignment of equipment with appreciable operating temperature difference after it has been run under actual operating conditions long enough to bring both units to operating temperatures.

31. SCREW THREADS

Screw threads used on fasteners or to mechanically transmit motion or power are made in conformance to established standards. These screw thread standards cover the cross-sectional shape of the thread, a range of diameters, and a specific number of threads per inch for each diameter. The most commonly used standard is the "Unified" standard for screw threads published by American National Standards Institute (ANSI). The Unified standard threads generally supersede the American National form (formerly known as the United States Standard) that was used for many years. Unified threads have substantially the same thread form and are mechanically interchangeable with the former American National threads of the same diameter and pitch. The principal differences between these standards relate to the application of allowances and tolerances. The general form of Unified threads shown in Fig. 31-1 is practically the same as the previous American standard.

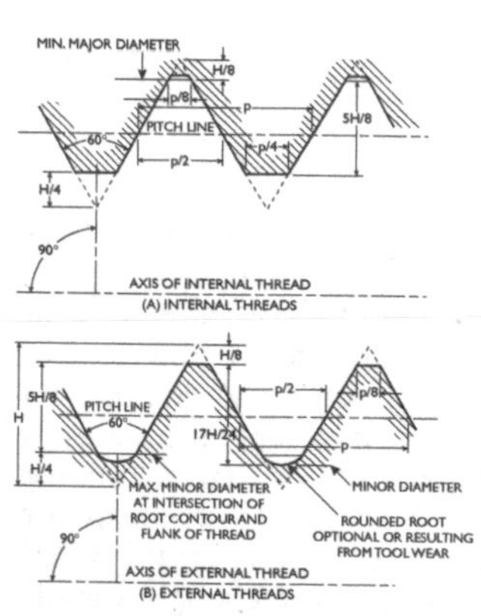

Fig. 31-1 Unified standard thread form.

Unified Standard Series

Unified standards are established by various thread series, which are groups of diameter-pitch combinations distinguished by the number of threads per inch applied to a specific diameter.

Coarse-Thread Series—Unified National Course (UNC)

Designated by the symbol *UNC*, this series is generally utilized for bolts, screws, nuts, and other general applications. It is applicable to rapid assembly or disassembly, or if corrosion or other slight damage might occur.

Fine-Thread Series—Unified National Fine (UNF)

Designated by the symbol *UNF*, this series is suitable for the production of bolts, screws, nuts, and other applications where a finer thread than that provided by the coarse series is required. External threads of this series have greater tensile stress area than comparable sizes of the coarse series. It is used where the length of engagement is short, where a smaller lead angle is desired, or where the wall thickness demands a fine pitch.

Extra-Fine Thread Series—Unified National Extra-Fine (UNEF)

Designated by the symbol *UNEF*, it is used where even finer pitches of threads are desirable for short lengths of engagement and for thin-walled tubes, nuts, ferrules, or couplings.

Constant-Pitch Series—Unified National Form (UN)

Designated by the symbol *UN*, various constant pitches are used on a variety of diameters. Preference is given wherever possible to use of the 8-, 12-, and 16-thread series.

Thread Classes

The Unified standard also establishes limits of tolerances called classes. Classes 1A, 2A, and 3A apply to external threads only, and Classes 1B, 2B, and 3B apply to internal threads only. Thread classes are distinguished from each other by the amounts of tolerance and allowance they provide. Classes 3A and 3B provide a minimum and Classes 1A and 1B a maximum.

Classes 2A and 2B are the most commonly used thread standards for general applications, including production of bolts, screws, nuts, and similar threaded fasteners. Classes

3A and 3B are used when close tolerances are desired. Classes 1A and 1B are used on threaded components where quick and easy assembly is necessary and a liberal allowance is required to permit ready assembly, even with slightly bruised or dirty threads.

Unified Thread Designation

When designating a screw thread, the standard method is to specify in sequence the nominal size, number of threads per inch, thread series symbol, and thread class symbol. For example, a ¾-in. Unified coarse-series thread for a common fastener would be designated as shown in Fig. 31-2.

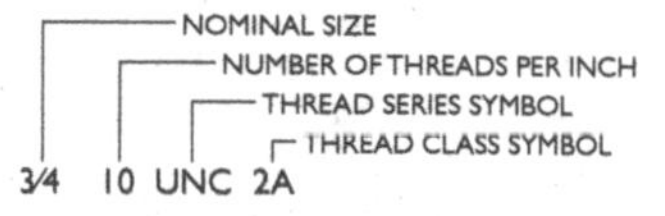

Fig. 31-2

Unless otherwise specified, threads are right-hand. A left-hand thread is designated by adding the letters LH after the thread class symbol.

The threads per inch for each UNC, UNF, and UNEF series from number 4 diameter to 3-in. diameter are listed in Table 31-1.

Table 31-1 Unified Standard Threads per Inch

Nominal Size	*UNC Threads per Inch*	*UNF Threads per Inch*	*UNEF Threads per Inch*
4	40	48	
5	40	44	
6	32	40	
8	32	36	
10	24	32	
12	24	28	32
¼	20	28	32
5⁄16	18	24	32
⅜	16	24	32
7⁄16	14	20	28
½	13	20	28
9⁄16	12	18	24
⅝	11	18	24
11⁄16			24
¾	10	16	20

Nominal Size	UNC Threads per Inch	UNF Threads per Inch	UNEF Threads per Inch
13/16			20
7/8	9	14	20
15/16			20
1	8	12	20
1 1/16			20
1 1/8	7	12	18
1 3/16			18
1 1/4	7	12	18
1 5/16			18
1 3/8	6	12	18
1 7/16			18
1 1/2	6	12	18
1 9/16			18
1 5/8			18
1 11/16			18
1 3/4	5		
2	4 1/2		
2 1/4	4 1/2		
2 1/2	4		
2 3/4	4		
3	4		

Screw Thread Terms

Only the more important screw thread terms are included in the following definitions. Many of these terms are illustrated in Fig. 31-1.

Major diameter. The largest diameter of a screw thread. The term *major diameter* applies to both internal and external threads. It replaces the term *outside diameter* as applied to the thread of a screw and the term *full diameter* as applied to the thread of a nut.

Minor diameter. The smallest diameter of a screw thread. The term *minor diameter* applies to both internal and external threads. It replaces the terms *root diameter* as applied to the thread of a screw and *inside diameter* as applied to the thread of a nut.

Pitch diameter. The diameter of an imaginary cylinder, the surface of which would pass through the threads at such points as to make equal the width of the threads and the width of the spaces cut by the surface of the cylinder.

Pitch. The distance from a point on the screw thread to a corresponding point on the next thread measured parallel to the axis.

Lead. The distance a screw advances axially in one turn. On a single-thread screw, the lead and pitch are the same; on a double-thread screw, the lead is twice the pitch; on a triple-thread screw, the lead is three times the pitch, and so on.

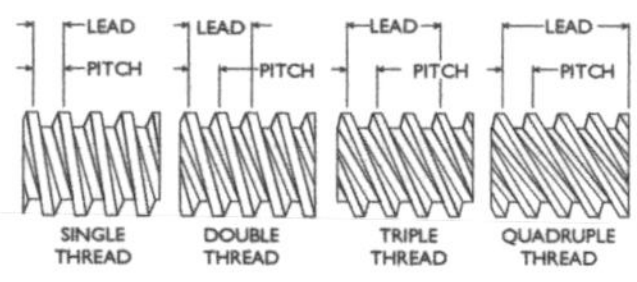

Fig. 31-3 Relation of lead and pitch of multiple threads.

Multiple thread. A screw thread that is formed of two or more single threads, as shown in Fig. 31-3. A double thread has two separate or single threads starting diametrically opposite or at points 180° apart. A triple thread has three single threads starting at points 120° apart. A quadruple thread has four single threads starting at points 90° apart. A multiple thread is used to increase the lead of a screw.

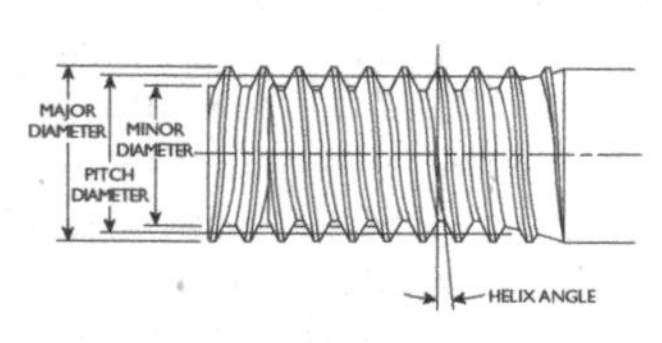

Fig. 31-4 Illustrating the helix angle.

Angle of thread. The angle included between the sides of the thread measured in an axial plane (Unified thread form is a 60° angle).

Helix angle. The angle made by the helix of the thread at the pitch diameter with a plane perpendicular to the axis, as shown in Fig. 31-4.

Depth of thread. The distance between the crest and the root of the thread measured normal to the axis.

Double Depth of Screw Threads

The double depth of a screw thread is equal to twice the thread depth (2 × depth). Its principal use is to determine the hole size required to produce an internal thread of a given pitch (threads per inch). Table 31-2 lists the double depth for various pitch Unified Form threads. Double depth may also be stated as the difference between the major and minor thread diameters. Therefore, minor thread diameter may be determined by subtracting the double depth from the major thread diameter. For example, hole size for a 2.000-in. × 20 threads per inch internal thread may be determined by subtracting the double depth for 20 threads per inch from 2.000 (2.000 – 0.064 = 1.935). Thread depth as well as major and minor diameters are illustrated in Fig. 31-5.

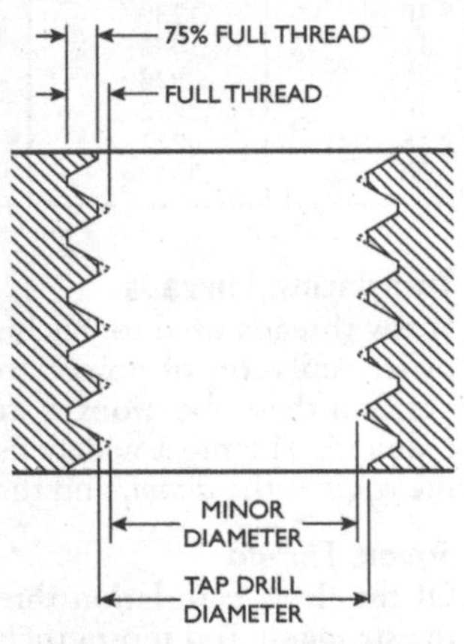

Fig. 31-5

A hole larger than the minor diameter is required to provide mating clearance. For precise machined threads this may be a very small amount. For tapped threads, a greater amount is allowed. Practice for tapped threads is to provide 25 percent clearance, resulting in a 75 percent full thread. Fig. 31-5 illustrates the relationship between full and 75 percent full threads.

Table 31-2 Double Depth of Threads

Threads per inch	Double depth	Threads per inch	Double depth
2	.6495	22	.0590
3	.4330	24	.0541
4	.3247	26	.0500
5	.2598	28	.0464
6	.2165	30	.0433
7	.1856	32	.0406
8	.1624	36	.0361
9	.1443	40	.0325
10	.1230	44	.0295
12	.1082	48	.0271
14	.0928	56	.0232
16	.0812	64	.0203
18	.0722	72	.0180
20	.0649	80	.0162

Translation Threads

Screw threads used to machine parts for adjustment, setting, or transmission of power are called *translation threads*. To perform these functions a stronger form than the V form is required. The most widely used translation thread forms are the *square*, the *acme*, and the *buttress*.

Square Thread

Of the three translation thread forms, the square thread is the strongest and most efficient, but it is also the most difficult to manufacture because of its parallel sides. The theoretical proportions of a square external thread are shown in Fig. 31-6. The mating nut must have a slightly larger thread space than the screw to allow a sliding fit. Similar clearance must also be provided on the major and minor diameters.

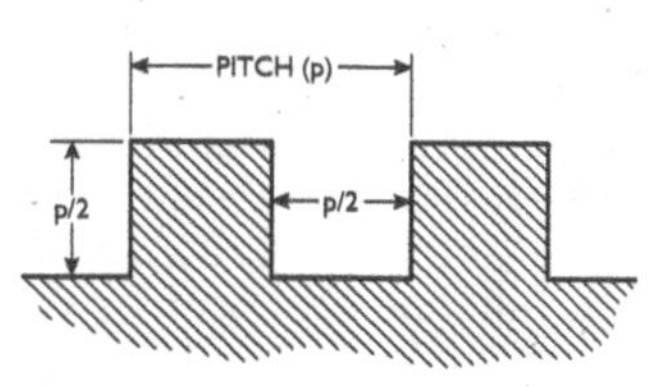

Fig. 31-6 Square thread.

Acme Thread

The acme form of thread has largely replaced the square thread for most applications. Although it is not quite as strong as the square thread, it is preferred for most applications because it is fairly easy to machine. The angle of an acme thread, measured in an axial plane, is 29°. The basic proportions of an acme thread are shown in Fig. 31-7. The *ANSI* for acme screw threads establishes thread series, fit classes, allowances and tolerances, and so on, for this thread form, similar to the standard for the Unified thread form.

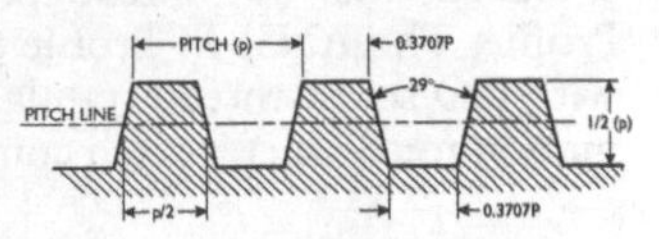

Fig. 31-7 Acme thread.

Buttress Thread

The buttress thread has one side cut approximately square and the other side slanting. It is used when a thread having great strength along the thread axis in one direction only is required. Because one side is cut nearly perpendicular to the thread axis, there is practically no radial thrust when the thread is tightened. This feature makes this thread form particularly applicable where relatively thin tubular members are screwed together. The basic thread form of a simple design of buttress thread in common use is shown in Fig. 31-8. Other buttress thread forms are more complex, with the load side of the thread inclined from perpendicular to facilitate machining. The *ANSI* buttress thread form angle of inclination is 7°.

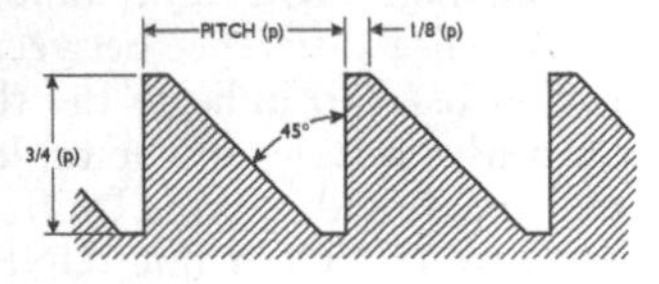

Fig. 31-8 Buttress thread.

Metric Screw Threads

In recent years, there has been a dramatic increase in the use of metric system threads in the area of mechanical fasteners. This can be attributed to the import and export business, most particularly to machine tool and automobile imports. Consequently, a need developed for metric thread information and domestic standardization. Efforts to establish standards resulted in the adoption by the American National Standards Institute (ANSI) of a 60° symmetrical screw thread with a basic International Organization for Standardization (ISO) profile. This is termed the ISO 68 profile. In the ANSI standard, this 60° thread profile is designated as the M-Profile. The ANSI M-Profile Standard is in basic agreement with ISO screw thread standards and features detailed information for diameter-pitch combinations selected as preferred standard sizes. The basic M-Profile is illustrated in Fig. 31-9.

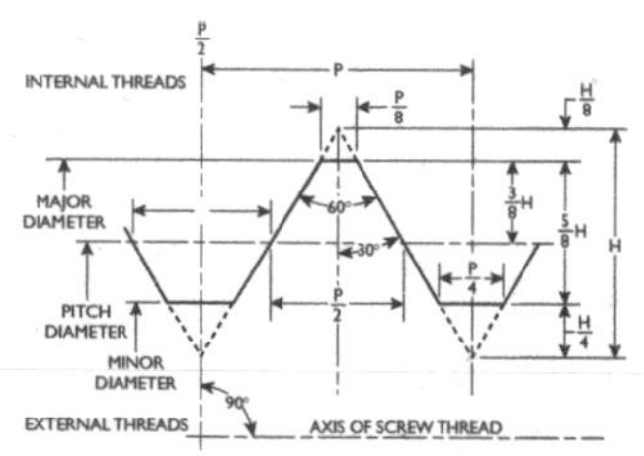

Fig. 31-9 Basic M thread profile (ISO basic profile).

The designations, terms, and so on used in the new metric standard differ in many respects from those used in the familiar Unified standard. Because the Unified standard incorporates terms and practices of long standing with which mechanics have knowledge and experience, an explanation of the M-Profile metric system may best be accomplished by comparing metric M-Profile standards to Unified standards.

A major difference between the standards is in the designation used to indicate the thread form. The Unified standard uses a series of capital letters, which not only indicate that the thread is of the Unified form, but also indicate if it is a coarse (UNC), a fine (UNF), an extra-fine (UNEF), or a constant pitch (UN), which is a series of pitches used on a variety of diameters. In the metric M-Profile standard the capital letter M is used to indicate the thread is metric M-Profile with no reference to pitch classification.

A second major difference is that the Unified standard designates the thread pitch in terms of the number of threads in one inch of thread length. The metric M-Profile standard states the specific pitch measurement—that is, the dimension in millimeters from the centerline of one thread to the centerline of the next thread.

A third major difference is that the Unified system establishes limits of tolerance called classes. Classes 1A, 2A, and 3A apply to external threads only; Classes 1B, 2B, and 3B apply to internal threads only. Classes 3A and 3B provide a minimum, and Classes 1A and 1B a maximum. In the metric M-Profile standard, the tolerance designation uses a number and a letter to indicate the pitch diameter tolerance, as well as a number and a letter to designate crest diameter tolerance. This results in four symbols for tolerance designation rather than the two used in the Unified system. In the metric M-Profile standard, the lowercase letter "g" is used for external threads and the capital letter "H" is used for internal threads. The numbers 4, 5, 6, 7, 8 are used to indicate degree of internal thread tolerance; the numbers 3, 4, 5, 6, 7, 8, 9 indicate the degree of external thread tolerance. Thus two nominally similar threads are designated as follows:

Unified	*Metric M*
3/8—16—UNC—2A	M10 × 1.5—6g6g

Common shop practice in designating Unified screw threads is to state only the nominal diameter and the threads per inch—for example, 3/8-16 for a coarse thread and 3/8-24 for a fine thread. Most mechanics recognize the thread series by the threads-per-inch number. As to the thread class symbol, it is usually assumed that if no class number is stated, the thread is a Class 2A or 2B. This is a safe assumption because Classes 2A and 2B are commonly used tolerances for general applications, including production of bolts, screws, nuts, and similar fasteners. When required to produce screw threads, most mechanics use commercial threading tools (that is, taps and dies), or if chasing threads, commercial gauging tools for size checking. These commonly used tools are made to Class 2A or 2B tolerances; therefore, the mechanic depends on tool manufacturers to maintain screw thread accuracy. Because manufacturers' products have extremely close tolerances, this practice is highly satisfactory and relieves the mechanic of practically all concern for maintaining sizes and fits within stated tolerances.

Reaching this same condition was one of the objectives during efforts to establish metric M-Profile standards. Because metric threads are in wide use throughout the world, a highly desirable condition would be one in which the standard is compatible with those in world use. As the International Organization for Standardization (ISO) metric standard was, for all practical purposes, recognized as the worldwide metric standard, the metric M-Profile standard is in large measure patterned after it. It has a profile in basic agreement with the ISO profile, and it features detailed information for diameter-pitch combinations selected as preferred standard sizes.

Metric M-Profile screw threads are identified by letter (M) for thread form, followed by the nominal diameter size and pitch expressed in millimeters, separated by the sign (×) and followed by the tolerance class separated by a dash (—) from the pitch. For example, a coarse-pitch metric M-Profile thread for a common fastener would be designated as shown in Fig. 31-10.

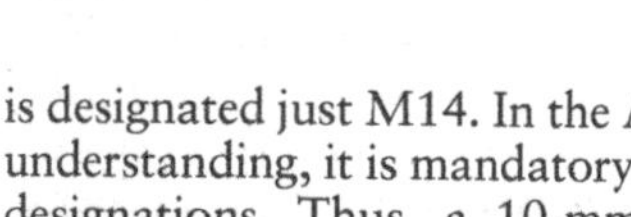

EXTERNAL THREAD M-PROFILE, RIGHT HAND
METRIC THREAD SYMBOL (ISO 68 METRIC THREAD)
NORMAL SIZE
PITCH
TOLERANCE CLASS
M 10 × 1.5 – 6g6g

Fig. 31-10

The simplified international practice for designating coarse-pitch ISO screw threads is to leave off the pitch. Thus, a M14 × 2 thread is designated just M14. In the ANSI standard, to prevent misunderstanding, it is mandatory to use the value for pitch in all designations. Thus, a 10-mm coarse thread is designated M10 × 1.5. When no tolerance classification is stated, it is assumed to be classification 6g6g (usually stated as simply 6g), which is equivalent to the United classification 2A, the commonly used classification for general applications.

The standard metric screw thread series for general-purpose equipment's threaded components and mechanical fasteners is a coarse-thread series. The diameter-pitch combinations selected as preferred standard sizes in the coarse-pitch series are listed in Table 31-3. These combinations are in basic agreement with ISO standards.

The metric M-Profile designation does not specify series of diameter/pitch combinations as does the Unified system (that is, coarse, fine, etc.). Although no indication of such grouping is given in the designation of a metric thread, series groupings are recommended. The coarse-pitch series of diameter/pitch combinations shown in Table 31-3 are described as standard metric screw thread series for general-purpose equipment's threaded components and mechanical fasteners.

A second series, called fine-pitch M-Profile metric screw threads and shown in Table 31-4, lists additional diameter/pitch combinations that are standard for general-purpose equipment's threaded components. These combinations, in some instances, list more than one pitch for a nominal diameter. As with the coarse-pitch series, they are in basic agreement with ISO screw standards.

Table 31-3 Standard Coarse-Pitch M-Profile General-Purpose and Mechanical Fastener Series

Nominal Size	*Pitch*	*Nominal Size*	*Pitch*
1.6	0.35	20	2.5
2	0.4	24	3.0
2.5	0.45	30	3.5
3	0.5	36	4.0
3.5	0.6	42	4.5
4	0.7	48	5.0
5	0.8	56	5.5
6	1.0	64	6.0
8	1.25	72	6.0
10	1.5	80	6.0
12	1.75	90	6.0
14	2.0	100	6.0
16	2.0		

Table 31-4 Standard Fine-Pitch M-Profile Screw Threads

Nominal Size	Pitch		Nominal Size	Pitch	
8	1		40	1.5	
10	0.75	1.25	42	2.0	
12	1	1.25	45	1.5	
14	1.25	1.5	48	2.0	
15	1		50	1.5	
16	1.5		55	1.5	
17	1		56	2.0	
18	1.5		60	1.5	
20	1		64	2.0	
22	1.5		65	1.5	
24	2		70	1.5	
25	1.5		72	2.0	
27	2		75	1.5	
30	1.5	2.0	80	1.5	2.0
33	2		85	2.0	
35	1.5		90	2.0	
36	2		95	2.0	
39	2		100	2.0	

Screw Thread Tapping

Cutting internal screw threads with a hand tap, called tapping, is an operation frequently performed by mechanics. Often this is a troublesome operation involving broken taps and time-consuming efforts to remove them. Some of these troubles may be avoided by a better understanding of the tapping operation and the tools involved.

Because threads must be made in many types of materials ranging from steel and cast iron to brass, aluminum, and plastics, procedures must be varied. Also, a thread that goes all the way through a part presents a different problem from a thread in a blind hole. Fig. 31-11 indicates the names of the various parts of a tap.

A tap basically consists of a shank, which has flats on the end to hold and drive it, and a threaded body, which does the thread cutting. The threaded body is composed of lands, which are the cutters, and flutes or channels to let the chips out and permit cutting fluid to reach the cutting edges. The threaded body is chamfered or tapered at the point to allow the tap to enter a hole and to spread the heavy cutting operation over several rings of lands or cutting edges. The radial thread relief shown in Fig. 31-11 refers to material having been removed from behind the cutting edges to provide clearance and reduce friction. Another form of clearance provided on commercial taps is termed "back taper." This is done by a very slight reduction in the thread diameter at the shank end. This provides thread relief so that the cutting is performed by the forward end of the tap and the remainder of the threads do not drag.

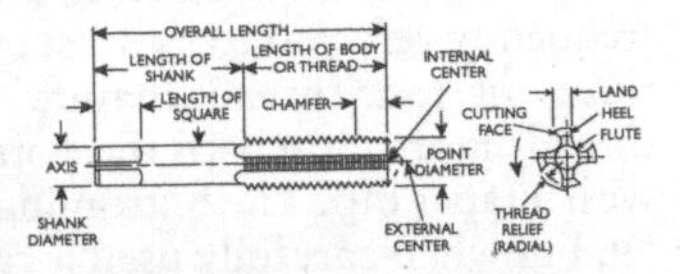

Fig. 31-11 Typical hand tap.

Because no single tap could possibly meet all the difficult tapping requirements, the manufacturers of threading tools modify the basic tap design in several ways, making tools that are especially suitable for particular tapping needs. The number of flutes may vary from two to as many as nine. Some taps have spiral flutes—winding either to the left or the right. Unless specific reference to the number of flutes is made, taps are supplied having the standard number of flutes for a given size and type. The most widely used sizes for general use are made with four flutes. The flutes serve to bring lubricant to the tap's cutting edges and provide chip clearance. For tapping especially tough materials, some taps are made with interrupted threads to provide even more chip accommodation and better cutting-edge lubrication. These interrupted thread taps have alternate cutting teeth omitted.

The "chamfer" on a tap refers to the angular reduction in diameter of the leading threads at its point. This chamfer allows the tap to enter the hole and the gradually increasing diameter of the threads serve to lead the threaded body of the tap into the hole as it is turned. The longer the chamfer, the smaller the chip each thread must cut, as the load is distributed over a greater number of cutting edges. Three different chamfer lengths are in common use; these are designated as "taper," "plug," and "bottoming."

Taper taps have the longest chamfer (8 to 10 threads) and are usually used for starting a tapped hole. The taper tap is frequently referred to as a "starting" tap. The plug tap has a three- or four-thread chamfer and is used to provide full threads more closely to the bottom of a hole than is possible with a taper tap. The bottoming tap has practically no chamfer, but when carefully used it can tap to the bottom of a hole if it is preceded by a plug tap. Taper taps should always be used for starting a tapped hole. While a tapped hole may be started with a plug tap if care is used, the load of the leading threads is extremely heavy because of the short chamfer. Do not attempt to start a tapped hole with a bottoming tap.

The first operation in internal threading with a hand tap is to drill the proper diameter hole. The usual method of selecting the drill size is to consult a "tap-drill size chart" as shown in Table 31-5 and Table 31-6. Note that the charts carry the statement *Based on Approximately 75 Percent Full Thread.*

This notation means that the drill diameter is larger than the minor diameter of the thread to be tapped. The drill size noted in the table will produce a hole enough oversize so that 25 percent of the crest of the internal thread will be missing.

An oversize tap hole is made to provide clearance between the tap hole and the minor diameter of the tap. If this is not done there will be no clearance and the tap will turn hard, tear the threads, and run a high risk of breakage. The 25 percent that is missing from the crest of the internal thread does not appreciably reduce its strength.

To tap a hole with a hand tap, begin with the taper tap. Start the tap by placing it in the hole and carefully turn it about one half turn until it starts to cut. Check the position of the tap by eye to keep it square with the work surface. Back it up to break the chip, and again turn it in about one half turn. Each time, check the tap to be sure it is square. After the tap is well started, check it with a square. If it is not true, pressure must be exerted in the direction to correct its position on the next several cuts. Use plenty of lubricant and occasionally completely remove the tap from the hole to clean out chips. Continue alternately cutting and reversing the tap to break the chips. When the tap is cutting properly, one can "feel" it as it is turned. If the tap resists turning and there is a springy feeling, it is not cutting properly and should be removed and inspected.

To tap a blind hole, use first the taper tap, then the plug tap, and finally the bottoming tap. Each tap must be carefully turned in until bottomed. Because the chips cannot fall through, special attention must be paid to chip removal. The tap must be removed more frequently than for a through hole, and the workpiece must be inverted and jarred to remove the chips. If the work cannot be inverted, blow the chips out with air or remove them with a magnet. In special cases, it might even be necessary to use a vacuum for chip removal.

Table 31-5 Tap-Drill Sizes—Based on Approximately 75 Percent Full Thread

National Coarse and Fine Threads				*Taper Pipe*		*Straight Pipe*	
Thread	*Drill*	*Thread*	*Drill*	*Thread*	*Drill*	*Thread*	*Drill*
0–80	3/64	7/16–14	U	⅛–27	R	⅛–27	S
1–64	No.53	7/16–20	25/64	¼–18	7/16	¼–18	29/64
1–72	No.53	½–12	27/64	⅜–18	37/64	⅜–18	19/32
2–56	No.50	½–13	27/64	½–14	23/32	½–14	47/64
2–64	No.50	½–20	29/64	¾–14	59/64	¾–14	15/16
3–48	No.47	9/16–12	31/64	1–11½	1 5/32	1–11½	1 3/16
3–56	No.45	9/16–18	33/64	1¼–11½	1½		1 33/64
4–40	No.43	⅝–11	17/32	1½–11½	1 47/64	1½–11½	1¾
4–48	No.42	⅝–18	37/64	2–11½	2 7/32	2–11½	2 7/32
5–40	No.38	¾–10	21/32	2½–8	2⅝	2½–8	2 21/32
5–44	No.37	¾–16	11/16	3–8	3¼	3–8	3 9/32

(continued)

Table 31-5 (continued)

National Coarse and Fine Threads				*Taper Pipe*		*Straight Pipe*	
Thread	***Drill***	***Thread***	***Drill***	***Thread***	***Drill***	***Thread***	***Drill***
6–32	No.36	7/8–9	49/64	3 1/2–8	3 3/4	3 1/2–8	3 25/32
6–40	No.33	7/8–14	13/16	4–8	4 1/4	4–8	4 9/32
8–32	No.29	1–8	7/8				
8–36	No.29	1–12	59/64				
10–24	No.25	1–14	59/64				
10–32	No.21	1 1/8–7	63/64				
12–24	No.16	1 1/8–12	1 3/64				
12–28	No.14	1 1/4–7	1 7/64				
1/4–20	No.7	1 1/4–12	1 11/64				
1/4–28	No.3	1 3/8–6	1 7/32				
5/16–18	F	1 3/8–12	1 19/64				
5/16–24	I	1 1/2–6	1 11/32				
3/8–16	5/16	1 1/2–12	1 27/64				
3/8–24	Q	1 3/4–5	1 9/16				

Table 31-6 Metric Tap Drill Chart—Based on Approximately 75 Percent Full Thread

Nominal Size	*Pitch*	*Tap Drill Size*	*Inch Decimal*
1.6	.35	1.25	.050
2	.4	1.6	.063
2.5	.45	2.05	.081
3	.5	2.5	.099
4	.7	3.3	.131
5	.8	4.2	.166
6	1.0	5.0	.198
8	1.0	7.0	.277
8	1.25	6.8	.267
10	.75	9.3	.365
10	1.25	8.8	.346
10	1.5	8.5	.336
12	1.0	11.0	.434
12	1.25	10.8	.425
12	1.75	10.3	.405
14	1.25	12.8	.503
14	1.5	12.5	.494
14	2.0	12.0	.474
16	1.5	14.5	.572

Nominal Size	Pitch	Tap Drill Size	Inch Decimal
16	2.0	14.0	.553
18	1.5	16.5	.651
20	1.0	19.0	.749
20	1.5	18.5	.730
20	2.5	17.5	.692
22	1.5	20.5	.809
24	2.0	22.0	.868
24	3.0	21.0	.830
25	1.5	23.5	.927

Millimeters × .03937 = inch decimals.

The values in the table result from the following formula:

Millimeter Drill Size for 75% Thread =
Major Diameter – (.974 × Pitch)

Using Heli-Coil Inserts

There are many reasons why internal threads on a machine assembly can become stripped or cross-threaded. Repeated use, over-torquing the wrench, or just one mistake by cocking a bolt while reassembling a machine, and the threads become useless. An old-timer's fix is to drill the hole out to allow tapping to the next size screw thread and use a larger bolt. While this certainly will produce a quality repair, this fix is not always possible because of space or assembly requirements. The Heli-Coil insert is a good maintenance fix.

The Heli-Coil insert is a precision-formed screw thread coil of stainless steel wire with a diamond-shaped cross section (Fig. 31-12). When they are installed into Heli-Coil tapped holes, they provide permanent conventional 60' internal screw threads that accommodate most standard bolts or machine screws. This procedure consists of choosing an insert that has the same thread size as the damaged thread. The old threads are drilled out and the hole is tapped with a Heli-Coil tap. Special tools are used to insert the proper Heli-Coil insert and to break off the driving tang. Now the hole can be reused for

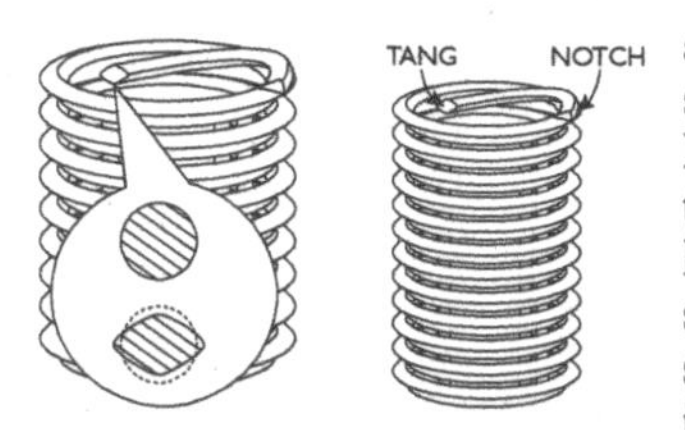

Fig. 31-12 The Heli-Coil insert. *Heli-Coil is a registered trademark of Emhart Teknologies, Inc.*

assembly using the original-sized bolt. In many cases, the Heli-Coil insert is stronger than the original threaded hole, and at the same time it significantly reduces the possibility of thread wear, seizing, and corrosion.

Each insert has a tang for installation purposes. It is notched for easy removal so that a through free-running threaded assembly results. The tang is broken off after installation by simply striking it with a piece of rod. In sizes over ½ in. and all spark plug applications, use long-nosed pliers, bending it up and down until it snaps off at the notch.

Heli-Coil inserts are retained in the hole with spring-like action. In the free state, they are larger in diameter than the tapped hole into which they are installed. In the assembly operation, the force applied to the tang reduces the diameter of the leading coil and permits it to enter the tapped thread. When the torque or rotation is stopped, the coils expand outward with a spring-like action, anchoring the insert permanently in place against the tapped hole. No staking, locking, swaging, key, or interference fits are required. Because the insert is made of wire, it automatically adjusts itself to any expansion or contraction of the parent material.

Repair Procedure

Heli-Coil inserts allow stripped threads to be fixed in three very easy steps, as outlined in Fig. 31-13.

1. Drill.
2. Tap.
3. Install.

If there are traces of thread left, they must be cleaned out before the hole can be prepared for the Heli-Coil insert. The hole must be drilled out to the size specified on the repair kit itself.

The hole is then threaded using the tap supplied in the kit. This is a tap specifically designed to prepare a hole for a Heli-Coil insert—it cannot be used for anything else, and no other tap can be used in its place for this purpose.

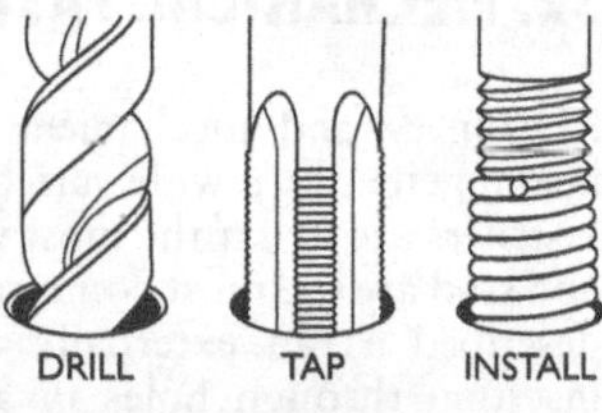

Fig. 31-13 Heli-Coil repair procedure. *Heli-Coil is a registered trademark of Emhart Teknologies, Inc.*

After tapping, the insert is wound into the hole using the installation tool supplied in the kit. When the insert has been fully installed (that is, when the top coil is ¼ to ½ turn below the surface), the tang is broken off and the result is a stainless steel thread that is the same size as the original. It is also stronger and more reusable than the original tapped hole.

External Threading

Most external screw threads are cut by means of dies because they cut rapidly and are capable of cutting threads to meet most commercial requirements for accuracy. There are two general classes of external threading dies: those removed from the thread by unscrewing, and those that may be opened so that the cutting edges clear the thread for removal. The non-opening type, usually split with provision for adjustment, commonly called "threading dies," is the type used by millwrights and mechanics. The die is placed in a "diestock," which provides both a means of holding the die and handles for leverage to turn it. External threading with hand dies is similar to tapping, in that care must be taken to start the die square, the same type of cutting "feel" is involved, the die must be occasionally reversed to break the chip, and adequate lubrication is necessary. Split dies are adjustable and may be opened so that the first cut may be more easily made or the external thread size varied to suit requirements.

32. MECHANICAL FASTENERS

Machinery and mechanical equipment are assembled and held together by a wide variety of fastening devices. Threaded fasteners are by far the most widely used, and the bolt, screw, and stud are the most common threaded fasteners. The bolt is described as an externally threaded fastener designed for insertion through holes in assembled parts. It is normally tightened and released by turning a mating nut. A screw differs from a bolt in that it is supposed to mate with an internal thread into which it is tightened or released by turning its head. Obviously, these descriptions do not always apply because bolts can be screwed into threaded holes and screws can be used with a nut. The third most common fastener, the stud, is simply a cylindrical rod threaded on either one or both ends or throughout its entire length. Fig. 32-1 shows some of the common bolt, screw, and stud types.

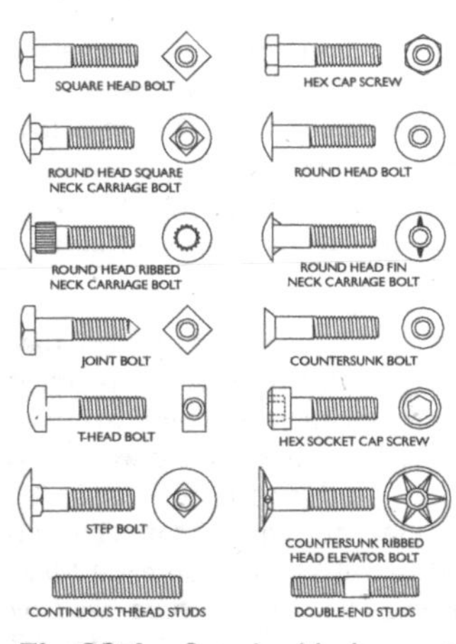

Fig. 32-1 Standard bolts, screws, and studs.

Threaded fasteners are furnished with either coarse threads conforming to Unified National Coarse (UNC) standards or Unified National Fine (UNF) threads.

Coarse Threads

For the majority of applications, fasteners with coarse threads provide these advantages:

- They assemble faster and more easily—providing a better start with less chance of cross-threading.

- Nicks and burrs from handling are less liable to affect assembly.
- They are less liable to seize in temperature applications and in joints where corrosion will form.
- They are less prone to strip when threaded into lower-strength materials.
- They may be more easily tapped in brittle materials and/or materials that crumble easily.

Fine Threads

Fine threads may provide superior fasteners for applications where specific strength or other qualities are required. Fine threads provide the following advantages:

- They are about 10 percent stronger than coarse threads due to greater cross-sectional area.
- In very hard materials, fine threads are easier to tap.
- They can be adjusted more precisely because of their smaller helix angle.
- Where length of engagement is limited, they provide greater strength.
- They may be used with thinner walls because of their smaller thread cross-section.

Washers

Most threaded fasteners are installed where vibration occurs. This mechanical motion tends to overcome the frictional force between the threads, causing the fastener to back off and loosen. Washers (Fig. 32-2) are usually placed beneath the fastener head to help maintain frictional resistance to loosening.

The primary function of a washer is to provide a surface against which the head of the fastener or the surface of the nut can bear. Flat washers provide this surface and spread the load over an increased holding area. Normally they do not provide much additional locking action to the

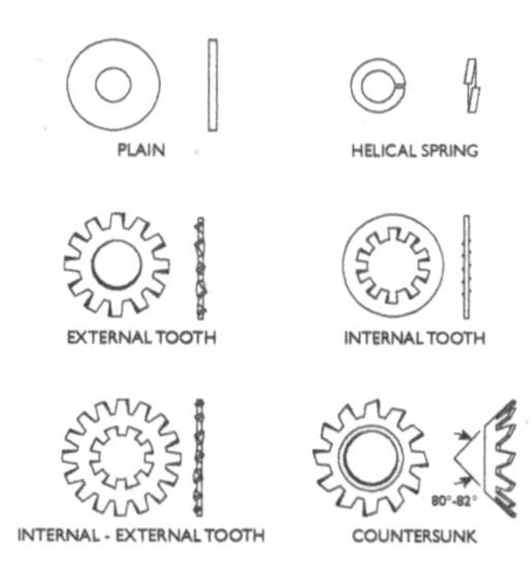

Fig. 32-2 Standard types of washers.

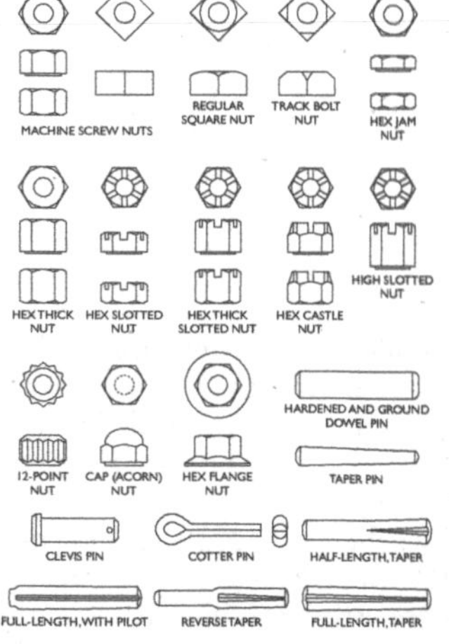

Fig. 32-3 Nuts and pins.

fastener. Lock washers tend to retard loosening of inadequately tightened fasteners. Theoretically, if fasteners were properly tightened, lock washers would not be necessary. Multiple-tooth locking washers provide a greater resistance to loosening because their teeth bite into the surface against which the head or nut bears. Their teeth are twisted to slide against the surface when tightened and hold when there is a tendency to loosen. Lock washers are most effective when the mating surface under the teeth is soft. A harder surface resists the digging action of the teeth.

Nuts and Pins

The nut is the mating unit used with bolt-type fasteners to produce tension by rotating and advancing on the bolt threads (Fig. 32-3). Nuts should be of an equal grade of metal with the bolt to provide satisfactory service.

Pins are used in conjunction with threaded fasteners in many applications. After the nut has been properly tightened on the fastener, slots in the nut are lined up with a hole drilled through the fastener body. Then some form of holding pin (Fig. 32-3) is inserted through the nut slots and fastener hole to prevent the nut from turning in relation to the fastener. The pins shown are not limited to use with fasteners because they are used for many other applications, such as shear pins, for locating and positioning parts, and hinge applications.

Mechanical Fastener Characteristics

Mechanical fasteners are manufactured in a variety of types and sizes too extensive to enumerate. Their design features or *characteristics* are standardized (Fig. 32-4). Fasteners with almost any combination of these characteristics are commercially available.

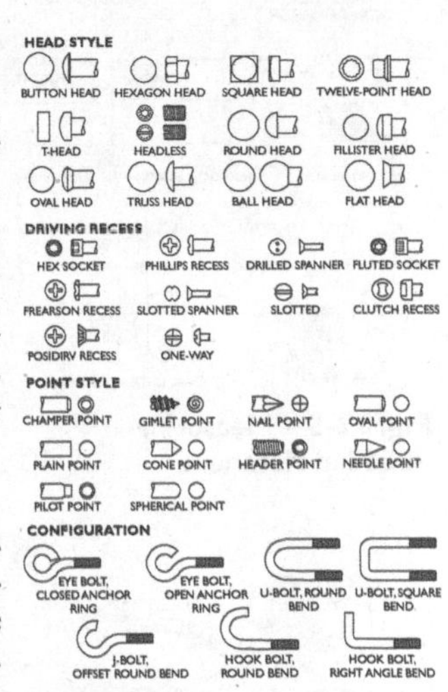

Fig. 32-4 Fastener characteristics.

Fastener Measurements

Threaded fasteners are identified by their nominal diameter and one or more of the measurements shown in Fig. 32-5.

Retaining Rings

While threaded fasteners are still the most common items found around a plant, metal "retaining rings" are replacing them in an ever-increasing number of applications. They may be used to replace screws and washers.

Because these rings are installed in easily cut grooves that can be machined simultaneously with other production processes, they eliminate threading, tapping, drilling, and other costly machining operations.

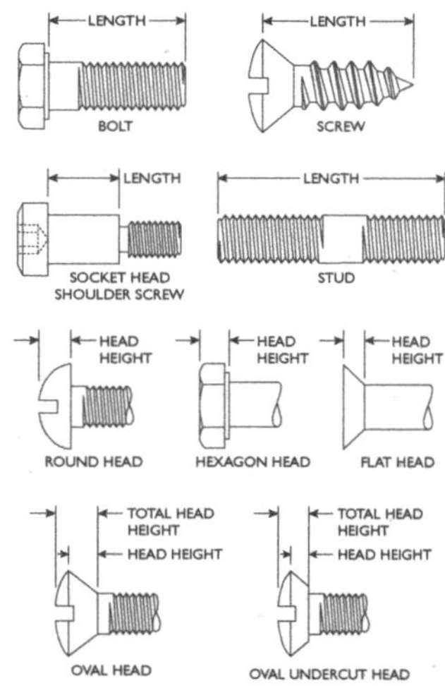

Fig. 32-5 Measuring threaded fasteners.

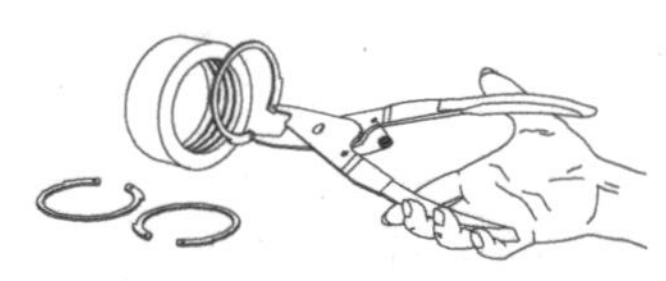

Fig. 32-6 Using retaining ring pliers.

The speed of assembly and disassembly with retaining rings may further reduce manufacturing costs. In addition to the greater economy, the rings often provide a more compact and functional design than other assembly methods. In many cases, they make possible assemblies that would be impractical with any other style of fastening device.

In wide use are a variety of retaining-ring styles designed to serve as shoulders for accurately locating, retaining, or locking components on shafts and in boxes and housing. They are assembled in an axial direction into precut grooves that ensure the precise location of parts. Assembly and disassembly are accomplished by expanding the external rings for assembly over a shaft or by compressing the internal rings for insertion into a bore or housing. The use of retaining ring pliers with an internal retaining ring is shown in Fig. 32-6.

Rivets

Rivets are non-threaded fasteners with heads and shanks manufactured from a malleable material. Two adjoining parts are assembled by drilling a hole through each of them, installing a rivet in the holes, and deforming the end of the shank to produce a second head—thereby locking the two pieces together.

While rivets are still in use by manufacturers, they are less often used in the maintenance of equipment. There is one exception—the use of POP rivets. These handy fasteners are made in three common diameters: ⅛ in., 5⁄32 in., and 3⁄16 in. Fig. 32-7 shows how they work, and Table 32-1 shows selection criteria.

POP rivets are "blind" rivets. That means they are inserted and set from the same side of the work. In Fig. 32-7, step 1 shows the POP rivet inserted into a hole in the work. Step 2 shows the jaws of the POP Rivetool gripping the mandrel of the rivet. In step 3 the setting tool is actuated, setting the rivet by pulling the mandrel head into the rivet body, expanding it, and forming a strong, reliable fastening.

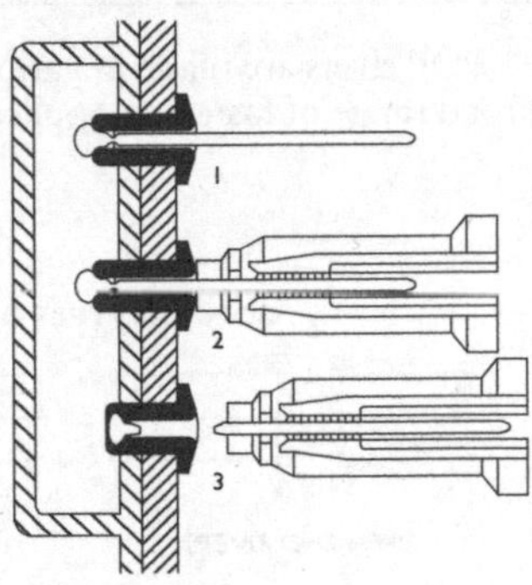

Fig. 32-7 How a POP rivet works. *POP is a registered trademark of Emhart Teknologies, Inc.*

Table 32-1 POP Rivet Selection

POP Rivet Type	*Application*
Open-end rivet	Most used for general applications. Covers a wide range of repairs or installations. Available with a dome head, countersunk head, or large flange head.
Closed-end sealing rivet	Prevents the leakage of vapor or liquid. One hundred percent mandrel retention. Provides greater tensile and shear strength than equivalent open-end rivet.
Multigrip rivet	Wide grip range. Optimum clamp-up force. Accommodates oversized and irregularly sized holes.

(continued)

Table 32-1 (continued)

POP Rivet Type	Application
T-rivet	Designed for structural and similar high-strength applications. High clamp loads. Exceptional pull-up properties.
Load-spreading rivet	Designed for plastics and other brittle or soft materials. Resists pull through or cracking.

POP rivets are made in various types and styles to handle a broad range of fastening applications, as shown in Fig. 32-8.

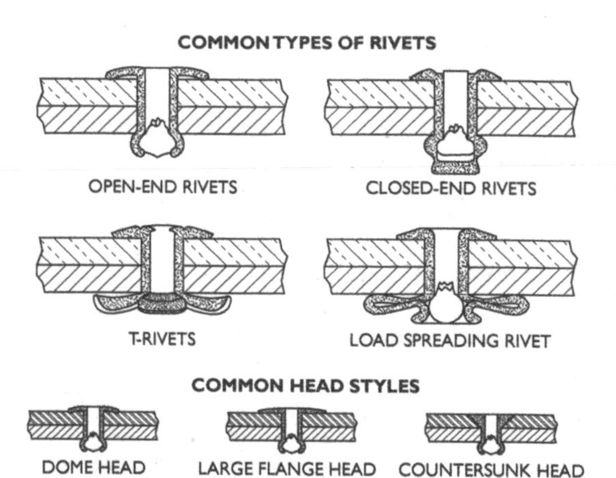

Fig. 32-8 Types and styles of POP rivets. *POP is a registered trademark of Emhart Teknologies, Inc.*

Engineering Adhesives

While adhesives have been around for a long time, engineering strides have been made to allow their use to permit fabrication of assemblies that are mechanically equal to or superior to threaded assemblies.

While the main use of adhesives is mechanical joining, they also seal and insulate. They may find good use in the assembly of lightweight metals, which can become distorted when using threaded fasteners.

Anaerobic adhesives are used for thread locking and bearing mounting applications. The word "anaerobic" means that the adhesive cures between two metal surfaces only when the air is excluded. The anaerobic adhesives available come in different grades, depending on the application. Obviously, a thread-locker must be strong enough to keep threads from loosening under vibration but not so strong that the bolt cannot be removed using a wrench. The anaerobic adhesives used to seal pipe threads or seal flanges are modified to produce various levels of holding or removal strength.

Other engineered adhesive products that are available are epoxies that have great resistance to chemicals or solvents, cyanoacrylates for joint spicing, and silicone adhesives for gasketing usage.

33. GASKETS, PACKING, AND MECHANICAL SEALS

Gaskets

A gasket is a material that is used to seal between two faces of a machine. Gaskets can be made of soft materials, such as asbestos or elastomers, or they can be made of harder materials, such as metal ring gaskets made of iron, steel, and other materials. Combinations of materials—such as spiral wound metal/asbestos-filled gaskets—are also common.

Gaskets are designed for compressibility and sealability. Compressibility is a measure of the gasket's ability to deflect and conform to the faces being sealed. Gasket compressibility compensates for surface irregularities, such as minor nicks, non-parallelism, corrosion, and variations in groove depth. Sealability is the measure of fluid leakage through and across both faces of a gasket. Most of the leakage for a properly installed face-to-face connection will occur through the gasket.

Types of Gaskets

Flat Gasket

Flat gaskets (Fig. 33-1) are gaskets cut from flat stock of gasket material. The material can be an elastomer, either natural (rubber) or synthetic (hypalon, viton, SBR). Fiber material is often formed with a binder into flat stock gasket material. Flat stock comes in standard thickness, ranging from 1/64 to 1/4 in. Flat stock gaskets are often cut into full-face gaskets, in which the gasket design incorporates the appropriate bolt hold pattern and the flange bolts are used to center the gasket in place. The full-face design is most often used with the flat-face flange design. Ring face gaskets are cut so that the outer diameter of the gasket rests inside the bolt pattern and the gasket is centered by resting it on the flange bolts.

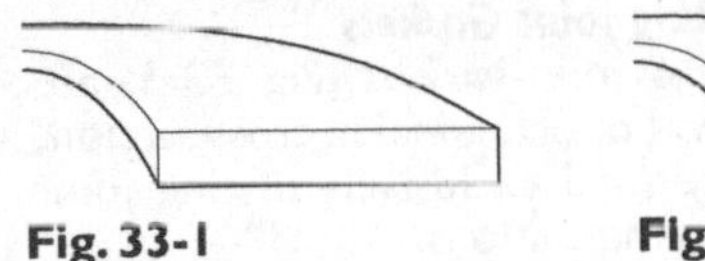

Fig. 33-1

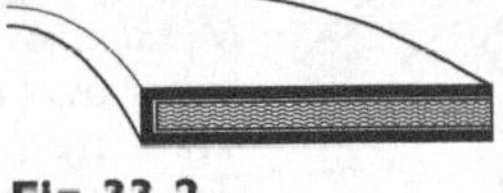

Fig. 33-2

Envelope Gasket

Envelope gaskets (Fig. 33-2) consist of an elastomeric material protected by another material as a jacket. TFE is a common envelope material because of its resistance to many chemicals. Envelope gaskets are usually ring-face sized so the gasket is centered by resting on the flange bolts. Different metals can be used as the envelope material.

Spiral-Wound Metal-Filled Gaskets

Spiral-wound metal-filled gaskets (Fig. 33-3) are the most common and popular. They consist of a thin metal spiral separated by different filler materials, depending on service conditions. They are available in full-face design, but they are most commonly employed as ring-face design. The spiral-wound section of the gasket exists only where the flange faces meet. The remainder of the gasket consists of a backing ring. Special spiral-wound gaskets with no backing rings are used in some tongue-and-groove and male-and-female joint designs.

Grooved Metal and Solid Flat Metal Gaskets

Grooved metal and *solid flat metal gaskets* (Fig. 33-4) are most often employed in tongue-and-groove and male-and-female joint design.

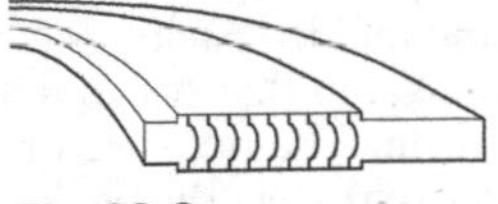

Fig. 33-3

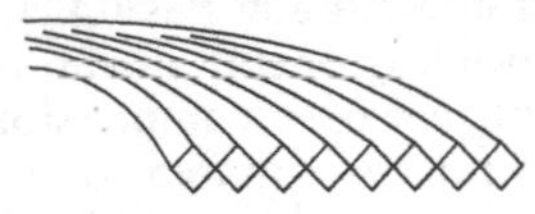

Fig. 33-4

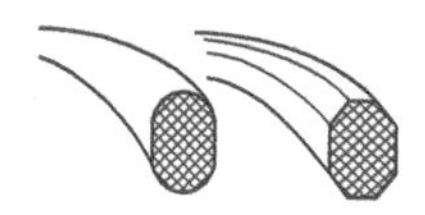

Fig. 33-5

Metal Ring Joint Gaskets

Metal ring joint gaskets (Fig. 33-5) are either oval or octagonal in cross section. They are used exclusively in ring joint flange configurations.

Choosing a Gasket

When choosing a gasket, remember that it must perform under the system temperature and pressure conditions. Gasket manufacturers commonly list the maximum temperature and maximum pressure ratings of their gasket materials. They also give the maximum pressure times temperature (P × T) rating. Table 33-1 gives P and T data for several common gaskets.

Table 33-1 Gasket Materials' Physical Properties

Material	*Temperature Max. °F*	*Pressure Max PSI*	*P × T, Max*
Natural rubber	200	100	15,000
SBR	200	100	15,000
Neoprene	250	150	20,000
Nitrile	250	150	20,000
EPDM	300	150	20,000
Asbestos/ rubber binder	900	3000	350,000
Asbestos/ SBR binder	750	1800	350,000
Asbestos/ neoprene binder	750	1500	350,000
Asbestos/ nitrile binder	750	1500	350,000

Formed and Molded Packings

The principle of operation of formed and molded packings is quite different from compression packings, in that no compression force is required to operate them. The pressure of the fluid being sealed provides the force that seats the packings against the mating surfaces. They are therefore often classed as automatic or hydraulic packings. Packings molded or formed in the shape of a "cup," "flange," "U-shape," or "V-shape" are classed as "lip"-type packings.

Lip-type packings are usually produced with lips slightly flared to provide automatic preload at installation. The fluid being sealed then acts against the lips, exerting the force that presses them against the mating surface. Lip-type packings are used almost exclusively for sealing during reciprocating motion. They must be installed in a manner that will give the lips freedom to respond to the fluid forces.

Packing Materials

Leather

Leather is one of the oldest packing materials and still the most satisfactory for rough and difficult applications. It has high tensile strength and resistance to extrusion. Because it absorbs fluids, it tends to be self-lubricating.

Fabricated

Fabricated materials are made by molding woven duck, asbestos cloth, and synthetic rubber. Fabric reinforcement gives strength to withstand high pressures and is resistant to acids, alkalis, and high temperature. It tends to wipe drier than leather packings, although the fabric will absorb some fluid and has slight lubricating action.

Homogenous

This type of packing is compounded from a wide variety of synthetic rubbers. It is characterized by low strength but high resistance to acids, alkalis, and high temperature. It requires a fine surface finish, close clearances, and clean operating conditions. Homogenous packing is non-absorbent, has no self-lubricating qualities, and usually wipes contact surfaces quite dry.

Plastic

This type of packing is molded from various kinds of plastics for special applications. It is inert to most chemicals and solvents but has little elasticity or flexibility. Some types have a slippery feel and resist adhesion to metal, but their friction under pressure is high.

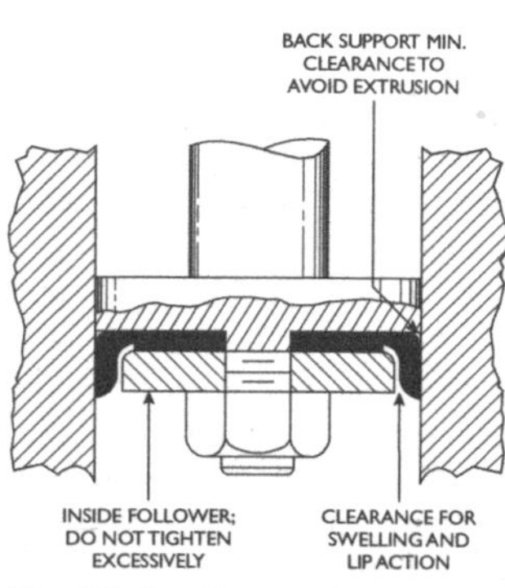

Fig. 33-6 Cup packing.

Cup Packings

Cup packings are one of the most widely used styles of packing, simple to install and highly satisfactory for plunger end applications. The inside follower plate must not be over-tightened, as the bottom of the packing will be crushed and cut through. The heel or shoulder is the point of greatest wear and usually the failure point. Clearances should be held to a minimum, and the lips should be protected from bumping (Fig. 33-6).

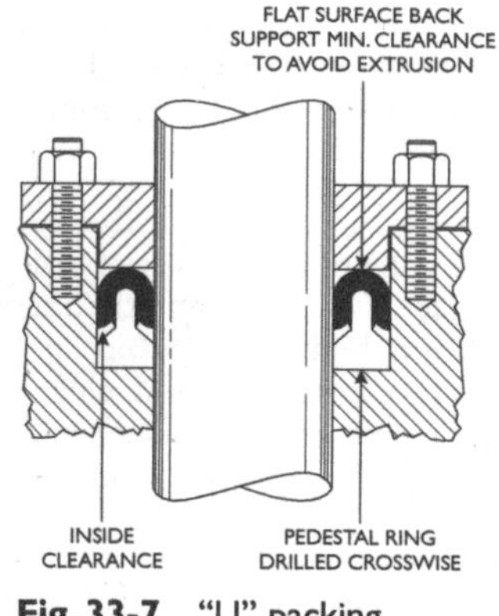

Fig. 33-7 "U" packing.

"U" Packings

This is a balanced packing, as sealing occurs on both the inside and the outside diameter surfaces (Fig. 33-7). To support the lips, the recess of the U is filled with flax, hemp, rubber, fiber, and so on. Metal rings called "pedestal" rings are also used for lip support. Fillers, if nonporous, must be provided with pressure-equalization openings (holes) to allow equalization of pressure on all inside surfaces. Clearance between pedestal rings and the inside packing wall is necessary to accommodate swelling and give the lips the freedom to respond to the actuating fluid.

"V" Packings

Installed in sets, each set of "V" packing consists of a number of V-rings and a male and female adapter. They have a small cross-section and are suitable for both high and low pressures. They operate as automatic packings, but they have the advantage of permitting taking up on the gland ring when excessive wear develops (Fig. 33-8).

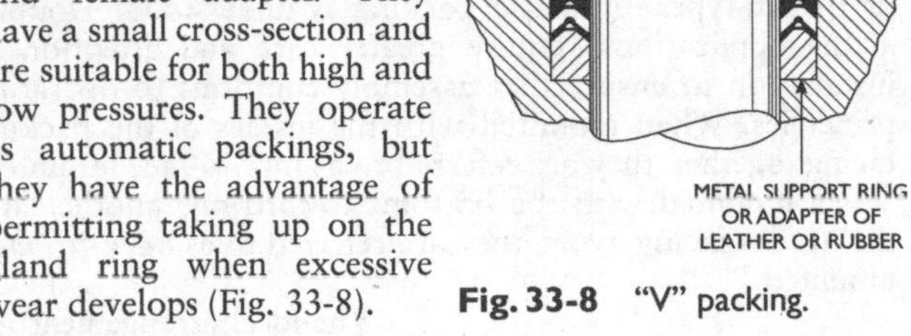

Fig. 33-8 "V" packing.

Flange Packings

This type of packing seals on the inside diameter only; it is generally restricted to low pressures. Base is sealed by an outside compressing force, usually a gland arrangement (Fig. 33-9). Lip action is the same as in cup packing, and it must be given the same consideration for clearance.

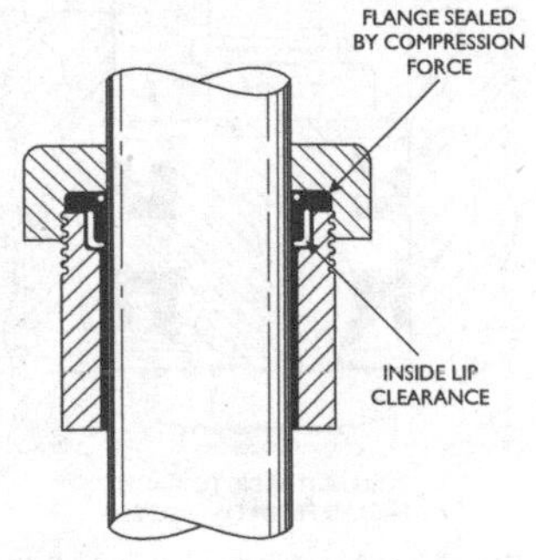

Fig. 33-9 Flange packing.

Packing Installation

The principle of operation of all lip-style packings is the same regardless of the type or material. They must be installed in a manner that will allow them to expand and contract freely. They should not be placed under high mechanical pressure, as this transforms them to compression packings. Over-tightening a lip-type packing improperly preloads it. Even though slight preload is needed for a tight fit and sealing at low pressure, it should occur automatically as a function of the size and shape of the packing.

Lip-type packings installed for one-directional scaling in glands or on pistons should be installed with the inside of the packing exposed to the actuating fluid. Proper installation of packings in this style of application is obvious if the principle of the lip-type automatic packing is understood. Double-acting applications require greater care and attention at installation to ensure that assembly conforms to operating principles. When mounted with the insides of the packing facing together, they are referred to as *face-to-face* mounted. When mounted with the bottoms toward one another and the insides facing away, they are referred to as *back-to-back* mounted.

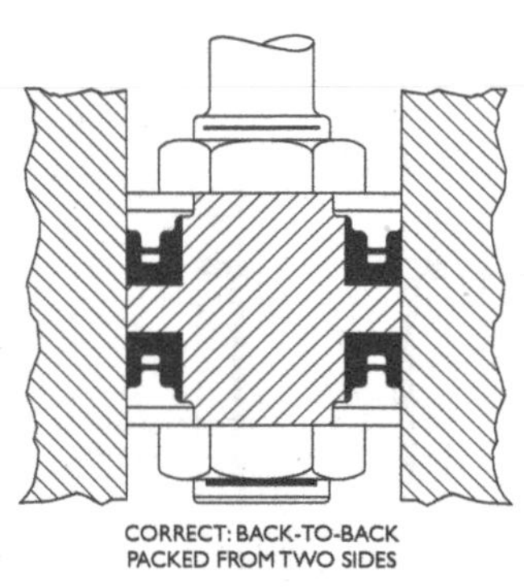

Fig. 33-10 Back-to-back "U" packing.

The ideal arrangement for double-acting packing assembly is back-to-back with solid shoulders for back support (Fig. 33-10). Each packing is fully supported, and no trapping of pressure between packings can occur. Because such an arrangement requires that packings be installed from both sides of the plunger head or end, it frequently is not practical. In such cases, face-to-face mounting is used (Fig. 33-11). While this allows packing from one end, it also introduces an undesirable condition. The actuating fluid must pass the first packing and open the second packing, which is facing toward the fluid. When the second packing is expanded by the fluid, pressure tends to back up into the first packing, expanding it and locking pressure between the two packings.

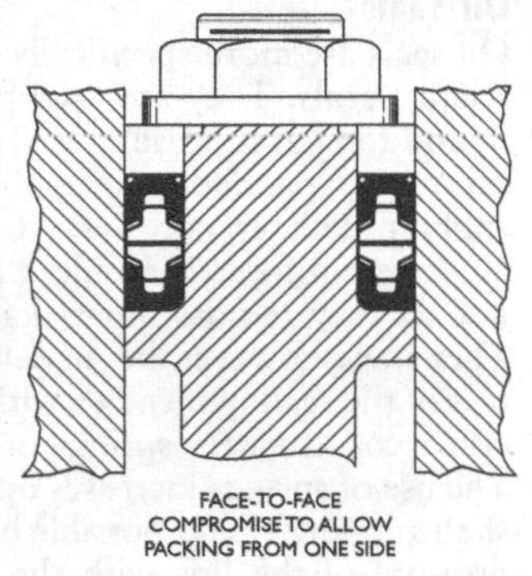

Fig. 33-11 Face-to-face "U" packing.

Clearance Seals

Clearance seals limit leakage between a rotating shaft and housing by maintaining a small, closely controlled annular clearance between shaft and housing. Because there is no rubbing contact, clearance seals offer many advantages: They do not wear, create heat due to friction, or consume power. The principal disadvantage of clearance seals is that they always have some leakage.

The two general types are *bushing* and *labyrinth*. The bushing seal is simply a close-fitting stationary sleeve. Leakage from the high-pressure end to the low-pressure end is limited by the throttling action of the limited clearance. The labyrinth seal is probably the most common type of clearance seal. It is composed of a series of lands on one member that mate closely with a series of grooves on the other member. These lands and grooves retard leakage because of the tortuous escape path they provide. The labyrinth seal is not an off-the-shelf piece of equipment, but is usually custom designed. Slingers, which might be called clearance seals, are also used for shaft sealing. They are usually discs or rings on the revolving member that interrupt lubricant or foreign material from traveling laterally. Centrifugal force tends to throw the lubricant or foreign material from the slinger to the housing surface, from which it is collected in a collector or sump.

Oil Seals

Oil seals are more specifically called radial lip seals or radial contact seals. They are used primarily for sealing lubricants in and foreign material out. They are constructed so the lips must be expanded at assembly, which ensures the positive rubbing contact necessary for their satisfactory operation. No external force is required as with compression packings, nor do they require internal fluid pressure to actuate them. They may, however, be made with a spring to improve contact of the sealing member with the shaft. The springs may be either coiled garter springs or bonded multiple leaf springs. The use of springs increases operating speed and allows some shaft runout. This is possible because of the increased contact pressure of the lips with the shaft that is provided by the springs. Oil seals are primarily used on rotating shaft applications. The use of this type of seal in reciprocating assemblies is very infrequent. The oil seal is made in a wide variety of types, designs, and materials, most of which can be classified as *bonded*, *cased*, or *composition* types.

Bonded seals are constructed of synthetic rubber or other elastic materials, bonded to a metal case or washer. They are seldom used where pressures are encountered. The metal element to which the elastic material is bonded provides the support and stiffness that is necessary for mounting. The lip of the seal is made of elastic material, the tip of which provides the flexible sealing ring that contacts the shaft surface. This flexible ring edge interrupts the lubricant or foreign material, preventing it from traveling laterally on the surface of the shaft.

Cased seals may use either elastic or leather sealing elements. The leather element is superior to elastic or synthetic rubber material where the seal is not constantly lubricated. Because leather is porous and will absorb lubricant, it will operate for fairly long periods without additional lubrication. Leather does not require as smooth a shaft finish for satisfactory operation as synthetic rubber materials do.

Leather, however, is limited to moderate temperatures and moderate surface speeds. Synthetic materials have greater resistance to acids and alkalis than leather and will withstand high temperatures. Another advantage of synthetic rubber is its ability to operate satisfactorily where runout or eccentricity is present. Cased seals require more space than bonded seals, as they are larger in dimension for a given shaft size.

The composition seal is similar in basic design to both cased and bonded seals. The composition seal is constructed with a rigid asbestos heel to which the elastic material is molded. This gives a rigid, hard backing to support the seal and allows tight mounting in the housing. The flexible lips are made of the same variety of materials as are used in either the bonded or cased seals. The principal advantages of this type of construction are improved sealing in the bore and the ability to conform to bores that may be somewhat out of round. Insofar as temperature and speed are concerned, the composition seal is satisfactory for the same conditions as bonded and cased seals of similar materials. Most split seals are of the composition-type construction. Split seals can be installed by placing them around the shaft, instead of sliding the seal over the shaft as is necessary with solid seals of the conventional type.

Combination or dual-element oil seals are designed for sealing where liquids are present on both sides of the seal. Other combinations provide a sealing action in one direction plus a wiping or an exclusion action in the reverse direction. The combination seal is similar in basic design to both cased and bonded seals except that it has double sealing lips, usually positioned in opposite directions. They are, however, made with both lips facing in the same direction for difficult sealing conditions or where leakage must be held to a minimum. The three functional installation modes of oil seal installation are *retention, exclusion,* and *combination* (Fig. 33-12).

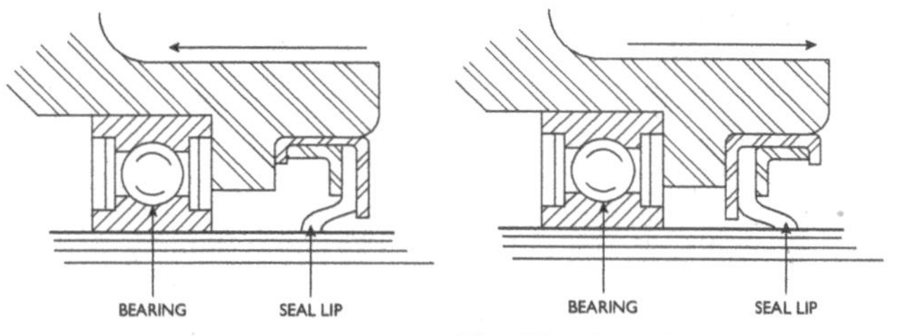

Fig. 33-12a Retention. **Fig. 33-12b** Exclusion.

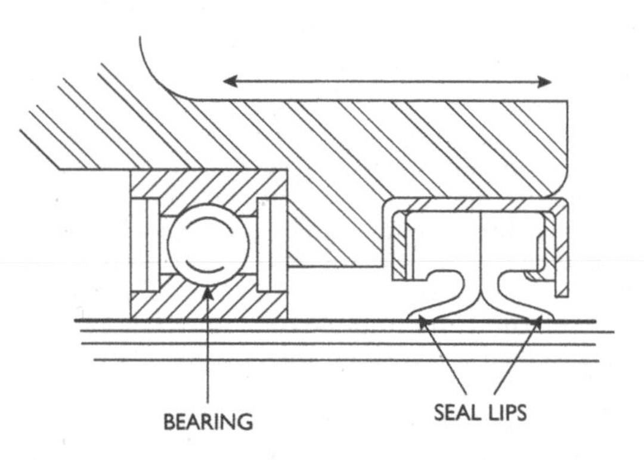

Fig. 33-12c Retention/Exclusion.

Oil Seal Installation

Check sizes carefully before starting installation. The seal should be a light press-fit into the housing, and the inside diameter of the seal such that its lips must be slightly expanded as they are assembled on the shaft. Surfaces should be checked carefully, as any sharp edges or roughness will soon damage the seal lips. The lips of the seal should be lubricated before assembly. This aids in installation and ensures that the seal will not run dry when the machine is started. Leather seals, though frequently pre-lubricated, should be oiled before installation. Synthetic seals can be dipped in oil before installing. This will also aid in pressing them into the housing.

In some cases, particularly with cased seals, it may be advisable to coat the outside diameter (OD) with some type of sealer. This will eliminate the possibility of leaks between the OD of the seal and the housing bore. Gasket cement, white lead, or some similar material may be used. In production operations special tools, such as thimbles to protect the seal lips and driving tools to apply the installation force properly, are usually provided. These are seldom available for field installation of oil seals. To protect the lips from sharp edges, such as keyways and splines, thin sheet material such as shimstock or plastic sheeting may be used to form a sleeve over which the seal can be assembled. In some cases, it may be possible to cover these edges with one of the many adhesive tapes that are available or even strong kraft paper.

The force to assemble the seal into the housing bore should be applied near the outer rim of the seal where it is stiff enough to withstand this treatment. If possible, a tubular tool or pipe that is slightly smaller in diameter than the OD of the seal should be used. Preferably this should be done in a press where the seal can be held square with the shaft.

In field installations where pressing is not possible and driving tools are not available, the seal may be tapped into the housing bore with a wooden block and hammer. Care must be taken that the seal is at right angles to the shaft and not cocked and bound as it is being tapped in. The block should be moved from side to side and around the seal to accomplish this, using light blows and care that the block is squarely against the outer rim of the seal.

Wipers and Rod Scrapers

Two seals, very similar in appearance to the oil seal, but designed for reciprocating applications, are the wiper and the rod scraper. The wiper may be used either as a retention or exclusion seal, and its function is accomplished in the same manner as an oil seal—that is, by pressure of the lip against the shaft surface. Wipers are easily confused with composition oil seals, as they are quite similar in appearance.

Rod scrapers are a special type of exclusion seal having a hard plastic or metallic lip or scraping element. They are used to scrape heavy or tenacious material from reciprocating shafts. Wipers are sometimes used behind them to catch any fine particles or fluid that pass the scraper. Some rod scrapers have polyurethane lip elements designed to exclude foreign materials such as dirt, mud, and metal chips.

O-Rings

The *O-ring* is a squeeze-type packing made from synthetic rubber compounds. It is manufactured in several shapes, the most common being the circular cross section from which it derives its name. The principle of operation of the O-ring can be described as controlled deformation. A slight deformation of the cross section, called a "mechanical squeeze," illustrated in Fig. 33-13a, deforms the ring and places the material in compression. The deformation squeeze flattens the ring into intimate contact with the confining surfaces, and the internal force squeezed into the material maintains this intimate contact.

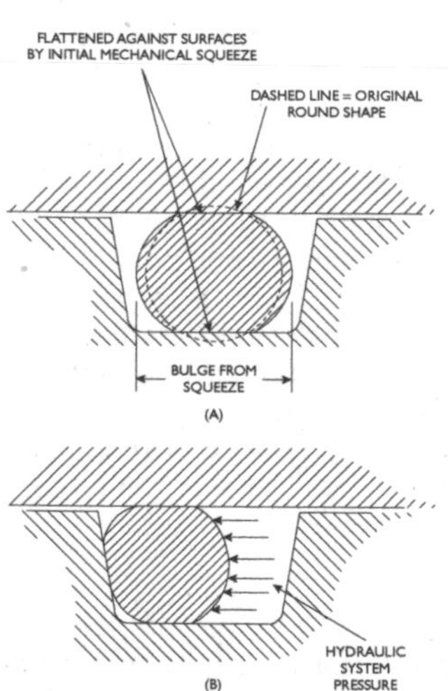

Fig. 33-13 How an O-ring develops a seal.

Additional deformation results from the pressure the confined fluid exerts on the surface of the material. This, in turn, increases the contact area and the contact pressure, as shown in Fig. 33-13b.

The initial mechanical squeeze of the O-ring at assembly should be equal to approximately 10 percent of its cross-sectional dimension. The general-purpose industrial O-ring is made to dimensions that, in effect, build the initial squeeze into the product. This is done by manufacturing the O-ring to a cross-sectional dimension 10 percent greater than its nominal size. The following are the nominal and actual cross-section diameter dimensions of the O-rings in general use:

Nominal	1/32	3/64	1/16	3/32	1/8	3/16	1/4
Actual	.040	.050	.070	.103	.139	.210	.275

Because the cross-section dimension of an O-ring is 10 percent oversize, the outside and inside dimensions of the ring must be proportionately larger and smaller. For example, a 2 × 2⅜ × 3/16 nominal size O-ring has actual dimensions of 1.975 × 2.395 × .210. Fig. 33-14 illustrates the relationship of such an O-ring to its groove.

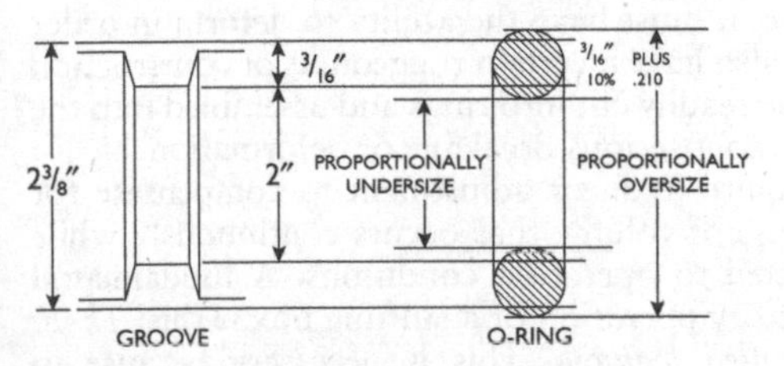

Fig. 33-14 Actual O-ring dimensions.

Stuffing Box Packings

The oldest and still one of the most widely used shaft seals is the mechanical arrangement called a *stuffing box*. It is used to control leakage along a shaft or rod. It is composed of three parts: the *packing chamber,* also called the *box;* the *packing rings;* and the *gland follower,* also called the *stuffing gland* (Fig. 33-15).

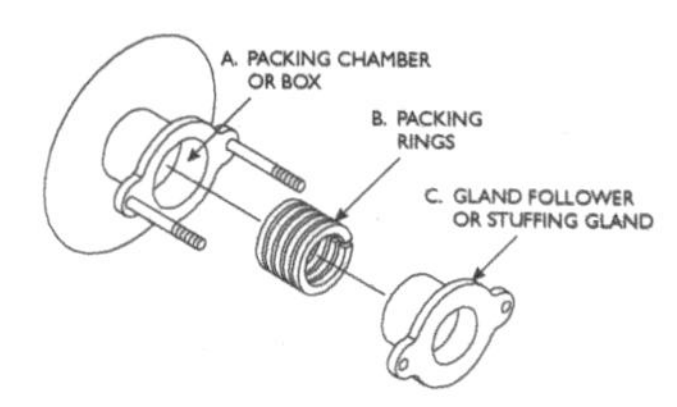

Fig. 33-15 Cutaway of stuffing gland.

Sealing is accomplished by squeezing the packing between the throat or bottom of the box and the gland. The packing is subjected to compressive forces that cause it to flow outward to seal against the bore of the box and inward to seal against the surface of the shaft or rod.

Leakage along the shaft is controlled by the intimate contact of the packings on the surface of the shaft. Leakage through the packing is prevented by the lubricant contained in the packing. The packing material is called "soft" or "compression" packing. It is manufactured from various forms of fibers and impregnated with binders and lubricant. The impregnated lubricant makes up about 30 percent of the total packing volume.

As the gland is tightened, the packing is compressed and wears; therefore, it must have the ability to deform in order to seal. It must also have a certain ruggedness of construction so that it may be readily cut into rings and assembled into the stuffing box without serious breakage or deformation.

Packings require frequent adjustment to compensate for the wear and loss of volume that occurs continuously while they are subjected to operation conditions. A fundamental rule for satisfactory operation of a stuffing box is this: *There must be controlled leakage.* This is necessary because in operation a stuffing box is a form of braking mechanism and generates heat. The frictional heat is held to a minimum by the use of smooth polished shaft surfaces and a continuous supply of packing lubricant to the shaft-packing interface. The purpose of leakage is to assist in lubrication and to carry off the generated heat. Maintaining packing pressures at the lowest possible level helps to keep heat generation to a minimum.

Stuffing Box Arrangement

The function of the multiple rings of packing in a stuffing box is to break down the pressure of the fluid being sealed so that it approaches zero atmospheric pressure at the follower end of the box. Theoretically, a single ring of packing, properly installed, will seal. In practice, the bottom ring in a properly installed set of packings does the major part of the sealing job. Because the bottom ring is the farthest from the follower, it has the least pressure exerted on it. Therefore, to perform its important function of a major pressure reduction, it is extremely important that it be properly installed.

Stuffing Box Packing Installation

While the packing of a stuffing box appears to be a relatively simple operation, it is often done improperly. It is generally a hot, dirty, uncomfortable job that is completed in the shortest possible time with the least possible effort. Short packing life and damage to shaft surfaces can usually be traced to improper practices, rather than deficiencies in material and equipment.

For example, a common improper practice is to lay out packing material on a flat surface and cut it to measured lengths with square ends, as shown in Fig. 33-16.

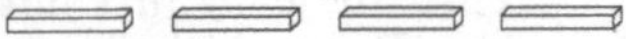

Fig. 33-16 Cut packing to length.

When lengths of packing with square ends are formed into rings they have a pie-shaped void at the ends (Fig. 33-17). The thicker the packing in relation to the shaft diameter, the more pronounced the void will be, as the outside circumference of a ring is greater than the inside circumference. Such voids cause unequal compression and distortion of the packing. Over-tightening is then required to accomplish sealing.

Correct practice is to cut the packing into rings while it is wrapped around a shaft or mandrel. A square-cut end to

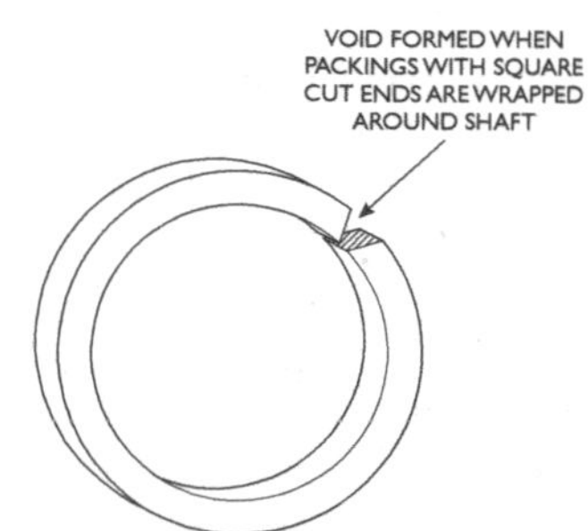

Fig. 33-17 Pie-shaped void—will leak.

form a plain butt joint is as satisfactory as step, angle, or scive joints.

The manner of handling and installing compression packing has a greater influence on its service life than any other factor. Therefore, maximum packing life can only be realized when correct packing practices are followed. The basic steps in correct packing installation are as follows:

1. Remove all old packing and thoroughly clean the box.
2. Cut packing rings on shaft or mandrel as shown in Fig. 33-18. Keep the job simple and cut rings with plain butt joints.
3. Form the first packing gently around the shaft and enter the ends first into the box. The installation of this first packing ring is the most critical step in packing installation. The first ring should be gently pushed forward, keeping the packing square with the shaft as it is being seated. It must be seated firmly against the bottom of the box with the butt ends together before additional packing rings are installed. *Never* put a few rings into the box and try to seat them with the follower. The outside rings will be damaged and the bottom rings will not be properly seated.

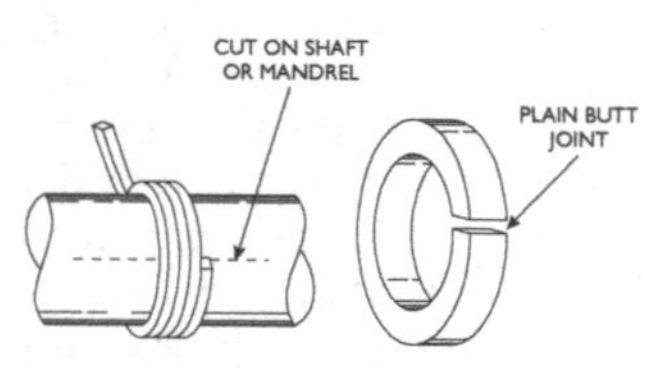

Fig. 33-18 Eliminating the pie-shaped void.

4. Insert additional rings individually, tamping each one firmly into position against the preceding ring. Stagger the joints to provide proper sealing and support.
5. Install the gland follower.
6. Tighten the follower snugly while rotating the shaft by hand. When done in this manner it is immediately apparent if the follower becomes jammed due to cocking or if the packing is over-tightened. Slack off and leave finger tight.
7. Open valves to allow fluid to enter equipment. Start equipment; fluid should leak from the stuffing box. If leakage is excessive, take up slightly on the gland follower. *Do not eliminate leakage entirely;* slight leakage is required for satisfactory service. During the first few hours of run-in operation the equipment should be checked periodically as additional adjustment may be required.

Stuffing Box Lantern Rings

Many stuffing box assemblies include a *lantern ring* or *seal cage*. Use of this device allows the introduction of additional lubricants or fluids directly to the interface of packing and shaft.

A common practice with pumps having a suction pressure below atmospheric pressure is to connect the pump discharge to the lantern ring. The fluid introduced through the lantern ring acts as both a seal to prevent air from being drawn into the pump and a lubricant for the packing and shaft.

Lantern rings are also commonly used in pumps handling slurries. In this case, clear liquid from an external source, at a higher pressure than the slurry, is introduced into the stuffing box through the lantern ring.

Stuffing boxes incorporating lantern rings require special attention when packing. The lantern ring must be positioned

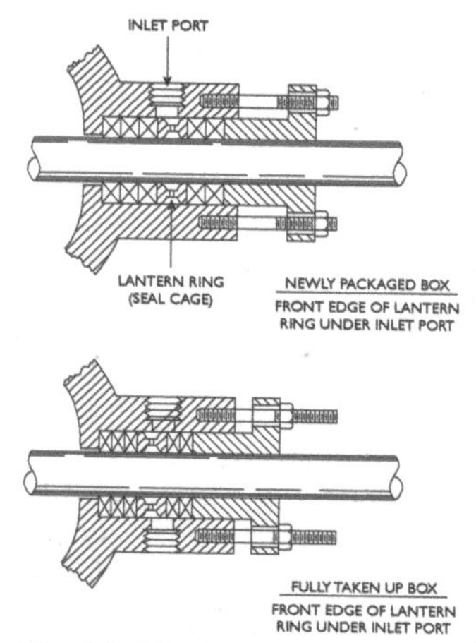

Fig. 33-19 Lantern ring positioning.

between the packing rings so its front edge is in line with the inlet port at installation. As the packings wear and the follower is tightened, the lantern ring will move forward under the inlet port. When the packings have been fully compressed, the lantern ring should still be in a position open to flow from the inlet port (Fig. 33-19).

While lantern rings may on occasion be troublesome to the mechanic, they should not be removed and discarded, as they are an important part of the stuffing box assembly.

Mechanical Seals

While the "stuffing box" seal is still widely used because of its simplicity and ability to operate under adverse conditions, it is used principally on applications where continuous slight leakage is not objectionable. The *mechanical seal* operates with practically no leakage and is replacing the stuffing box seal in an ever-increasing number of applications.

Principle of Operation

The mechanical seal is an *end-face* type designed to provide rotary seal faces that can operate with practically no leakage. This design uses two replaceable anti-friction mating rings—one rotating, the other stationary—to provide sealing surfaces at the point of relative motion. These rings are statically sealed, one to the shaft, the other to the stationary housing. The mechanical seal therefore is made up of three individual seals; two are static, having no relative movement, and the third is the rotary or dynamic seal at the end faces of the mating rings.

The mechanical seal also incorporates some form of self-contained force to hold the mating faces together. This force is usually provided by a spring-loaded apparatus such as a single coil spring, multiple springs, or wave springs, which are thin spring washers into which waves have been formed. The basic design and operating principle of the "inside" mechanical seal is illustrated in Fig. 33-20.

Mechanical-Seal Types

While there are many design variations and numerous adaptations, there are only three basic types of mechanical seals. These are the *inside, outside,* and *double* mechanical seals.

Inside Seal

The principle of the "inside" type is shown in Fig. 33-20. Its rotating unit is inside the chamber or box, thus its name. Because the fluid pressure inside the box acts on the parts and adds to the force holding the faces together, the total force on the faces of an inside seal will increase as the pressure of the fluid increases. If the force built into the seal, plus the hydraulic force, is high enough to squeeze out the lubricating film between the mating faces, the seal will fail.

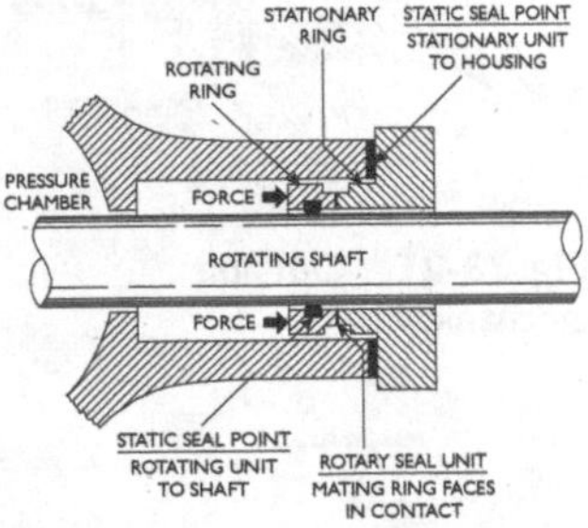

Fig. 33-20 Inside mechanical seal.

While the mechanical seal is commonly considered to be a positive seal with no leakage, this is not true. Its successful operation requires that a lubricating film be present between the mating faces. For such a film to be present there must be a very slight leakage of the fluid across the faces. While this leakage may be so slight that it is hardly visible, if it does not occur the seal will fail. This is why a mechanical seal must never run dry. The stuffing box must be completely filled and the seal submerged in fluid before the equipment is started, and always while operating.

Outside Seal

This style of seal is also named for the location of the rotating unit, in this case outside of the stuffing box, as shown in Fig. 33-21. Because all the rotating parts are removed from the liquid being handled, it is superior for applications where corrosive or abrasive materials are present. Because the hydraulic pressure of the fluid is imposed on the sealing faces, tending to open them by overcoming the self-contained force, it is limited to moderate pressures.

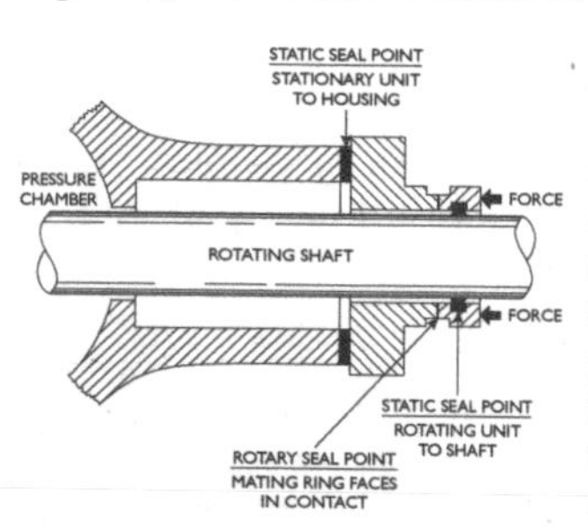

Fig. 33-21 Outside mechanical seal.

Double Seal

A double seal is basically an arrangement of two single inside seals placed back to back inside a stuffing box. The double seal provides a high degree of safety when handling hazardous liquids. This is accomplished by circulating a non-hazardous liquid inside the box at higher pressure than the material being sealed. Any leakage therefore will be the non-hazardous

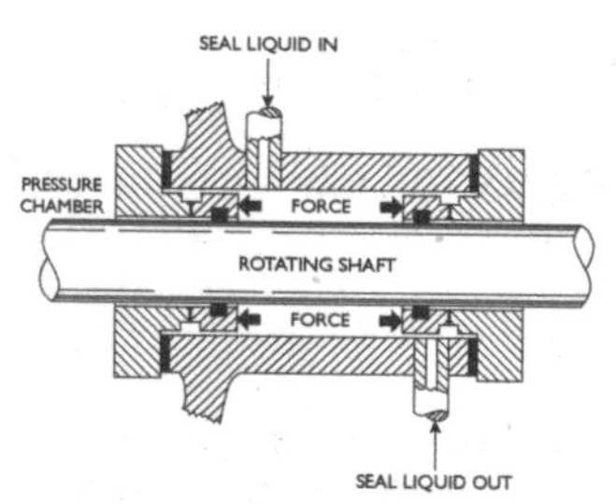

Fig. 33-22 Double mechanical seal.

lubricating fluid inward, rather than the hazardous material that is being sealed leaking outward. The basic principle of the double mechanical seal is illustrated in Fig. 33-22.

Lubrication of Seal Faces

The major advantage of the mechanical seal is that it slows the leakage rate. This is so low there is virtually no visible leakage. Operating satisfactorily in this manner requires that the film of lubricant between the seal faces be extremely thin and uninterrupted. Maintenance of this extremely thin film is made possible by machining the seal faces to very high tolerances in respect to flatness and surface finish. To protect this high quality, precise surface finish requires that seal parts be carefully handled and protected. Mating faces should never be placed in contact without lubrication.

Operation of the mechanical seal depends on this thin film of lubricant furnished by leakage. If there is insufficient leakage to provide the lubricating film the faces will quickly overheat and fail. Liquid must always be present during operation, as running dry for a matter of seconds can destroy the seal faces.

Mechanical-Seal Installation

No one method or procedure for the installation of mechanical seals can be outlined because of the variety of styles and designs. In cases where the seal is automatically located in correct position by the shape and dimensions of the parts, installation is relatively simple and straightforward. In many cases, however, the location of the seal parts must be determined at installation. In such cases, the location of these parts is critical, as their location determines the amount of force that will be applied to the seal faces. This force is a major factor in seal performance, as excessive face pressure results in early seal failure. Parts must be located to apply sufficient force to hold the mating rings together without exerting excessive face pressure.

The procedure or method of location and installation of outside seals is usually relatively obvious and easily accomplished. The inside style of mechanical seal is more difficult to install, as some parts must be located and attached while the equipment is disassembled. The location of these parts must be such that the proper force will be applied to the seal faces when the assembly is complete.

Seal designs and styles vary with manufacturers; however, the same basic principles apply to all when locating and installing inside mechanical seals. The following general procedure is applicable to most styles in common use.

Step I

Determine the compressed length of the seal component incorporating the force mechanism. This is its overall length when it is in operating position (springs properly compressed). Two widely used seal designs are shown in Fig. 33-23; one incorporates multiple springs and the other a single helical spring. In either case, the spring or springs must be compressed the amount recommended by the seal manufacturer before the measurement to determine that the compressed length is made. This is vitally important because the force exerted on the seal faces is controlled by the amount the springs are compressed.

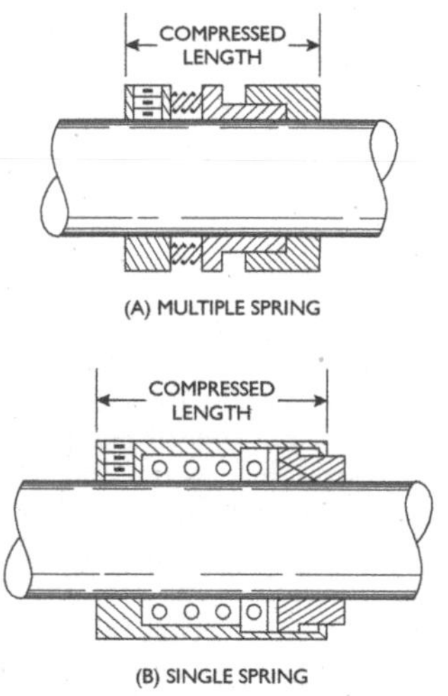

Fig. 33-23 Typical inside seal designs.

Manufacturers' practices vary in the method of determining correct spring compression. In some cases, the springs should be compressed to obtain a specific gap or space between sections of the seal assembly. In other cases, it is recommended that the spring or springs be compressed until alignment of lines or marks is accomplished. In any case, consult manufacturers' instructions, and be sure that the method used to determine compressed length is correct for the make and model of seal being installed.

Step 2

Determine the insert projection of the mating seal ring. This is the distance the seal face will project into the stuffing box when it is assembled into position. Care must be exercised to be sure the static seal gasket is in position when this measurement is made. Obviously, the amount of projection can be varied by modifying the thickness of the gasket (Fig. 33-24).

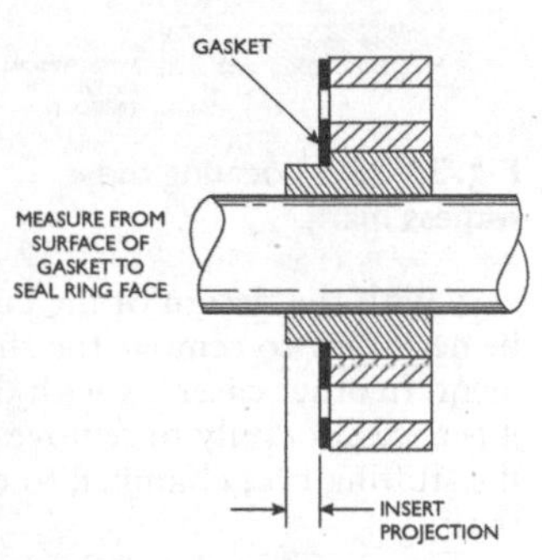

Fig. 33-24 Insert projection of a seal.

Step 3

Determine the "location dimension." This is done by simply adding the "compression-length" dimension found in *step 1* to the "insert-projection" dimension found in *step 2.*

Step 4

"Witness-mark" the shaft in line with the face of the stuffing box (Fig. 33-25). A good practice is to blue the shaft surface in the area where the mark is to be made. A flat piece of hardened steel such as a tool bit, ground on one side only to a sharp edge, makes an excellent marking tool. The marker should be held flat against the face of the box and the shaft rotated in contact with it. This will provide a sharp clear

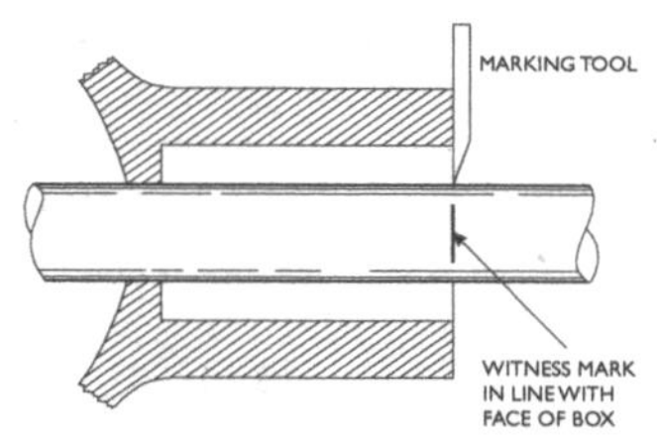

Fig. 33-25 Locating the witness mark.

witness-mark line that is exactly in line with the face of the box.

Step 5

At this point the equipment must be disassembled in a manner to expose the area of the shaft where the rotary unit of the seal is to be installed. The amount and method of disassembly will vary with the design of the equipment. In some cases it may be necessary to remove the shaft completely from the equipment. In other cases, as with the back pull-out design pumps, it is necessary only to remove the back cover, which contains the stuffing box chamber, to expose the required area of the shaft.

Step 6

With the shaft either removed or exposed, blue the area where the back face of the rotary unit will be located. From the witness mark, which was placed on the shaft in *step 4*, measure the location-dimension distance and place a second mark on the shaft. This is called the "location mark," as it marks the point at which the back face of the rotary unit is to be located. The location dimension is the sum of the compressed length plus the insert projection (Fig. 33-26).

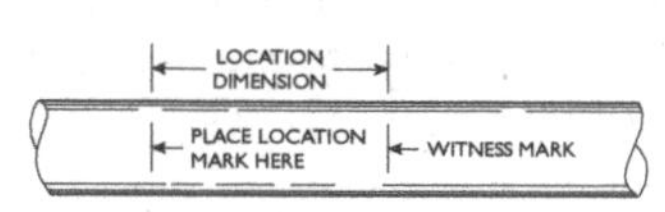

Fig. 33-26 Placing location mark on shaft.

Step 7

Assemble the rotary unit on the shaft with its back face on the location mark. Fasten the unit securely to the shaft at this location. Some seal designs allow separation of the rotary unit components. In such cases the back collar may be installed at this time and the other rotary unit parts later. Illustrated in Fig. 33-27 is a single-spring type rotary unit assembled on the shaft with its back face on the location mark. The spring is extended *and* will be compressed by tightening the stationary ring at assembly.

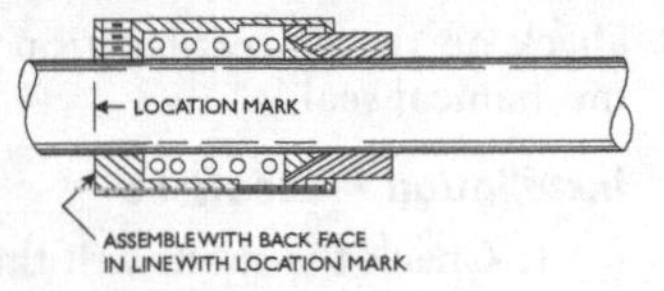

Fig. 33-27 Locating seal on shaft.

Step 8

Reassemble the equipment with the rotary unit on the shaft inside the stuffing box chamber. Complete the seal assembly, and *check that the seal faces are lubricated.*

The sequence of parts assembled depends on the type and design of the seal and the equipment. Fig. 33-28 shows a completely installed inside seal of the multiple spring design.

The final assembly operation will be the tightening of the gland follower bolts or nuts. When this is done the lubricated seal faces should be brought into contact very carefully. When the faces initially contact there should be a space between the face of the box and the follower gland gasket (Fig. 33-28). This space should be the same amount as the springs were compressed in *step 1,* when the compressed length of the rotary unit was determined. This should be very carefully observed, as it is a positive final

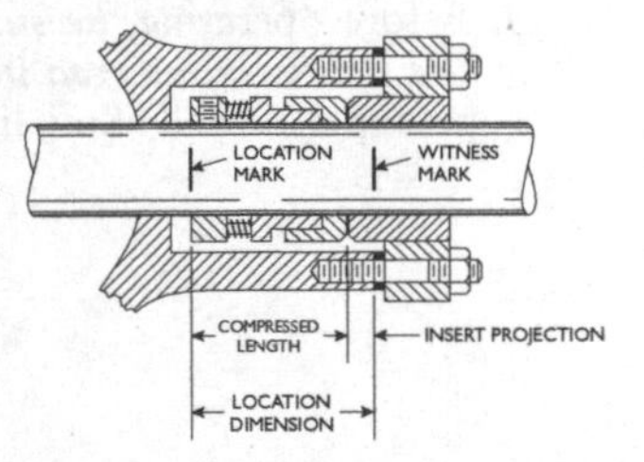

Fig. 33-28 Inside seal completely installed.

check on the correct location of the rotary unit of an inside mechanical seal.

Installation Precautions

1. Check the shaft with the indicator for runout and end play. Maximum TIR allowable is .005 in.
2. All parts must be clean and free of sharp edges and burrs.
3. All parts must fit properly without binding.
4. Inspect seal faces carefully. No nicks or scratches are allowed.
5. Never allow faces to make dry contact. Lubricate them with a good grade of oil or with the liquid to be sealed.
6. Protect all static seals such as O-rings, V-rings, V-cups, wedges, and so on from damage on sharp edges during assembly.
7. Before operating, *be sure proper valves are open and the seal is submerged in liquid.* If necessary, vent box to expel air and allow liquid to surround seal rings.

34. PUMPS

Pumps are broadly classified with respect to their construction or the service for which they are designed. The three groups into which most pumps in common use fall are *centrifugal, reciprocating,* and *rotary.*

Centrifugal Pumps

Centrifugal force, from which this pump takes its name, acts on a body moving in a circular path, tending to force it farther from the center of the circle (Fig. 34-1).

Inside the body of a centrifugal pump, the impeller forces the liquid to revolve and generate centrifugal force. The impeller blades (on the right side of Fig. 34-1) are usually curved backward with reference to the direction of rotation. The liquid is drawn in through the center or "eye" of the impeller, is whirled around by the blades, is thrown outward by the centrifugal force, and passes through the discharge outlet.

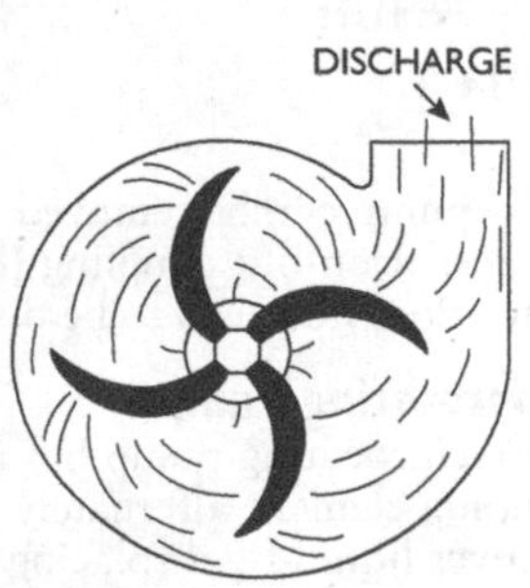

Fig. 34-1

Many centrifugal pumps employ a volute casing design (Fig. 34-2). The volute casing has an increasing radial space between the impeller and casing as the liquid approaches the discharge end of the pump. This design increases the efficiency of the pump by minimizing the amount of heat generated. The volute design minimizes turbulence in the casing, which would heat up the liquid and waste energy.

There are two principal classes of centrifugal pumps, single-stage and multistage. The single-stage pump has a single impeller. By arranging a number of centrifugal pumps in series so that the discharge of one leads to the suction of the succeeding pump, the head or pressure may be multiplied as required. Multistage pumps are made with a common housing, and internal passages are so arranged that liquid flows from the discharge of one stage to the inlet of the next.

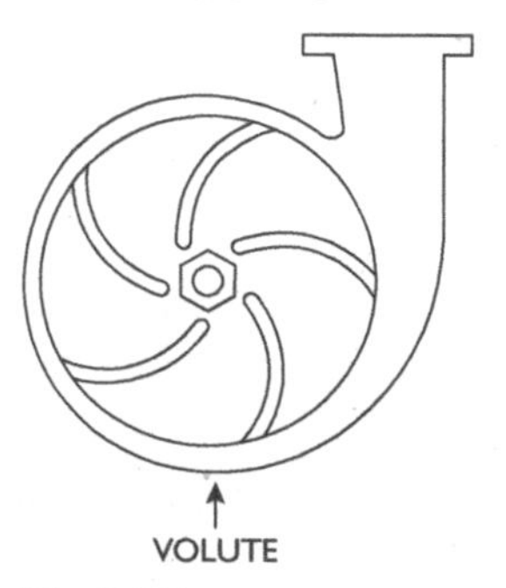

Fig. 34-2

Many pump manufacturers offer a pump design known as a standard ANSI back pull-out design (Fig. 34-3). The volute casing is designed so that it can remain installed to the piping flanges, while the power end of the pump can be removed for service. If the assembly uses a spacer drop-out coupling (between pump and motor), the motor does not need to be disconnected or moved either.

Reciprocating Pumps

The reciprocating pump has a back-and-forth motion as the pumping element alternately moves forward and backward. It moves liquid by displacing the liquid with a solid, usually a piston or a plunger. The principle of operation is called *positive displacement.*

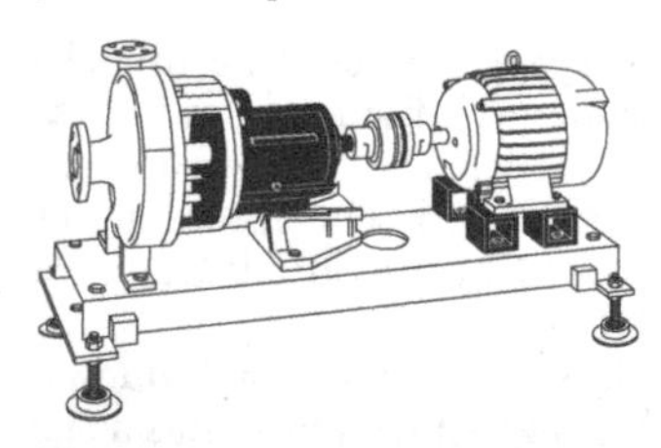
Fig. 34-3 ANSI back pull-out design.

The piston-pumping element is a relatively short cylindrical part that is moved back and forth in the pump chamber, or cylinder. The distance that the piston travels back and forth, called the *stroke,* is generally greater than the length of the piston. Leakage past the piston is

Table 34-1 Centrifugal Pump Troubleshooting Chart

Problem or Symptom	*Cause*	*Correction*
No liquid being pumped	Suction line plugged or clogged	Clear obstructions
	Pump not primed	Reprime pump
	Impeller clogged	Try back flushing to clear impeller
	Incorrect rotation proper rotation	Reverse motor leads to achieve
Pump not producing rated flow or pressure	Air leak at stuffing box	Tighten packing or replace mechanical seal
	Air leak at gasket	Replace with new gasket and proper gasket sealant if required.
	Impeller partly clogged	Back flush
	Low suction head	Open suction valve all the way and check suction piping
Hot bearings	Improper lubrication	Check for proper grease or oil and relubricate
	Improper alignment	Realign pump and motor

(continued)

Table 34-1 (continued)

Problem or Symptom	Cause	Correction
Pump vibrates	Poor alignment	Align motor and pump shafts
	Worn bearings	Replace bearings
	Pump is cavitating	Recalculate pump or check system attributes
	Unbalanced impeller	If due to foreign material, back flush; if impeller is worn, replace
	Broken parts on impeller or shaft	Replace impeller
Pump begins to pump and then stops	Air pocket in suction line	Change piping to eliminate air pocketing
	Air leak in suction line	Plug the leak
	Pump not primed properly	Reprime pump
Motor amperage draw is excessive	Rotating parts are binding	Check rebuilding procedures and correct interfering parts
	Pump is pumping too much liquid	Machine impeller to smaller size
	Stuffing box packing is too tight	Readjust
Leakage at stuffing box	Worn mechanical seal parts	Replace seal
	Shaft sleeve (if used) is scored or cut	Replace sleeve with new, or remachine and replace
	Packing improperly adjusted	Tighten or replace packing material

usually controlled by packings or piston rings. The piston in normal operation moves back and forth within the cylinder.

A *plunger* pumping element is generally longer than the stroke of the pump. In operation the plunger moves into and withdraws from the cylinder. To prevent leakage past the plunger, packings are contained in the end of the cylinder through which the plunger moves.

As the pumping element in a reciprocating pump travels to and fro, liquid is alternately moved into the pump chamber and moved out. The period during which the element is withdrawing from the chamber and liquid is entering is called the *suction,* or *intake*, stroke. Travel in the opposite direction, during which the element displaces the liquid, is called the *discharge* stroke. Check valves are placed in the suction and discharge passages to prevent backflow of the liquid. The valve in the suction passage is opened and the discharge passage valve is closed during the suction stroke. Reversal of liquid flow on the discharge stroke causes the suction valve to close and the discharge valve to open.

Fig. 34-4 illustrates the position of the valves during travel in each direction. At one end of each cylinder the suction valves are open to admit liquid and the discharge valves closed to prevent backflow from the discharge passage. On reversal of direction, the suction valves are closed to prevent backflow into the suction passage, and liquid moves out through the open discharge valves.

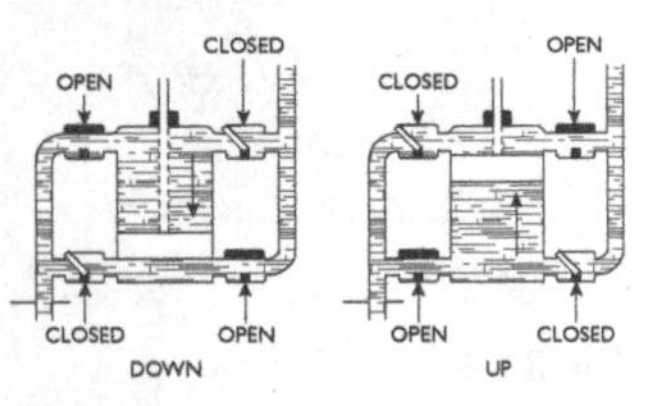

Fig. 34-4

Rotary Pumps

Rotary pumps are also positive-displacement type of pumps in operation. As their flow is continuous in one direction, no check valves are required. Different designs make use of such elements as vanes, gears, lobes, cams, and so on, to move the material. The principle of operation is similar with all of these

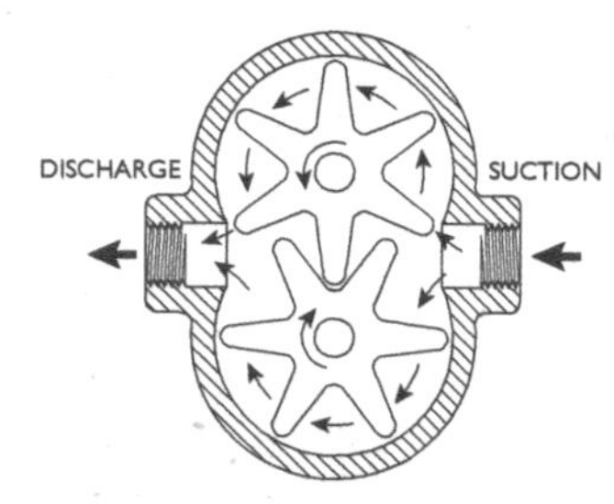

Fig. 34-5 Operating principle of a rotary pump.

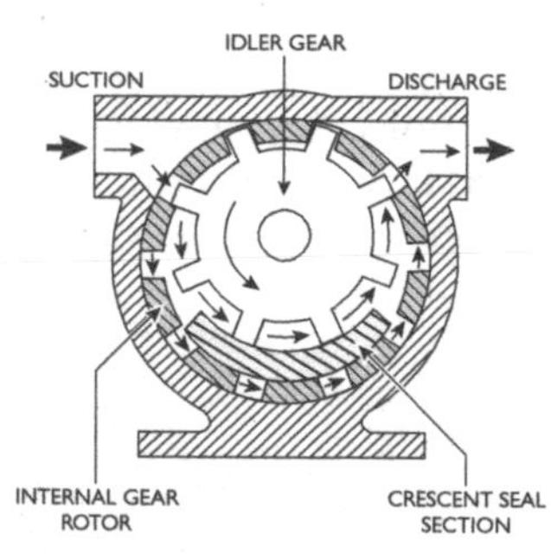

Fig. 34-6

elements, in that the element rotates within a close-fitting casing that contains the suction and discharge connections (Fig. 34-5).

At the pump suction port the liquid enters chambers formed by spaces in the elements, or between the surface of the elements and the internal chamber surface. The liquid is carried with the elements as they rotate, and it is literally squeezed out the discharge as the elements mesh or the volume of the chambers is reduced to practically zero (Fig. 34-6).

Rotary pumps have close running clearances and generally are self-priming. In operation they produce a very even continuous flow with almost no pulsation. The delivery capacity is constant regardless of pressure, within the limits of operating clearances and power.

Troubleshooting Guide for Rotary Pumps

No Liquid Delivered

1. Stop pump immediately.
2. If pump is not primed, prime according to instructions.
3. Lift may be too high. Check this factor with a vacuum gauge on the inlet. If the lift is too high, lower the

position of the pump and increase the size of the inlet pipe; check the inlet pipe for air leaks.

4. Check for incorrect direction of rotation.

Insufficient Liquid Delivered

1. Check for air leak in the inlet line or through stuffing box. Oil and tighten the stuffing-box gland. Paint the inlet pipe joints with shellac or use RTV rubber to seal.
2. Speed is too slow. Check the RPM with manual tach or strobe light. The driver may be overloaded, or the cause may be low voltage or low steam pressure.
3. Lift may be too high. Check with vacuum gauge. Small fractions in some liquids vaporize easily and occupy a portion of the pump displacement.
4. There is too much lift for hot liquids.
5. Pump may be worn.
6. Foot valve may not be deep enough (not required on many pumps).
7. Foot valve may be either too small or obstructed.
8. Piping is improperly installed, permitting air or gas to pocket inside the pump.
9. There are mechanical defects, such as defective packing or damaged pump.

Pump Delivers for Short Time and Quits

1. There is a leak in the inlet.
2. The end of the inlet valve is not deep enough.
3. There is air or gas in the inlet.
4. Supply is exhausted.
5. Vaporization of the liquid in the inlet line has occurred. Check with vacuum gauge to be sure the pressure in the pump is greater than the vapor pressure of the liquid.

6. There are air or gas pockets in the inlet line.
7. Pump is cut by the presence of sand or other abrasives in the liquid.

Rapid Wear

1. Grit or dirt is in the liquid that is being pumped. Install a fine-mesh strainer or filter on the inlet line.
2. Pipe strain on the pump casing causes working parts to bind. The pipe connections can be released and the alignment checked to determine whether this factor is a cause of rapid wear.
3. Pump is operating against excessive pressure.
4. Corrosion roughens surfaces.
5. Pump runs dry or with insufficient liquid.

Pump Requires Too Much Power

1. Speed is too fast.
2. Liquid is either heavier or more viscous than water.
3. Mechanical defects occur, such as bent shaft, binding of the rotating element, stuffing box packing too tight, misalignment of pump and driver, misalignment caused by improper or sprung connections to piping.

Noisy Operation

1. Supply is insufficient. Correct by lowering pump and increasing size of inlet pipe.
2. Air leaks in inlet pipe cause a cracking noise in pump.
3. There is an air or gas pocket in the inlet.
4. Pump is out of alignment, causing metallic contact between rotor and casing.
5. Pump is operating against excessive pressure.
6. Coupling is out of balance.

35. FANS

Any type of device that moves air or a gas is called a *fan* or *blower.* The shop name for a centrifugal fan is a "squirrel cage" fan. A fan is usually considered for low-pressure applications below 1 psi, while blowers are in use for pressures up to 10 psi.

Types of Fans

Fans are divided into two major classes: *centrifugal* and *axial*. Airflow through a centrifugal fan is circular. Air enters along the axis of rotation of a centrifugal fan and discharges in a radial direction, as shown in Fig. 35-1. Airflow in an axial fan is straight. While a centrifugal fan builds pressure by moving the air in a circular motion, the axial fan adds energy to the air by pushing it through the fan.

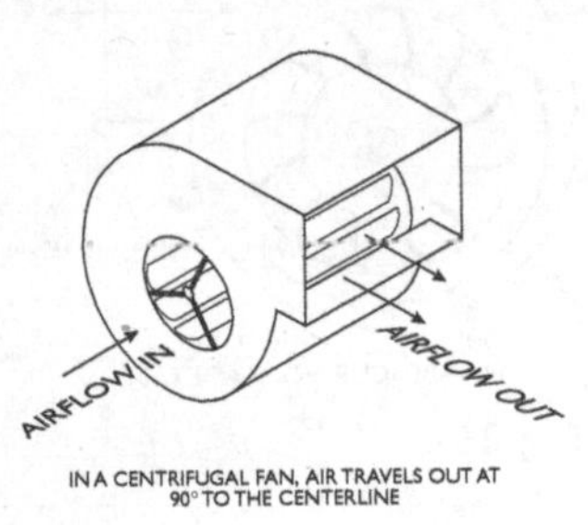

IN A CENTRIFUGAL FAN, AIR TRAVELS OUT AT 90° TO THE CENTERLINE

Fig. 35-1

Centrifugal Fans

The *fan housing* or *scroll* is the metal casing that encloses the fan. The *cut-off* is a piece of metal that keeps air that is discharged from the wheel from reentering the scroll and reducing fan efficiency. The *fan wheel* includes the blades, backpiece, and any framework that holds these components in position.

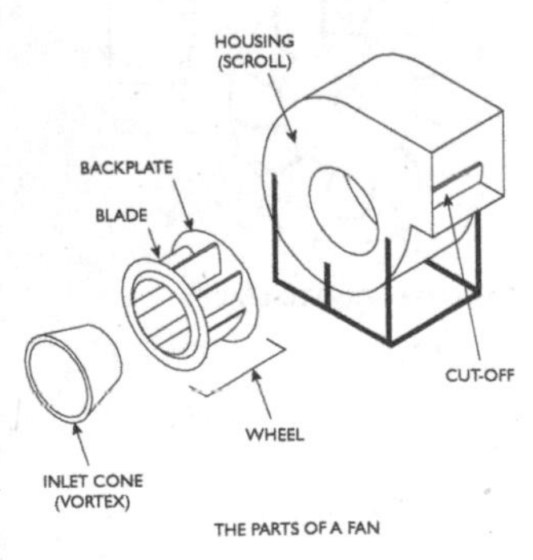

THE PARTS OF A FAN

Fig. 35-2

Blade design varies and is dependent on the design purpose of the fan. The *inlet* or *vortex* is usually a conical component, which allows efficient draw of the intake air into the rotating wheel.

There are numerous subcategories of centrifugal fans, but the most common ones are identified by the type of blade used to move the air:

Forward-curved. Used for small systems, often in residences.

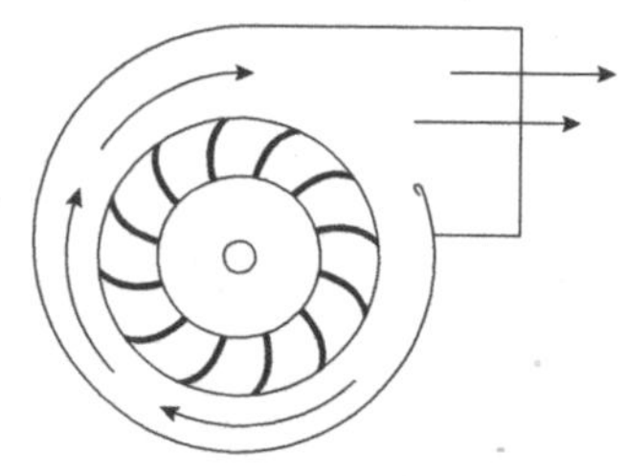

FORWARD-CURVED-BLADE FAN

Fig. 35-3

Backward-curved. Blades slant away from the direction of rotation—curved blades.

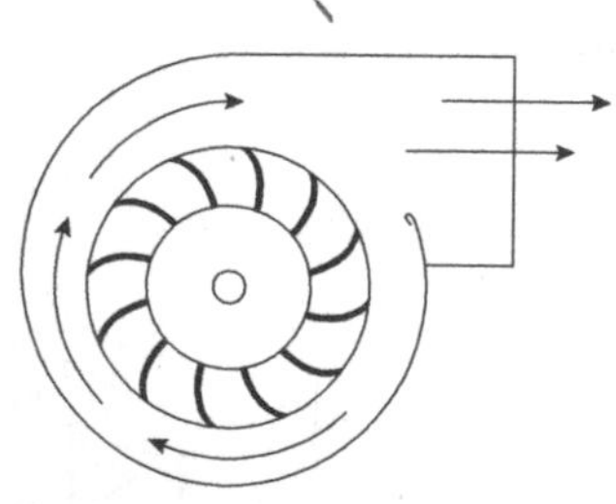

BACKWARD-CURVED BLADES

Fig. 35-4

Backward-inclined. Blades slant away from the direction of rotation—straight blades.

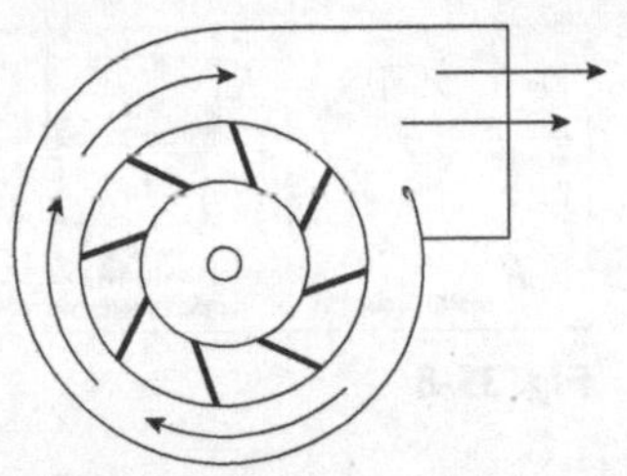

Fig. 35-5

Radial blade. Used for material handling.

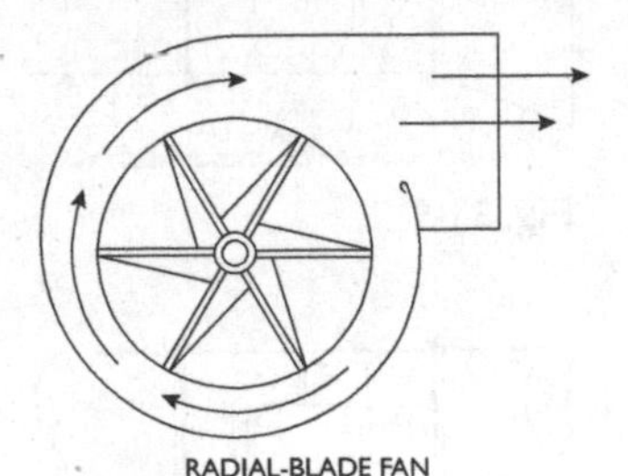

Fig. 35-6

Airfoil. Cutaway of blade looks like airplane wing—most efficient fan.

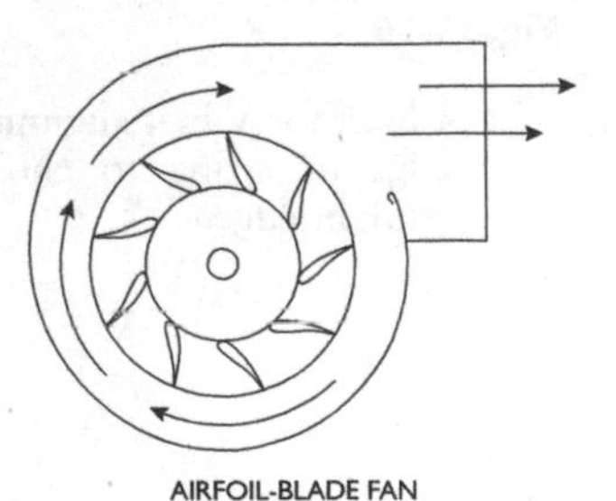

Fig. 35-7

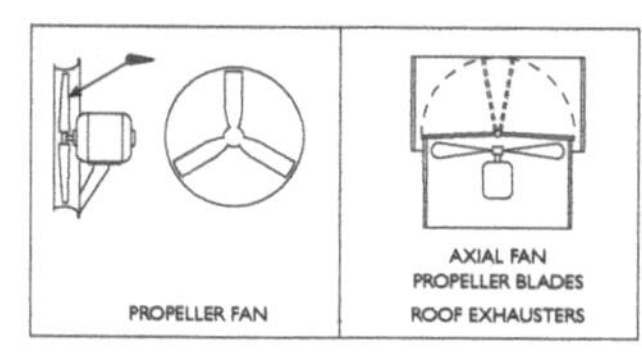

Fig. 35-8

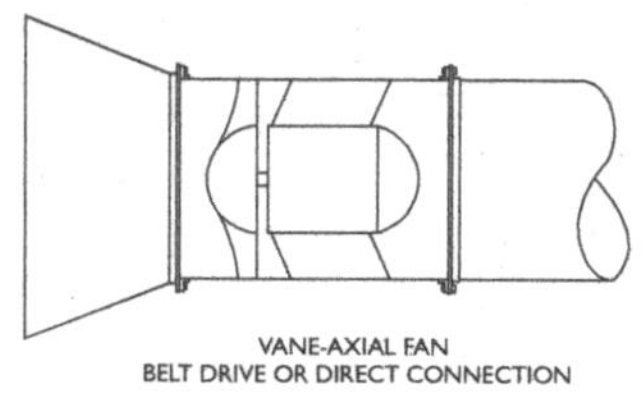

Fig. 35-9

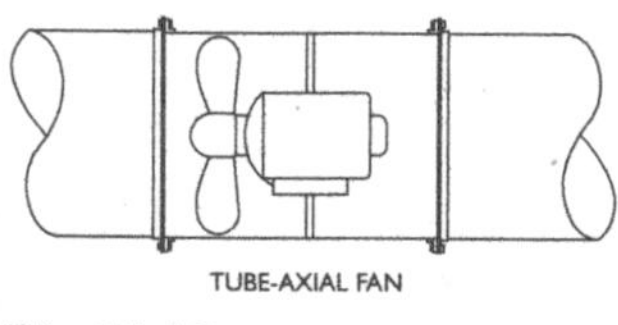

Fig. 35-10

Axial Fans

Axial fans are also categorized by blade style:

Propeller. Resembles an aircraft propeller.

Vane-axial. Uses short propeller-type blades mounted on a large wheel and enclosed in a tube with guide vanes.

Tube-axial. Similar to propeller, but enclosed in housing with no guide vanes.

Inspection of Fans

A checklist for fan maintenance and inspection would include the following items.

V-Belt Drive

- Inspect for sheave wear. Look at side wall of sheave for grooving and curving.
- Inspect for belt wear. Look for fraying, cracking, or slapping belts.
- Check for V-belt alignment and tension. Use a straightedge or string to check sheave alignment or long straight edge.

Fan Wheel

- Look for corrosion, rust buildup, and corrosion cracking on blades.
- Check for loss of balancing weights or clips.
- Look for dirt buildup on blades. Use a wire brush or water blast to clean.
- Check for missing bolts or set screws that attach wheel to shaft.
- Look for hairline cracks at welds.

Lubrication

- Clean fan bearing grease fitting, and lubricate properly.
- Clean motor bearing and grease fitting, and lubricate properly.
- Lubricate damper assembly.

Noises or Knocks

- Listen for unusual noises, rattles, rubbing, grinding, or knocking.
- Feel and listen for air escaping—whistling—indicating leakage.
- Listen for fluctuating air noise. Check for possible loose damper.

Vibration

- Test for excessive vibration using vibration meter.
- Look at any baseplate spring isolators and see if any are broken or completely collapsed.
- Check for broken welds or cracks in expansion joints in ductwork.

Troubleshooting Fan Problems

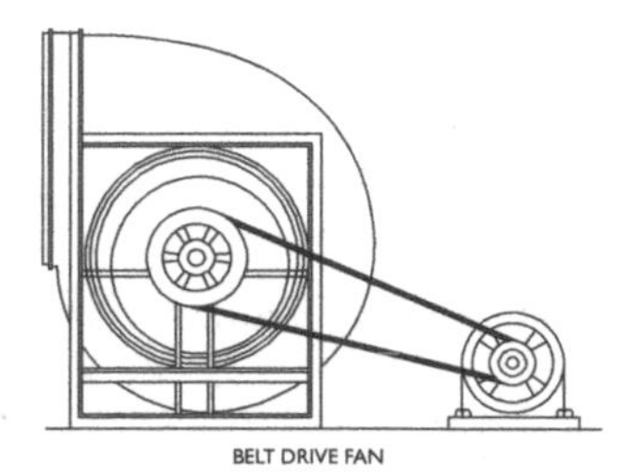

Fig. 35-11

Vibration

Check for the following:

1. Loose bolts in bearings and pedestals or improper mounting.
2. Poor alignment of bearings or shaft couplings.
3. Excessively worn or defective bearings.
4. Unbalanced wheel due to dirt buildup or abrasive wear to blades.
5. Cracked welds.
6. Improper wheel clearance between wheel and inlet vortex.
7. Loose or slapping V-belts.
8. Reversed wheel rotation.
9. Loose setscrews on wheel to shaft hub.
10. Bent shaft due to high-temperature shutdown without proper cooling.
11. Beat frequency with other fans on common base.
12. Motor or fan is causing structural base to resonate. If so, base must be stiffened or motor/fan speed must be changed.

Hot Motor

Check for the following:

1. Improper ventilation to motor or blocking of the cooling air.
2. High ambient temperature.
3. High current draw by checking amperage.
4. Power problems caused by brownouts or other causes of low voltage.
5. Motor is wrong rotation for cooling fan to give proper cooling.

High Bearing Temperature

Check for the following:

1. Over-lubrication of bearings.
2. Improper lubrication, such as mixed lubes or contaminated lubrication.
3. V-belts too tight.
4. Defective or misaligned bearings after recent overhaul.
5. Lack of lubrication to bearings.
6. Heat flinger missing on fan shaft.
7. Floating bearing endplay is restricted.

Excessive Starting Time

Check for the following:

1. Improper sizing of motor to fan.
2. Failure to close inlet dampers during startup.
3. Low voltage at motor terminals.
4. Improper selected time-delay starting circuit.

Air or Duct Noise

Check for the following:

1. Duct thinner than housing.
2. Flattened, cracked, or compressed expansion joints.
3. Poor duct design.
4. Rusted or cracked ducting.
5. Foreign material in fan housing.

36. BASIC ELECTRICITY

Electrical Safety

Because of the ever-present dangers of electrical energy, a basic requirement when working with electricity is that there is no guesswork or risk taking. Activities in this area should be restricted to those things with which you have experience or about which you have specific knowledge or understanding.

Electrical Terms

Electromotive force. The force that causes electricity to flow when there is a difference of potential between two points. The unit of measurement is the *volt* (V).

Direct current. The flow of electricity in one direction. This is commonly associated with continuous direct current, which is non-pulsating, as from a storage battery.

Alternating current. The flow of electricity that is continuously reversing or alternating in direction, resulting in a regularly pulsating flow.

Voltage. The value of the electromotive force in an electrical system. It may be compared to pressure in a hydraulic system.

Amperage. The quantity and rate of flow in an electrical system. It may be compared to the volume of flow in a hydraulic system. The unit of measure is the *amp* (short for ampere).

Resistance. The resistance offered by materials to the movement of electrons, commonly referred to as the flow of electricity. The unit of measurement commonly used is the *ohm.*

Cycle. The interval or period during which alternating current (using zero as a starting point) increases to maximum force in a positive direction, reverses and decreases to zero, then increases to maximum force in a negative direction, then reverses again and decreases to zero value (Fig. 36-1).

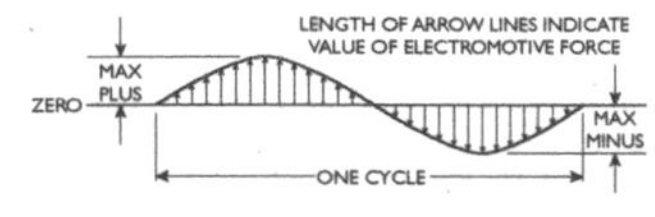

Fig. 36-1 Single phase alternating current.

Frequency. The number of complete cycles per second of the alternating current flow. The most widely used alternating current frequency is 60 cycles per second. This is the number of complete cycles per second; thus, the pulsation rate is twice this, or 120 pulses per second. Frequency is now specified as so many hertz. The term *hertz* (Hz) is defined as cycles per second.

Phase. The word "phase" applies to the number of current surges that flow simultaneously in an electrical circuit. Fig. 36-1 is a graphic representation of single-phase alternating current. The single line represents a current flow that is continuously increasing or decreasing in value.

Three-phase current has three separate surges of current flowing together. In any given instant, however, their values differ, as the peaks and valleys of the pulsations are spaced equally apart (Fig. 36-2). The waveforms are lettered A, B, and C to represent the alternating current flow for each phase during a complete cycle. In three-phase current flow, any one current pulse is always one-third of a cycle out of matching with another.

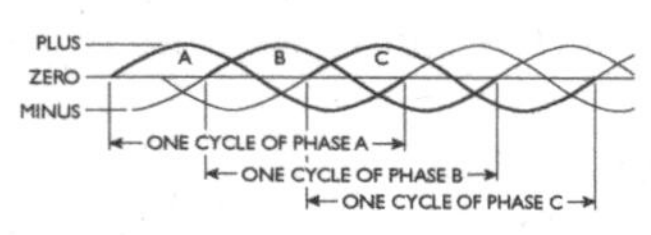

Fig. 36-2 Three-phase alternating current.

Watt. The *watt* (W) is the electrical unit of power, or the rate of doing work. One watt represents the power that is used when one ampere of current flows in an electrical circuit with a voltage or pressure of one volt.

Watt hour. The *watt hour* expresses watts in time measurement of hours. For example, if a 100 W lamp is in operation for a two-hour period, it will consume 200 watt hours of electrical energy.

Kilowatt hour. One *kilowatt* is equal to 1000 W. One kilowatt hour is the electrical energy expended at the rate of one kilowatt (1000 W) over a period of one hour.

Electrical Calculations

Most simple electrical calculations associated with common electrical power circuits involve the use of two basic formulas. These are the *Ohm's Law* formula and the basic *electrical power* formula. By substitution of known values into these formulas, and their rearrangements, unknown values may be easily determined.

Ohm's Law

This is the universally used electrical law stating the relationship of current, voltage, and resistance. This is done mathematically by the formula shown here. Current is stated in *amperes* and abbreviated *I*. Resistance is stated in *ohms* and abbreviated *R*, and voltage in *volts* and abbreviated *E*.

$$\text{Current} = \frac{\text{voltage}}{\text{resistance}} \text{ or } I = \frac{E}{R}$$

The arrangement of values gives two other forms of the same equation:

$$R = \frac{E}{I} \text{ and } E = I \times R$$

Example

An ammeter placed in a 110 V circuit indicates a current flow of 5 amps; what is the resistance of the circuit?

$$R = \frac{E}{I} \text{ or } R = \frac{110}{5} \text{ or } R = 22 \text{ ohms}$$

Power Formula

This formula indicates the rate at any given instant at which work is being done by current moving through a circuit. Voltage and amperes are abbreviated E and I as in Ohm's Law, and watts are abbreviated W.

$$\text{Watts} = \text{voltage} \times \text{amperes or } W = E \times I$$

The two other forms of the formula, obtained by rearrangement of the values, are these:

$$E = \frac{W}{I} \text{ or } I = \frac{W}{E}$$

Example

Using the same values as used in the preceding example, 5 amps flowing in a 110 V circuit, how much power is consumed?

$$W = E \times I \text{ or } W = 110 \times 5 \text{ or } W = 550 \text{ watts}$$

Example

A 110 V appliance is rated at 2000 W; can this appliance be plugged into a circuit fused at 15 amps?

$$I = \frac{W}{E} \text{ or } I = \frac{2000}{110} \text{ or I} = 18.18 \text{ amperes}$$

Obviously, the fuse would blow if this appliance were plugged into the circuit.

Circuit Basics

An electrical circuit is composed of conductors or conducting devices such as lamps, switches, motors, resistors, wires, cables, batteries, or other voltage sources. Lines and symbols are used to represent the elements of a circuit on paper. These are called *schematic diagrams*. The symbols used to represent the circuit elements, including the voltage source, are standardized. Table 36-1 shows the symbols commonly used in industrial applications.

Table 36-1 Electrical Symbols

Symbol	*Meaning*	*Symbol*	*Meaning*
	Crossing of conductors not connected		Knife switch
	Crossing of conductors connected		Double-throw switch
	Joining of conductors not crossing		Cable termination
	Grounding connection		Resistor
	Plug connection		Reactor or coil
	Contact normally open		Transformer
	Contact normally closed		Battery
	Fuse	A	Ammeter

(continued)

Table 36-1 (continued)

Symbol	Meaning	Symbol	Meaning
	Air circuit breaker		Voltmeter
	Oil circuit breaker		

Electrical circuits may be classified as *series* circuits, *parallel* circuits, or a combination of series and parallel circuits. A series circuit is one where all parts of the circuit are electrically connected end to end. The current flows from one terminal of the power source through each element and to the other power-source terminal. The same amount of current flows in each part of the circuit. Fig. 36-3 shows an example of a series circuit.

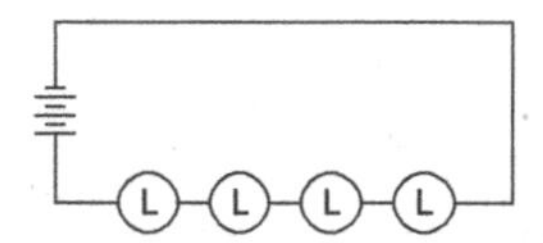

Fig. 36-3 Series circuit.

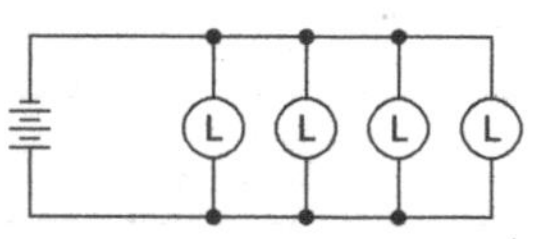

Fig. 36-4 Parallel circuit.

In a parallel circuit, each element is so connected that it has direct flow to both terminals of the power source. The voltage across any element in a parallel circuit is equal to the voltage of the source, or power supply. Fig. 36-4 shows an example of a parallel circuit.

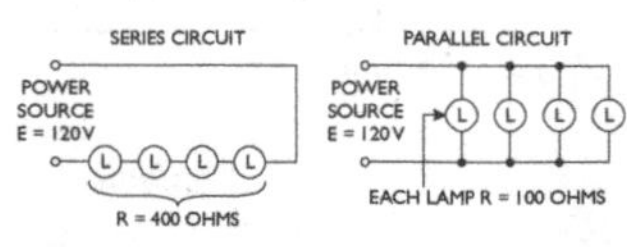

Fig. 36-5

The relationship of values in series and parallel circuits using Ohm's Law and the power formula are illustrated in Fig. 36-5 and compared in the following examples:

Current Flow through Series Circuit	$I = \frac{E}{R} = \frac{120}{400} = .3$ amp
Current Flow through Parallel Circuit	$I = \frac{E}{R} = \frac{120}{25} = 4.8$ amps
Voltage across One Lamp	$E = IR = .3 \times 100 = 30$ volts
Current Flow through One Lamp	$I = \frac{E}{R} = \frac{120}{100} = 1.2$ amps
Current Flow through One Lamp	$I = \frac{E}{R} = \frac{30}{100} = .3$ amp
Voltage across One Lamp	$E = IR = 1.2 \times 100 = 120$ volts
Power Used by One Lamp	$W = EI = 30 \times .3 = 9$ watts
Power Used by One Lamp	$W = EI = 120 \times 1.2 = 144$ watts
Power Used by Circuit	$W = EI = 120 \times .3 = 36$ watts
Power Used by Circuit	$W = EI = 120 \times 4.8 = 576$ watts

Electrical Wiring

The term *electrical wiring* is applied to the installation and assembly of electrical conductors. The size of the wire used for electrical conductors is specified by gauge number according to the American Wire Gauge (AWG) system. The usual manner of designation is by the abbreviation AWG. Table 36-2 lists the AWG numbers and the corresponding specifications using the *mil* unit to designate a 0.001-in. measurement.

Table 36-2 AWG Table

Size of Wire, AWG	Diameter of Wire, mils	Cross Section, Circular, mils	Resistance, Ohms/1000 ft at 68°F (20°C)	Weight, pounds per 1000 ft
0000	460	212,000	0.0500	641
000	410	168,000	0.062	508
00	365	133,000	0.078	403
0	325	106,000	0.098	319
1	289	83,700	0.124	253
2	258	66,400	0.156	201
3	229	52,600	0.197	159
4	204	41,700	0.248	126
5	182	33,100	0.313	100
6	162	26,300	0.395	79.5
7	144	20,800	0.498	63.0
8	128	16,500	0.628	50.0
9	144	13,100	0.792	39.6
10	102	10,400	0.998	31.4
11	91	8,230	1.26	24.9
12	81	6,530	1.59	19.8
13	72	5,180	2.00	15.7
14	64	4,110	2.53	12.4
15	57	3,260	3.18	9.86
16	51	2,580	4.02	7.82
17	45	2,050	5.06	6.20
18	40	1,620	6.39	4.92
19	36	1,290	8.05	3.90
20	32	1,020	10.15	3.09
21	28.5	810	12.80	2.45
22	25.3	642	16.14	1.94
23	22.6	509	20.36	1.54
24	20.1	404	25.67	1.22
25	17.9	320	32.37	0.970
26	15.9	254	40.81	0.769
27	14.2	202	51.47	0.610
28	12.6	160	64.90	0.484
29	11.3	127	81.83	0.384
30	10.0	101	103.2	0.304
31	8.9	79.7	130.1	0.241
32	8.0	63.2	164.1	0.191

Table 36-3 Current Capacities

Wire Size	In Conduit or Cable: Type RHW*	In Conduit or Cable: Type TW, R*	In Free Air: Type RHW*	In Free Air: Type TW, R*	Weatherproof Wire
14	15	15	20	20	30
12	20	20	25	25	40
10	30	30	40	40	55
8	45	40	65	55	70
6	65	55	95	80	100
4	85	70	125	105	130
3	100	80	145	120	150
2	115	95	170	140	175
1	130	110	195	165	205
0	150	125	230	195	235
00	175	145	265	225	275
000	200	165	310	260	320

** Types RHW, TW, or R are identified by markings on outer cover.*

Table 36-4 Adequate Wire Sizes

Load in Building, A	Distance, in ft, from Pole to Building	Recommended* Size of Feeder Wire for Job
Up to 25 A, 120 V	Up to 50	No. 10
	50 to 80	No. 8
	80 to 125	No. 6
20 to 30 A, 240 V	Up to 80	No. 10
	80 to 125	No. 8
	125 to 200	No. 6
	200 to 350	No. 4
30 to 50 A, 240 V	Up to 80	No. 8
	80 to 125	No. 6
	125 to 200	No. 4
	200 to 300	No. 2
	300 to 400	No. 1

** These sizes are recommended to reduce "voltage drop" to a minimum.*

Table 36-5 Circuit Wire Sizes for Individual Single-Phase Motors

Horsepower of Motor	Volts	Approximate Starting Current, A	Approximate Full Load Current, A	Length of Run, in ft, from Main Switch to Motor								
				Feet	25	50	75	160	150	200	300	400
1/4	120	20	5	Wire Size	14	14	14	12	10	10	8	6
1/3	120	20	5.5	Wire Size	14	14	14	12	10	8	6	6
1/2	120	22	7	Wire Size	14	14	12	12	10	8	6	6
3/4	120	28	9.5	Wire Size	14	12	12	10	8	6	4	4
1/4	240	10	2.5	Wire Size	14	14	14	14	14	14	12	12
1/3	240	10	3	Wire Size	14	14	14	14	14	14	12	10
1/2	240	11	3.5	Wire Size	14	14	14	14	14	12	12	10
3/4	240	14	4.7	Wire Size	14	14	14	14	14	12	10	10
1	240	16	5.5	Wire Size	14	14	14	14	14	12	10	10
1 1/2	240	22	7.6	Wire Size	14	14	14	14	12	10	8	8
2	240	30	10	Wire Size	14	14	14	12	10	10	8	6
3	240	42	14	Wire Size	14	12	12	12	10	8	6	6
5	240	69	23	Wire Size	10	10	10	8	8	6	4	4
7 1/2	240	100	34	Wire Size	8	8	8	8	6	4	2	2
10	240	130	43	Wire Size	6	6	6	6	4	4	2	1

Table 36-6 Types and Usage of Extension Cords

	Type	*Wire Size*	*Use*
Ordinary Lamp Cord	POSJ SPT	No. 16 or 18	In residences for lamps or small appliances
Heavy duty—with thicker covering	S, SJ or SJT	No. 10, 12, 14, or 16	In shops, and outdoors for larger motors, lawn mowers, outdoor lighting, etc.

Table 36-7 Cord Carrying Capacity

Wire Size	*Type*	*Normal Load*	*Capacity Load*
No. 18	S, SJ, SJT or POSJ	5.0 A (600 W)	7 A (840 W)
No. 16	S, SJ, SJT or POSJ	8.3 A (1000 W)	10 A (1200 W)
No. 14	S	12.5 A (1500 W)	15 A (1800 W)
No. 12	S	16.6 A (1900 W)	20 A (2400 W)

Table 36-8 Length of Cord Set

Light Load (to 7 A)	*Medium Load (7–10 A)*	*Heavy Load (10–15 A)*
To 25 ft— Use No. 18	To 25 ft— Use No. 16	To 25 ft— Use No. 14
To 50 ft— Use No. 16	To 50 ft— Use No. 14	To 50 ft— Use No. 12
To 100 ft— Use No. 14	To 100 ft— Use No. 12	To 100 ft— Use No. 10

Note: As a safety precaution, be sure to use only cords that are listed by Underwriters' Laboratories. Look for the Underwriters' seal when you make your purchase.

The *circular mil* unit used in the table is a measurement of cross-sectional area based on a circle one mil in diameter.

Switches are the most widely used of all electric wiring devices. They are connected in series with the devices they control, and they allow current to flow when closed and interrupt current flow when open. One of the most common of switch applications is the control of one or more lamps from a single location. The schematic diagram for such a circuit is illustrated in Fig. 36-6, and a sketch of actual wiring connections is shown in Fig. 36-7.

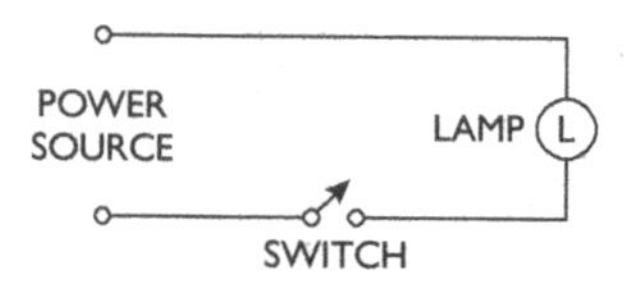

Fig. 36-6 Schematic diagram—lighting circuit.

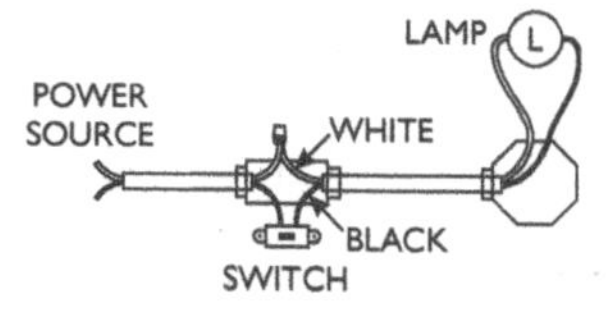

Fig. 36-7 Actual wiring—lighting circuit.

Controlling lamps from two points requires a switch called a three-way switch. It has three terminals, one of which is so arranged that current is carried through it to either of the other two. Its function is to connect one wire to either of two other wires. The diagram in Fig. 36-8 shows a lamp circuit controlled from two points, using three-way switches. The actual connection boxes and the wires in the cable between the boxes for the three-way switch circuit are shown in Fig. 36-9.

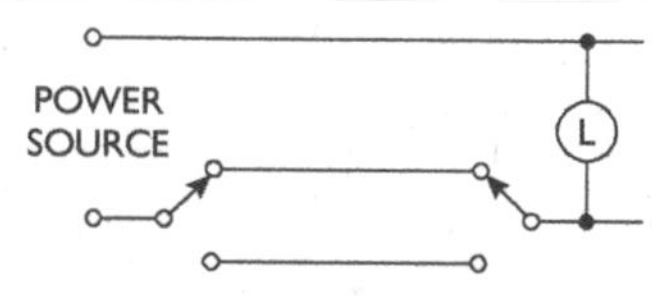

Fig. 36-8 Schematic—three-way switching.

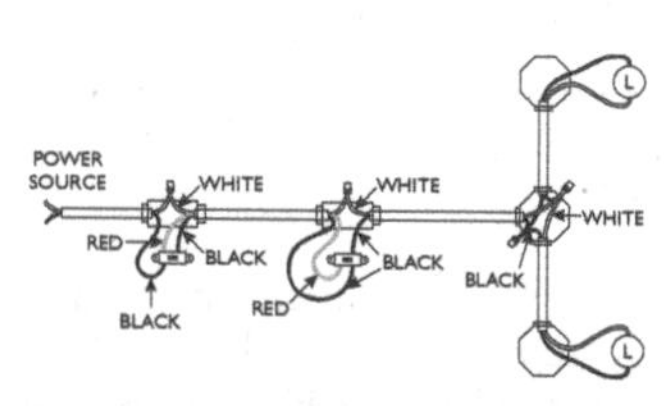

Fig. 36-9 Actual—three-way switching.

Circuit Breakers

A circuit breaker looks like a toggle switch and has a handle to turn power on and off

(Fig. 36-10). Inside a circuit breaker is a mechanism that trips the breaker if an electrical overload exists. When the breaker trips, the load is disconnected. In most cases, resetting the breaker consists of forcing the handle beyond the OFF position and then returning it to the ON position. All circuit breakers have a time delay built into them. The breaker will carry 80 percent of its rated load indefinitely. The breaker can carry a small overload for a short period of time before tripping, but it trips quickly if a large overload is present. A circuit breaker installed in a motor circuit has enough time delay that the motor can start without tripping the breaker.

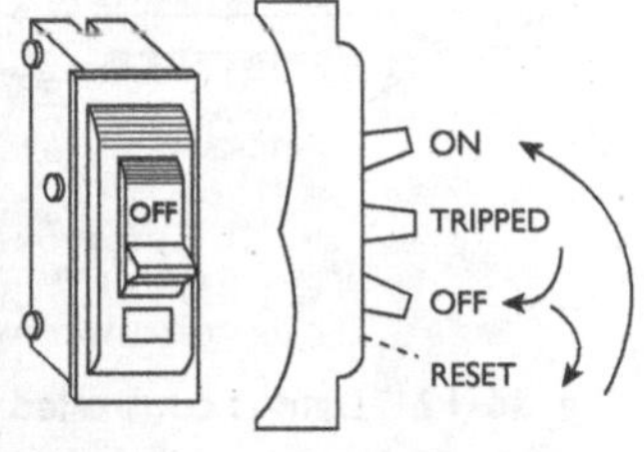

Fig. 36-10 Circuit breaker.

Switching

In factory or residential circuits it is often important to control lighting (or other devices) from more than one switching location. Use of two-way, three-way, and four-way switches can be a bit confusing. While some simple schematics were shown previously in this section, the following diagrams are useful to compare the circuits and the connections. Fig. 36-11, Fig. 36-12, and Fig. 36-13 depict schematic diagrams for the use of these switching devices and show the differences in the connections required to affect the various types of control.

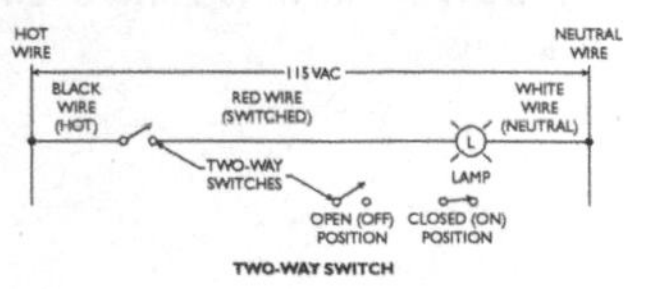

Fig. 36-11 Light is controlled from one switch location.

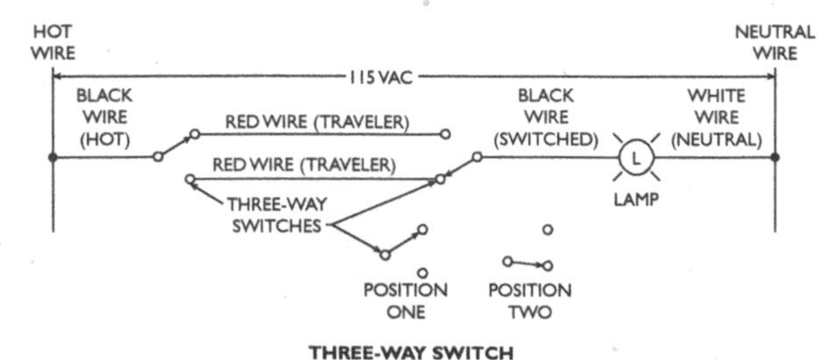

Fig. 36-12 Light is controlled from two switch locations.

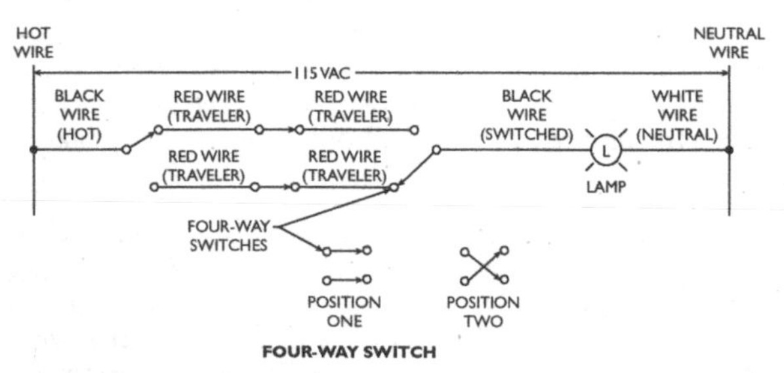

Fig. 36-13 Light is controlled from three or more switch locations.

37. AC MOTORS

Magnetism makes motors work. All motors include a series of electromagnets that are energized to produce a force that allows the motor to run. Electromagnets are simple devices. They consist of a coil of wire wound around an iron core (Fig. 37-1). When a current passes through the wire, a magnetic field is produced. At one end of the iron core is a north pole, at the other a south—and the poles can be reversed by changing the direction of current in the coil.

While there are many types of electric motors, the basic principle of all motors can be shown using a permanent magnet and two electromagnets. Fig. 37-2a shows a current passed in coil 1 and coil 2 in a direction to cause north and south poles to occur next to the permanent magnet. In this simple example, the permanent magnet represents the moving portion of the motor, spinning around an axis. The laws of physics show that like unlike poles attract and like poles repel. The permanent magnet starts to spin because the poles at each end are repelling. As the permanent magnet gets half way around, the force of attraction between unlike poles keeps the permanent magnet rotating. In Fig. 37-2b the rotating magnet turns all the way around until the unlike poles line up (attraction) and the rotor would be expected to stop. Suddenly

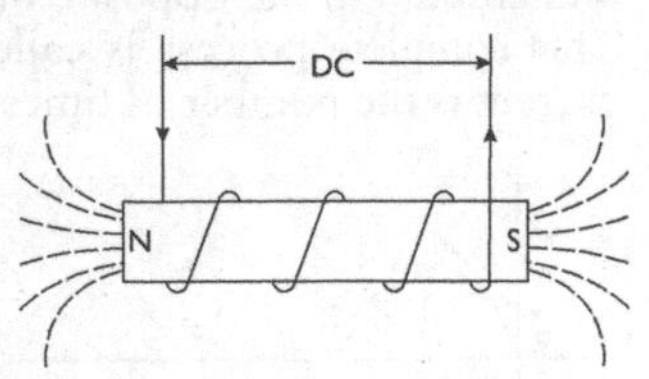

Fig. 37-1 Coil and iron core.

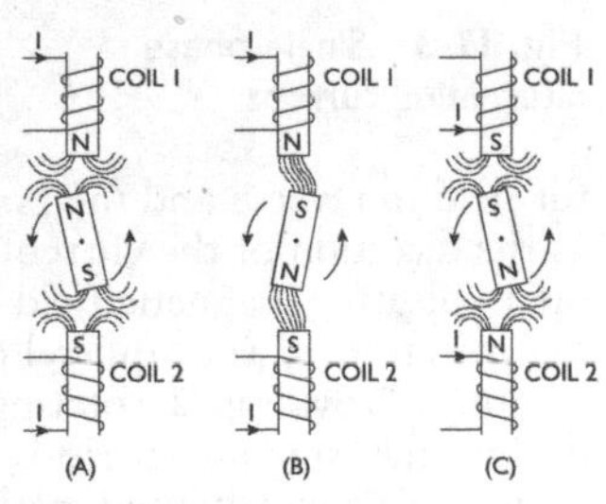

Fig. 37-2 Simple example of how motors work.

the current is reversed, and the poles of the electromagnets also change and reverse. Now there is a condition where the poles are the same and repulsion takes place. The rotor continues to spin, as shown in Fig. 37-2c. In short, if the current in the coils is reversed every time the permanent magnet turns half way around, then the magnet (rotor) would continue to operate. This simple example shows the principle by which all motors work. Motors may be more complex, but the idea is the same.

Single-Phase Induction Motors

Alternating current (AC) alters its direction (or reverses) many times in each second. The amplitude of current increases to a maximum in one direction, drops off to zero, and increases to a maximum in the opposite direction, as shown in Fig. 37-3. This complete process is called a *cycle*. The *frequency* of a current is the number of times that this process (cycle) occurs in one second and is normally expressed in cycles per second or hertz (Hz). Most AC power in the United States operates at a frequency of 60 cycles per second—60 Hz.

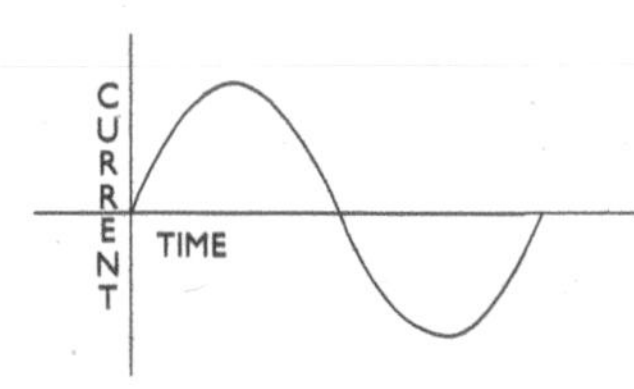

Fig. 37-3 Single-phase alternating current.

Single-phase current does not set up a rotating magnetic field in the stator. Single-phase current flow causes the polarity of the stator field to reverse and the poles to alternate back and forth as the direction of the current changes. This alternation sets up a pulsating magnetic field in the stator, which can maintain rotation of the squirrel-cage rotor but cannot start it rotating. Providing a rotating magnetic field in the stator during the starting period starts single-phase induction motors. Two widely used methods of doing this are by phase splitting and by capacity.

Split-Phase Motors

Split-phase motors range in size from ⅛ to ¾ horsepower, are relatively low in cost, and are widely used on appliances and other light applications. The split-phase motor uses a squirrel-cage rotor with two separate sets of windings in the stator. One set of windings is for running; the other set is for starting only. The starting-coil windings are displaced 45° from the running-coil windings, as shown in Fig. 37-4.

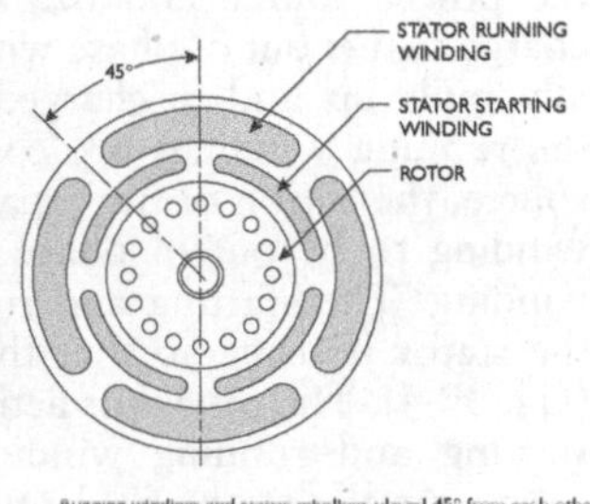

Fig. 37-4 Split-phase motor.

The starting windings have finer wire, offering more resistance than the heavy wire on the running windings. The greater resistance in the starting windings delays the current enough to let it peak at a different time from the current in the running windings. The alternating current flowing through coils of differing resistance sets up rotation of the stator magnetic field. Similar to the three-phase motor operation, the rotating magnetic field of the stator induces current in the rotor. This induced current sets up a magnetic field in the rotor, and it begins to rotate. When the rotor reaches about ¾ speed, the magnetic field of the starting windings is no longer needed, and the power source is automatically disconnected. The pulsating field of the stator, with its reversal of polarity, provides the necessary force of rotation for the rotor, and the motor maintains a constant running speed.

Capacitor-Start Motors

The *capacitor-start* motor ranges from ⅛ to about 7½ horsepower and is used to run small equipment such as compressors that require high starting power. The capacitor-start motor also uses a squirrel-cage rotor and extra starting windings in the stator. As with the split-phase motor, once the motor is running, the starting windings are not needed. To start the motor, a capacitor is put in the circuit between the power source and the starting windings. An electric charge that is out of phase with the power source is continually built up and discharged from the capacitor. Because the running windings are connected directly to the power source, the action of the capacitor causes the current in one winding to be out of phase with the current in the other winding. The starting and running windings are placed in the stator housing so that their poles are about 45° apart (Fig. 37-4). The capacitor action and the arrangement of the starting and running windings cause the stator field to rotate. As the stator field rotates, current is induced in the squirrel-cage rotor and the magnetic field of the rotor is attracted by the magnetic field of the stator, causing rotation. At about ¾ speed the starting windings are disconnected and the motor uses the running windings to operate.

Polyphase Induction Motors

The effects of magnetism are used in an electric motor to rotate the shaft and convert electrical energy to mechanical power. Most widely used for industrial application is the three-phase induction-type electric motor. Some of its advantages are fairly constant speed under load, simple construction, a wide range of horsepower, and an ability to start under load. The two basic parts of an induction motor are the *stator* and the *rotor* (Fig. 37-5). The stator is a set of electromagnets contained in a frame and arranged to form a cylinder. The rotor is an assembly of conductor bars in the form of a cage (called a *squirrel cage*) embedded in an iron core and free to rotate inside the stator.

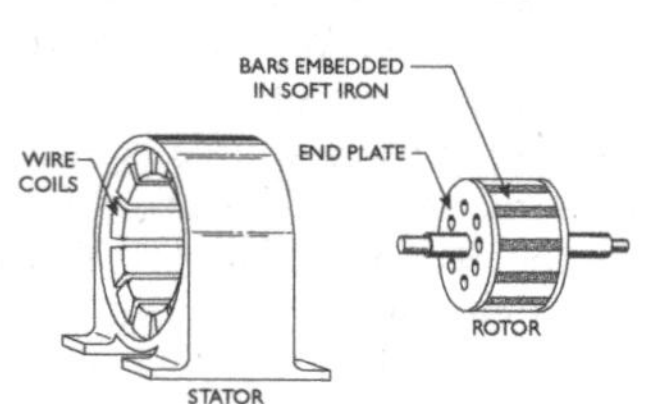

Fig. 37-5 Induction motor—stator and rotor.

The ends of the cage of conductor bars in the rotor are connected to plates at each end of the soft iron core. This core, which is usually constructed of many thin iron plates called laminations, acts to concentrate the strength of the magnetic field of the rotor. The end plates act as conductors and connect the bars to form the cage within the iron core. Power to energize the electromagnets in the stator is supplied from an outside source.

When three-phase electric current flows to the stator magnets arranged in a cylinder, magnetic fields are set up that are increasing, decreasing, and reversing as the current alternates. This alternation of current causes the polarity of the magnetic fields in the stator to move or rotate (Fig. 37-6).

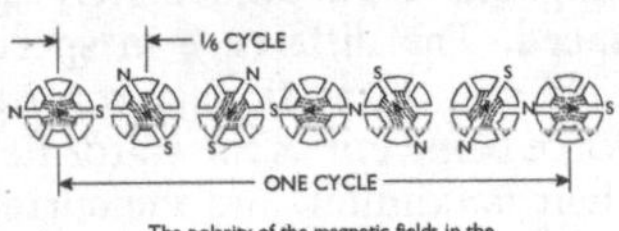

Fig. 37-6 Rotating magnetic fields.

This rotation of the magnetic field does not occur when single-phase alternating current is supplied to the stator. A pulsation of the magnetic field is all that occurs because there is only a single flow of power. With three-phase current, there are three separate surges of power. The maximum intensity of each (either positive or negative) is separated by one-sixth of a cycle. This results in the two magnetic poles shifting steadily or, in effect, rotating always in the same direction. While the electromagnets in the stator windings are stationary, their magnetic fields rotate because of the regular alternation of both current strength and polarity of each phase of the three-phase current. The speed of this rotation, which in turn determines the speed of the motor, is established by the frequency of the electric current and by the construction of the stator windings.

Because the rotor is completely surrounded by the stator electromagnets, it is within the magnetic field of the stator. The effect of the magnetic flux lines of the stator field moving past the conducting bars of the rotor is the same as moving a conductor through a magnetic field. Electric current is induced in the conducting bars of the rotor, which in turn sets up magnetic fields in the rotor. A north pole in the stator field sets up a south pole in the rotor, and a stator south pole sets up a rotor north pole. Because unlike poles attract, the moving north pole of the stator attracts the south pole of the rotor, and vice versa. In effect, the rotor becomes one large bar magnet whose poles are attracted by the rotating poles of the stator, and so it rotates to follow the rotating stator field.

At starting, when the rotor is stationary and an outside power source is connected to the stator windings, the stator's magnetic field immediately starts to rotate at a constant speed. The difference in speed between the stator rotating field and the stationary rotor is at a maximum. The lines of force being cut as the stator field crosses the rotor bars are at their maximum, and the current induced in the rotor is also at its maximum. Therefore, during the starting period, an induction motor develops a strong rotational force or torque. As the speed of the rotor increases, fewer lines of force are cut and less current is induced. If the rotor were to reach the same rotational speed as the magnetic field, no current would be induced in the rotor because the conductor bars would not be cutting any magnetic flux lines. To induce current in the rotor, it must always turn more slowly than the stator field.

This required difference in speed between the stator field and the rotor is called *slip*. Without this slip, a squirrel-cage induction motor cannot run. As loads increase on an induction motor, slip increases, which induces higher rotor current and causes a stronger magnetic field in the rotor. The increase in the strength of the magnetic field increases horsepower output to handle the increased load. Although under different loads there are different amounts of slip, the overall speed of squirrel-cage induction motors is fairly constant. The direction of rotation of the rotor in the squirrel-cage induction motor is the same as that of the stator's rotating magnetic field. The direction of a three-phase motor can be reversed by interchanging the connections of any two supply leads. This interchange will reverse the sequence of phases in the stator and the direction of the rotor rotation. The interchanging of leads may be made at any point before the power reaches the motor, and the effect will be the same. For this reason, it is important that when disconnecting and reconnecting supply voltage to a three-phase motor, care must be taken not to interchange leads. If this is not practical, check the motor direction carefully after reconnection.

Motor Control

Some form of control equipment is required to operate all electric motors. In some cases, this need be only a device to connect the motor to the power source and to disconnect it as needed. Or it may be a more complicated control to stop, start, reverse, or vary the rotational speed. Other controls may be necessary to start and stop automatically at time intervals or in synchronization with related equipment. Two general terms are used to describe motor controls—*starter* and *controller*. The term *starter* is commonly used when referring to simple start-and-stop devices and *controller* when referring to more complex equipment that may perform several functions.

For fractional-horsepower, low-voltage motors, a simple on–off contact switch may adequately serve as a starter. In this case, the line fuses provide the only overload protection. The electromagnetic full-voltage (across-the-line), starter is widely used in industry for stop–start control of larger AC motors. Generally, this type of starter employs a solenoid to close the contacts electromagnetically when the control's start button is pushed. A simplified diagram of a simple full-voltage starter is shown in Fig. 37-7.

When the control's start button is pushed, current flows to the solenoid, which closes the line switch and connects the power source to the motor. In this simplified diagram the power to energize the solenoid comes from the same power source as the motor. For safety, the control circuit for motors

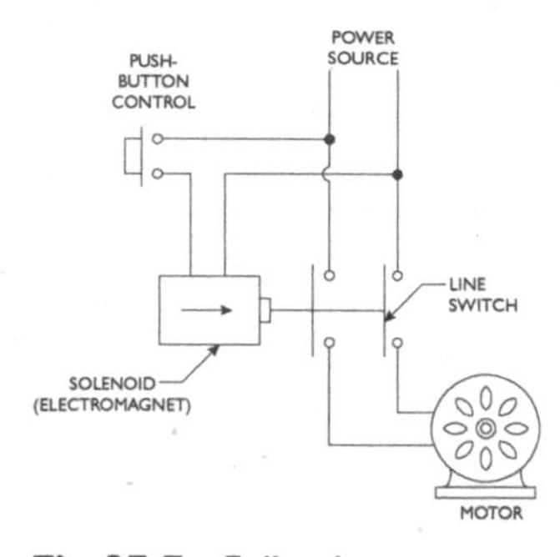

Fig. 37-7 Full-voltage starter circuit.

operating at high voltage is supplied with a lower-voltage current. The control switches touched directly by the operator do not carry the high voltage supplied to the motor.

A starter for control of a three-phase motor is, in effect, three simultaneously acting switches—because all three power lines are controlled. In addition, a fourth switch to supply current to a holding coil is a part of the starter mechanism. The three main sets of contacts control the current flow to the motor. The fourth switch keeps current flowing to the holding coil after the start button is released. When the stop button is pushed, the current flow to the holding coil is interrupted, the electromagnet loses its magnetism, and the starter opens. Fig. 37-8 shows a three-phase starter and push-button stations with the single-phase control circuit power supplied from the main power line.

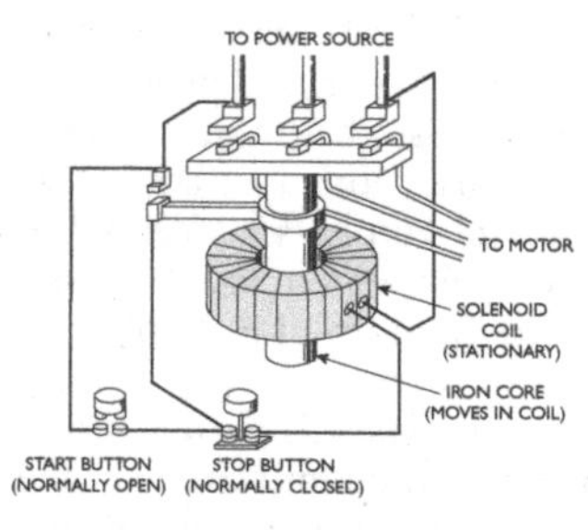

Fig. 37-8 Three-phase starter—single-phase control.

The solenoid used to close the contacts on a starter electromagnetically is made up of a stationary coil, called a holding coil, and a movable soft iron core. Due to the weight of the core and the parts attached to it, the core drops down out of the center of the coil. Many solenoids employ a spring to supply the force to move the core out of the center of the coil. When current is supplied to the holding coil, its core is magnetically pulled into the center of the coil, thus closing a switch. When current is cut off, gravity or force supplied by the spring causes the core to move out of the center of the coil and open the switch.

Overload devices to protect the motor from damage are also placed in the control circuit. These are usually actuated by the heat that results from excessive current flow. The heat causes the overload contacts to open, interrupting the current in the control circuit in the same manner as when the stop button is pushed. Because the protective device can be made ineffective if the start button is held in contact, it should be pressed only momentarily. The current to the holding coil will then flow through the magnetically operated contacts, and the circuit can be interrupted by either the stop button or the overload device.

Troubleshooting Polyphase AC Motors

Table 37-1 Polyphase Motor Troubleshooting Chart

Symptom	*Possible Cause*	*Solution*
Motor will not start	Overload tripped	Wait for overload to cool. Reset and try again.
	Power not connected	Connect power to control and control to motor. Check clip contacts.
	Faulty fuses	Test fuses.
	Low voltage	Check motor nameplate values against power supply. Also make sure motor wire size is adequate.
	Wrong control connections	Check connections against wiring diagram.
	Loose terminals	Tighten connections.
	Driven machine locked	Disconnect motor from load. If motor starts properly, check driven machine for blockage or damage.

(continued)

Table 37-1 (continued)

Symptom	*Possible Cause*	*Solution*
Motor will not start	Open circuit in stator or rotor winding	Check for open circuits.
	Short circuit in stator winding	Check for shorted coil.
	Winding grounded	Test for grounded winding.
	Bearings stiff	Free bearings or replace.
	Grease too stiff	Use different lubricant.
	Faulty control	Check control wiring.
	Overload	Reduce load on motor.
Motor noisy	Motor running single-phase	Stop motor, try to start again. If it is single phasing, it will not start.
	Electrical load unbalance	Check current balance with ammeter.
	Shaft bumping (sleeve bearing motor)	Check alignment and condition of belt drive. Check end play and axial centering of motor.
	Vibration	Driven machine may be unbalanced. Remove motor from load. If motor is still noisy, rebalance motor rotor.
	Air gap not uniform or rotor rubbing on stator	Center the rotor, and if necessary replace bearings.
	Noisy ball bearings	Check lubrication. Replace bearings if necessary.
	Loose parts or loose rotor on shaft	Tighten all holding bolts.
	Object caught between fan and end shields	Disassemble motor and clean.
	Motor loose on foundation	Tighten hold-down bolts. Motor might have to be realigned.
	Coupling loose	Tighten coupling bolts. Realign if necessary.

Symptom	Possible Cause	Solution
Motor temperature too high	Overload	Measure motor loading with ammeter. Reduce load.
	Electrical load unbalance	Check for voltage unbalance or single phasing. Check for an opening in one of the lines or circuits.
	Restricted ventilation	Clean air passages and windings.
	Incorrect voltage and frequency	Check motor nameplate values with power supply. Also check voltage at motor terminals under load.
	Motor stalled by driven machine or tight bearings	Remove power from motor. Check machine for cause of stalling.
	Stator windings with loose connections	Test windings for short circuit or ground.
	Motor used for rapid reversing service	Replace with motor designed for this service.
	Belt too tight	Remove tension and excessive pressure on bearings.
Bearings hot	End bells loose or not replaced properly	Make sure end bells fit squarely and are properly tightened.
	Excessive belt tension or excessive gear slide thrust	Reduce belt tension or gear pressure, and realign shafting.
	Bent shaft	Straighten shaft or replace.
Sleeve bearings hot	Insufficient oil	Add oil. If oil supply is very low, drain, flush, and refill.
	Foreign material in oil or poor grade of oil	Drain oil, flush, and relubricate with proper lubricant.
	Oil rings rotating slowly or not rotating at all	Oil is too heavy. Drain and replace.

(continued)

Table 37-1 (continued)

Symptom	*Possible Cause*	*Solution*
Sleeve bearings hot	Motor tilted too far	Level motor or reduce tilt and realign, if necessary.
	Oil rings bent or otherwise damaged in reassembling	Replace oil rings.
	Oil ring out of slot	Adjust or replace retaining clip.
	Motor tilted, causing end thrust	Level motor, reduce thrust, or use motor designed for thrust.
	Defective bearings or rough shaft	Replace bearings. Resurface shaft.
Ball bearings hot	Too much grease	Remove relief plug and let motor run. If excess grease does not come out, flush and relubricate.
	Wrong grade of grease	Add proper grease.
	Insufficient grease	Remove relief plug and regrease bearing
	Foreign material in grease	Flush bearings and relubricate; make sure the grease supply is clean. Keep can covered when not in use.
	Bearings misaligned	Align motor and check bearing housing assembly. See that bearing races are exactly 90° with the shaft.
	Bearings damaged	Replace bearings.

Table supplied by Maintenance Troubleshooting — Newark, Delaware, USA.

38. PREVENTIVE MAINTENANCE OF EQUIPMENT

Preventive maintenance (PM) is a vital and necessary step in keeping equipment in good operating condition. PM follows some logical steps:

1. Get a process drawing, building equipment location drawing, or other "map" of the plant or process that lists all the equipment and shows the location.
2. Make a list of equipment that will require PM. It is beneficial to group the equipment by name, number, location, and type. If equipment has never been numbered, it might be a smart idea to "invent" an intelligent numbering system that helps in sorting or locating the equipment.

   Examples of intelligent numbering are these:

   PU-2004. A PUmp located on the second deck and the fourth one in the process train.

   100-EX-405. A heat EXchanger in building 100 on the fourth floor and the fifth item in the process train.

   100-AC-405. An AC motor used to drive the gear pump on the heat exchanger shown in the previous entry.

3. Perform a field visit to each piece of equipment. Make sure it exists and that is corresponds to the information previously obtained. Take down all nameplate information to determine the vendor, model number, serial number, and other identification characteristics. A digital camera can speed up this process.
4. Locate all the maintenance manuals for the equipment. Be thorough—check the engineering offices, foreman's desk, and mechanics' benches. If there are

no manuals available, use the nameplate information to find the local sales office or manufacturer's location to obtain a contact person or phone number. The Internet is also a good source of manufacturer information. Contact the manufacturer, supply the nameplate information, and request a maintenance manual for each piece of equipment. In some cases there may be a nominal charge. Make sure to request manuals with "cut sheet" or "exploded diagrams" to make spare part identification easier when working on the equipment.

5. Create a set of PM (or equipment) files. Use the number for the equipment, and set up an empty manila folder for each piece of equipment. Use this file to store nameplate information, vendor contact information, maintenance manuals, or other reference materials for each particular piece of equipment.
6. Use a copy machine to copy the PM page for the particular equipment from the equipment manual. Look over the manufacturer's recommended PM activities. Mentally group the activities into lubrication tasks, visual observation tasks, shutdown inspection tasks, teardown inspection tasks, and any other grouping that puts like activities together.
7. Set up a meeting with key personnel who are responsible for the operation of the equipment—both maintenance and operations. Review the manufacturers' information with others and see if the recommended PM makes sense. Some manufacturers will use an overkill approach and suggest much more PM than is necessary. In addition, pay attention to the suggested frequencies of PM that are mentioned. Read closely. A manufacturer suggests that the equipment be serviced every month, but the recommendation is based on continuous usage. If the equipment is run for only one week in the month, then the PM frequency needs to be adjusted.

8. Begin setting up the PM activities. Pay attention to common types of equipment (all the gearboxes, all the motors) or common activities (greasing, oil changes, chain inspection).
   - It may be more expeditious to assign PM by tasks—that is, one person follows a route through the facility and performs all the lubrication.
   - It may be more expeditious to assign PM by equipment type—that is, a work order for PM calls for PM inspection of all gearboxes during an outage.
   - It may be more expeditious to assign PM by area—that is, all the PM for a certain floor or process area is done by one person.

   Usually a combination of these methods will produce a workable PM system.
9. Assign frequencies to the PM activities (daily, weekly, hours of operation, etc.).
10. Set up the PM routes. Schedule things to be accomplished so that people use their time properly. Make sure personnel do not have to retrace their steps or back up to accomplish the PM tasks.
11. Begin the PM system. In some cases, it is easier to take a small section of the plant or building as a model to see how well the route is laid out. If the model meets with success, then expand it. If the model has problems, correct them before going further.
12. Incorporate the entire PM system into the daily planning and scheduling of personnel.
13. Develop a feedback loop to adjust tasks, frequency, and duration of activities. This feedback loop should allow tailoring of the PM system to adjust for changes in production, changes in machinery loading, or changes from aging of equipment.

14. Continue to audit the PM program every three months. Look for problems. Look for solutions. Effect changes.
15. Modify the program to change tasks, change frequencies, add equipment, remove equipment, change routes, and change personnel based on results of the PM audit.

Visual Inspection Checklist

Visual inspection of machinery or machine components should be quantified. Using a checklist with specific activities and some method of measurement is preferred.

Example

Gearbox inspection (partial list shown):

1. Make sure the gearbox is shut down and locked out.
2. Clean off the area around the inspection cover.
3. Remove the inspection cover and using a flashlight check for gear wear.
   - [ ] Gears show no wear
   - [ ] Gears show some signs of pitting
   - [ ] Gears show mushrooming and have sharp edges when touched
   - [ ] Gears have cracked teeth
   - [ ] Gears have missing teeth
4. Inspect each oil seal on gearbox.
   - [ ] Input shaft seal—drive side leaking
     - [ ] small [ ] medium [ ] extreme
   - [ ] Input seal—drive side not leaking
   - [ ] Input shaft—opposite drive side leaking
     - [ ] small [ ] medium [ ] extreme
   - [ ] Input shaft—opposite drive side not leaking

[ ] Output shaft on drive side leaking
    [ ] small [ ] medium [ ] extreme
[ ] Output shaft on drive side not leaking
[ ] Output shaft—opposite drive side leaking
    [ ] small [ ] medium [ ] extreme
[ ] Output shaft on opposite drive side not leaking

Quantifying things is important. A well-designed checklist gives the mechanic an exact procedure and a place to record observations easily without a lot of writing, which may be difficult to do in the field.

Instrument or Gauge Inspection

Some basic instruments or checking gauges make PM inspections more meaningful. A strobe light can be used to perform a slow-motion study of a shaft or coupling while it is rotating. Cracks, loose parts, and movement are readily observable using this method.

A mechanical stethoscope is inexpensive and a great asset to a PM mechanic. If the PM checklist calls for listening for noises it makes good sense to give the mechanic a tool to pinpoint the source.

Sheave inspection on the V-belt drive can be done with a sheave inspection tool (Fig. 38-1). Belt tensioning can be checked using a tension tester.

A pocket scale or tape measure allows the check of chain elongation to determine wear.

Any device that allows a mechanic to measure for wear or obtain information without shutting the equipment down can greatly enhance a PM system.

Fig. 38-1 Inspection using a sheave gauge. *Courtesy of Maintenance Troubleshooting.*

39. STRUCTURAL STEEL

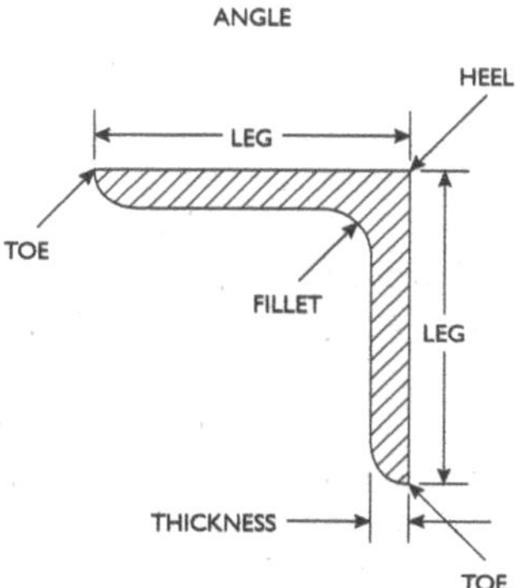

Fig. 39-1 Standard angle.

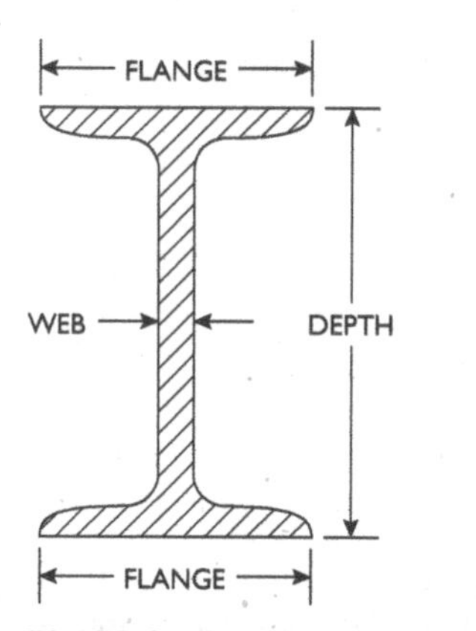

Fig. 39-2 Standard I-beam.

American Standard Angles

The symbol used to indicate an angle shape is (∠). The usual method of billing is to state the symbol, then the long leg, the short leg, the thickness, and finally the length—for example, ∠ 6 × 4 × ⅜ × 12'4". When the legs are equal, both lengths are stated (Fig. 39-1).

American Standard Beams

The symbol used to indicate the beam shape is the letter (I) because of the beam's resemblance to the capital letter "I." The usual method of billing is to state the depth, the symbol, the weight per foot, and finally the length—for example, 15 I 42.9 × 18'4½" (Fig. 39-2).

American Standard Channels

A channel may be compared to an I-beam that has been trimmed on one side to give a flat back web. The symbol used to indicate the standard channel is ([). The usual method of billing is to state the depth, the symbol, the weight per foot, and finally the length. For example, 10 [15.3 – 16'16" (Fig. 39-3).

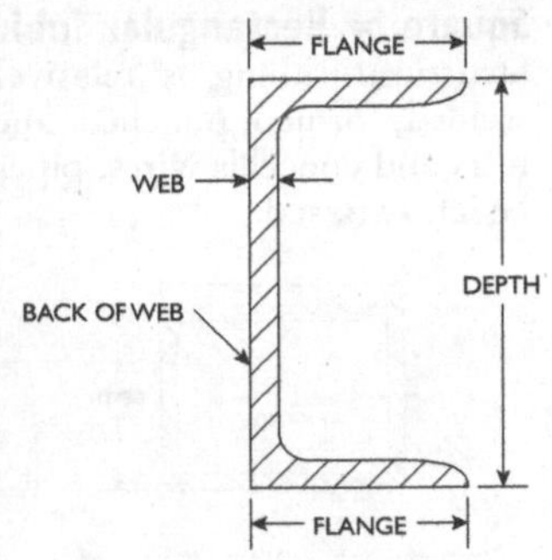

Fig. 39-3 Standard channel.

American Standard Wide Flange Beams or Columns

Wide flange beams are also referred to as "B," "CB," or "H" shapes. They may also be compared to an I-beam with extra wide flanges. The symbol used to indicate the standard wide flange shape is (WF). The usual method of billing is to state the depth, the symbol, the weight per foot, and finally the length—for example, 12 (WF) 45—24'8" (Fig. 39-4).

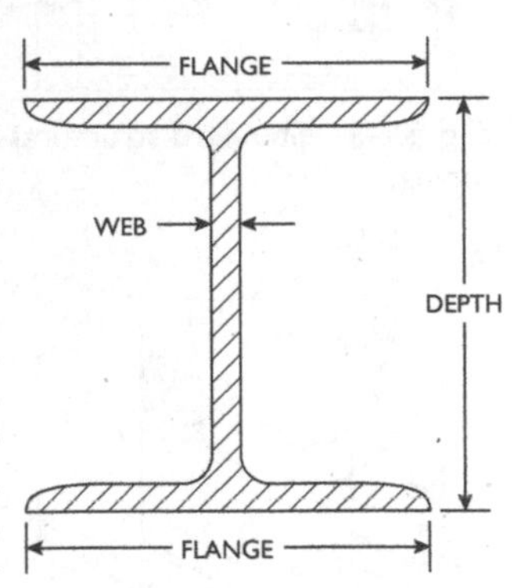

Fig. 39-4 Standard wide flange.

Structural steel is produced at rolling mills in a wide variety of standard shapes and sizes. In this form, it is referred to as "plain material." In addition to the plates and bar stock, the four shapes illustrated previously are the most widely used. Other standard shapes produced are *Tee's, Zee's, Rails,* and various special shapes. All standard structural shapes are made to a standardized series of nominal sizes. Within each size group there is a wide range of weights and dimensions.

Square or Rectangular Tubing (Structural Tubing)

Structural tubing is relatively low in cost and is easily welded, formed, punched, and drilled. Its hollow shape protects and conceals wires, pipes, and moving parts; and it can be left exposed.

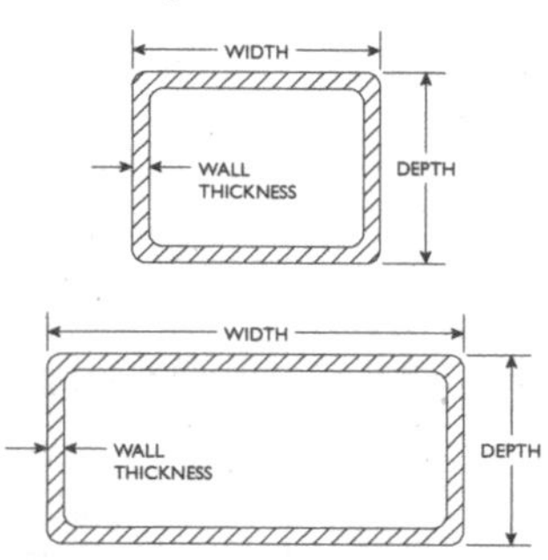

Fig. 39-5 Standard structural tubing.

The structural square or rectangular tubing often facilitates the incorporation of hydraulic lines or control wiring within the framework of a machine—accomplishing two purposes at the same time.

Simple Square-Framed Beams

Square-framed beams are, as the name implies, beams that intersect or connect at right angles. This is the most common type of steel construction. Two types of connections may be used in framed construction—*framed* and *seated*. In the framed type, shown in Fig. 39-6a, the beam is connected by means of fittings (generally a pair of angle irons) attached to its web. With the seated connection (Fig. 39-6b), the end of the beam rests on a ledge or seat.

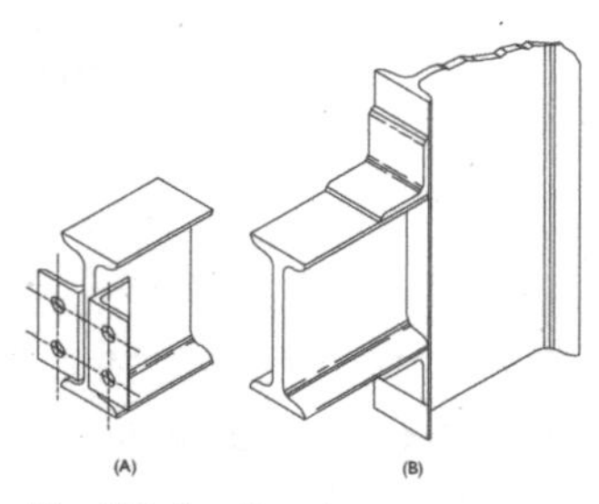

Fig. 39-6 Simple square framed beams.

Clearance Cuts

When connecting one member to another, it is often necessary to notch or cut away both flanges of the entering member to avoid flange interference. Such a notch is called a *cope,* a *block,* or a *cut.* The term "cope" is usually used if the cut is to follow closely the shape of the member into which it will fit. When the cut is rectangular in shape with generous clearance, it is usually called a "block-out." Unless there is some reason for a close-matching fit, the block-out is recommended, as it is the easiest and most economical notch to make (Fig. 39-7).

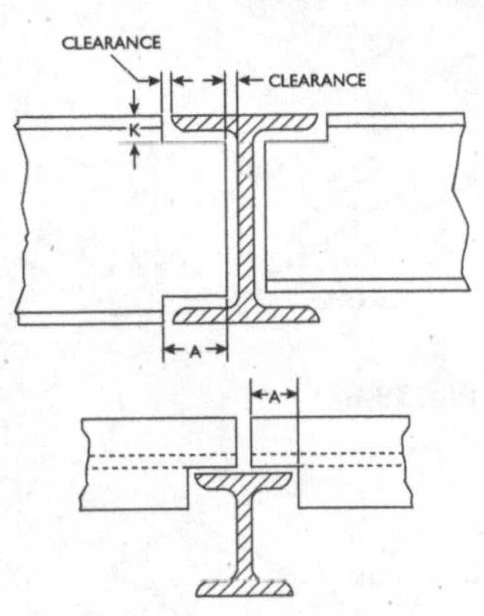

Fig. 39-7 Typical cutouts on a square framed beam.

When making block-out cuts, the dimensions of the rectangular notch may be obtained from tables of structural steel dimensions. Tables 39-3, 39-4, and 39-5 list the "K" and "a" dimensions for various size and weight members. The "K" and "a" values determine the dimensions of the cut, as they indicate the maximum points of interference. While the notch is made in the entering member, the values from the table for the supporting member determine the notch dimensions. The steel must be cut to length before the block-out cut is made.

Square-Framed Connections (Two-Angle Type)

"Standard" connections should be used when fabricating structural steel members to ensure proper assembly with supporting members at installation.

> **Spread.** The distance between hole centers in the web of the supporting member and the holes in the *attached* connection angles is called the *spread* of the holes. The spread dimension is standardized at 5½ in., as shown in Fig. 39-8.

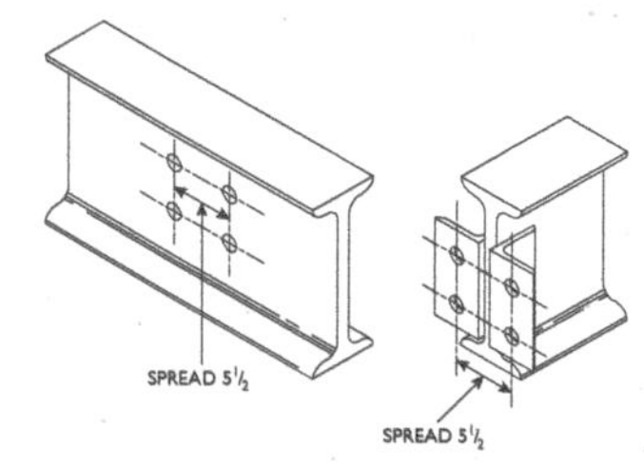

Fig. 39-8

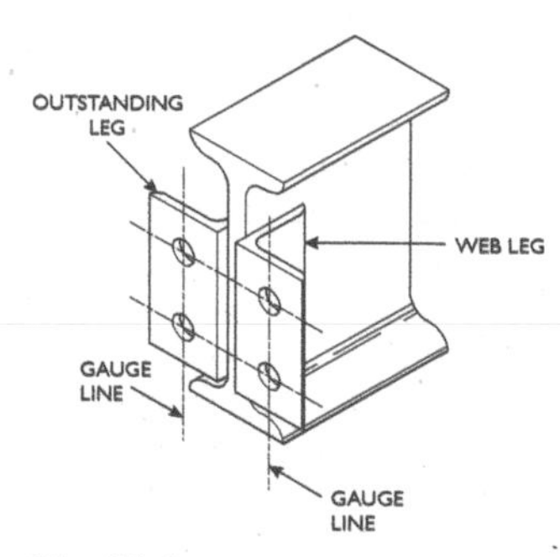

Fig. 39-9

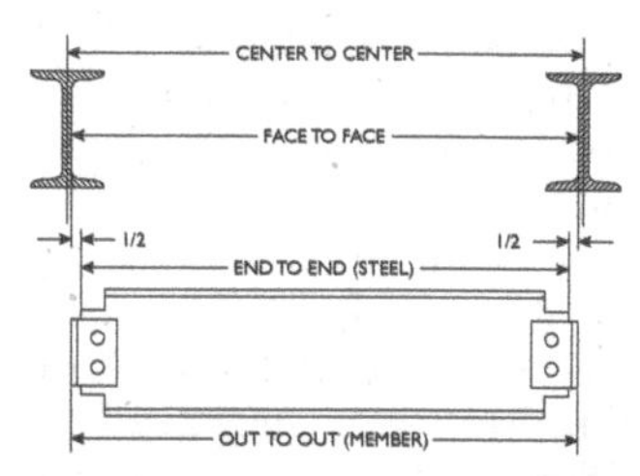

Fig. 39-10

Angle Connection Legs

The legs of the angles used as connections are specified according to the surface to which they are connected, as shown in Fig. 39-9. The legs that attach to the entering steel to make the connection are termed *web* legs. The legs of the angles that attach to the supporting member are termed *outstanding* legs. The lines on which the holes are placed are called *gauge* lines. The distances between gauge lines, or from a gauge line to a known edge, are called *gauges*.

Fabrication Terms

Commonly used structural steel fabrication terms are illustrated in Fig. 39-10. Using these terms aids in understanding steel fabrication and reduces the probability of errors.

Steel. Various structural steel shapes and forms.

Member. An assembly of a length of steel and its connection fittings.

Center-to-center. The distance from the centerline of one member to the centerline of another member.

Face-to-face. The distance between the facing web surfaces of two members. It is the center-to-center distance minus the ½-web thickness of each member.

End-to-end. The overall length of the steel. It should be 1 in. shorter than the opening into which it will be placed. The 1-in. clearance is provided for assembly and as an allowance for inaccuracies.

Out-to-out. The overall length of the assembled member. To provide assembly clearance, the connection angles are positioned on the member so that the out-to-out distance before assembly is slightly less than the face-to-face distance.

Connection Hole Locations

The terms and the constant dimension for standard square-framed two-angle type connections are illustrated in Fig. 39-11.

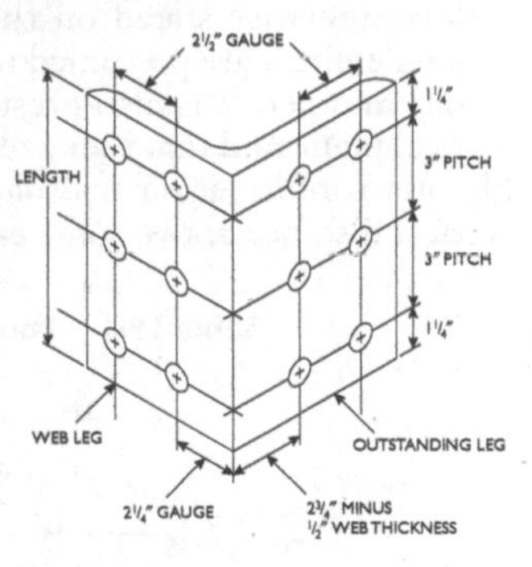

Fig. 39-11

Web-Leg Gauge

The distance from the heel of the angle to the first gauge line on the web leg is called *web-leg gauge*. This dimension is constant, as it is standardized at 2¼ in.

Outstanding-Leg Gauge

The distance from the heel of the angle to the first gauge line on the outstanding leg is called the *outstanding-leg gauge*. This dimension varies, as the thickness of the web of the member varies in order to maintain a constant 5½-in. spread dimension. The outstanding leg-gauge dimension is determined by subtracting the web thickness from 5½ in. and dividing by 2. A simpler way to determine the outstanding-leg dimension is to subtract the ½-in. web thickness from 2¾ in., which is one-half the spread.

Gauges. The distances between gauge lines, or from a gauge line to a known edge, are called "gauges." When more than one row of holes is used, the gauge is 2½ in. This dimension is constant.

Pitch. The distance between holes on any gauge line is called *pitch*. This dimension is standardized at 3 in.

End Distance. The end distance is equal to one-half of the remainder left after subtracting the sum of all pitches from the length of the angle. By common practice the angle length is selected to give a 1¼-in. end distance.

Steel Elevation

Unless otherwise stated on the drawing, all square-framed beam members are presumed to be parallel or at right angles to one another. It is also presumed that their webs are in a vertical plane and that they are in a level position end to end. Elevation information is usually given by note, stating the vertical distance above some established horizontal plane.

Table 39-1 Standard Channels

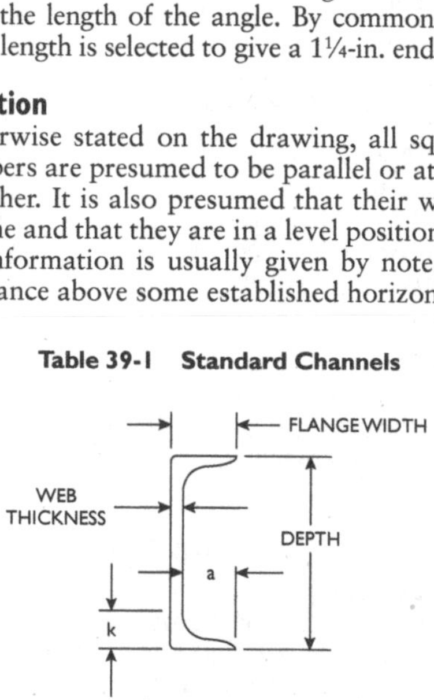

Depth (Inches)	*Weight per Foot*	*Flange Width*	*Web Thickness*	*a*	*k*
3	4.1	1⅜	3/16	1¼	⅝
	5.0	1½	¼	1¼	⅝
	6.0	1⅝	⅜	1¼	⅝

Depth (Inches)	Weight per Foot	Flange Width	Web Thickness	a	k
4	5.4	1 3/8	3/1	1 3/8	5/8
	7.25	1 3/46	5/16	1 3/8	5/8
5	6.7	1 3/4	3/16	1 1/2	11/16
	9.0	1 7/8	5/16	1 1/2	11/16
6	8.2	1 7/8	3/16	1 3/4	1 3/4
	10.5	2	5/16	1 3/4	1 3/4
	13.0	2 1/8	7/16	1 3/4	1 3/4
7	9.8	2 1/8	1/4	1 7/8	13/16
	12.25	2 1/4	5/16	1 7/8	13/16
	14.75	2 1/4	7/16	1 7/8	13/16
8	11.5	2 1/4	1/4	2	13/16
	13.75	2 3/8	5/16	2	13/16
	18.75	2 1/2	1/2	2	13/16
9	13.4	2 3/8	1/4	2 1/4	7/8
	15.0	2 1/2	5/16	2 1/4	7/8
	20.0	2 5/8	7/16	2 1/4	7/8
10	15.3	2 5/8	1/4	2 3/8	5/16
	20.0	2 3/4	3/8	2 3/8	5/16
	25.0	2 7/8	9/16	2 3/8	5/16
	30.0	3	11/16	2 3/8	5/16
12	20.7	3	5/16	2 5/8	1 1/16
	25.0	3	3/8	2 5/8	1 1/16
	30.0	3 1/8	1/2	2 5/8	1 1/16
15	33.9	3 3/8	7/16	3	1 5/16
	40.0	3 1/2	9/16	3	1 5/16
	50.0	3 3/4	3/4	3	1 5/16
18	42.7	4	17/16	3 11/2	1 5/16
	45.8	4	11/2	3 11/2	1 5/16
	51.9	4 1/8	15/8	3 11/2	1 5/16
	58.0	4 1/4	111/16	3 11/2	1 5/16

Table 39-2 Standard Beams

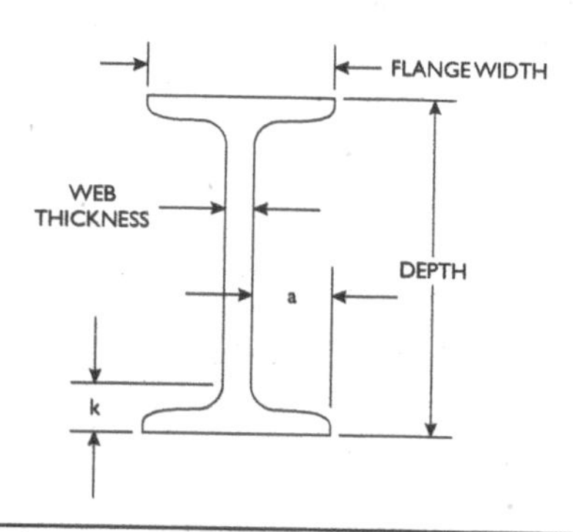

Depth (Inches)	Weight per Foot	Flange Width	Web Thickness	a	k
3	5.7	2 3/8	3/16	1 1/a	9/16
	7.5	2 1/2	3/8	1 1/a	9/16
4	7.7	2 5/8	3/16	1 1/4	5/8
	9.5	2 3/4	5/16	1 1/4	5/8
5	10.0	3	1/4	1 3/8	11/16
	14.75	3 1/4	1/2	1 3/8	11/16
6	12.5	3 3/8	1/4	1 1/2	3/4
	17.75	3 5/8	1/2	1 1/2	3/4
7	15.3	3 5/8	1/4	1 3/4	13/16
	20.0	3 7/8	7/16	1 3/4	13/16
8	18.4	4	5/16	1 7/8	7/8
	23.0	4 1/8	7/16	1 7/8	7/8
10	25.4	4 5/8	5/16	2 1/8	1
	35.0	5	5/8	2 1/8	1
12	31.8	5	3/8	2 3/8	1 1/8
	35.0	5 1/8	7/16	2 3/8	1 1/8
12	40.8	5 1/4	1/2	2 3/8	1 5/16
	50.0	5 1/2	11/16	2 3/8	1 5/16

Depth (Inches)	Weight per Foot	Flange Width	Web Thickness	a	k
15	42.9	5½	7/16	2½	1¼
	50.0	5⅝	9/16	2½	1¼
18	54.7	6	½	2¾	1⅜
	70.0	6¼	½	2⅞	1 9/16
20	65.4	6¼	⅝	2⅞	1 9/16
	75.0	6⅜	¾	2¾	1⅜
20	85.0	7	11/16	3¼	1¾
	95.0	7¼	13/16	3¼	1¾

Table 39-3 Wide Flange — CB Sections

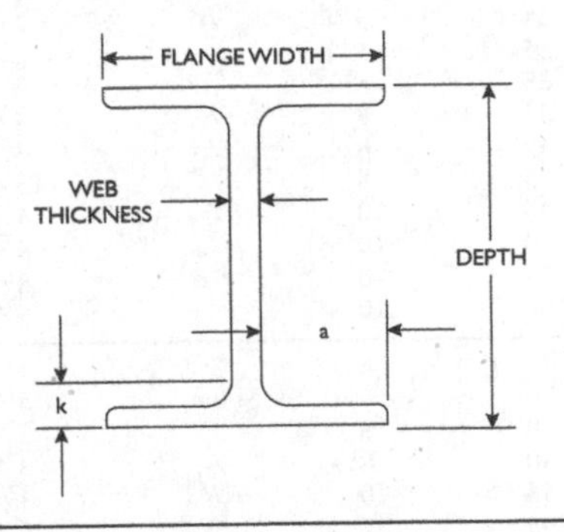

Nom. Depth (Inches)	Weight per Foot	Flange Width	Web Thickness	a	k
4	10	4	¼	1⅞	7/16
5	16	5	¼	2⅜	⅝
6	12	4	¼	1⅞	9/16
	15.5	6	¼	2⅜	9/16

(continued)

Table 39-3 *(continued)*

Nom. Depth (Inches)	Weight per Foot	Flange Width	Web Thickness	a	k
8	13	4	1/4	1 7/8	9/16
	17	5 1/4	1/4	2 1/2	5/8
	20	5 1/4	1/4	2 1/2	11/16
	24	6 1/2	1/4	3 1/8	13/16
	28	6 1/2	5/16	3 1/8	13/16
	31	8	5/16	3 7/8	13/16
	35	8	5/16	3 7/8	7/8
	40	8 1/8	3/8	3 7/8	15/16
	48	8 1/8	7/16	3 7/8	1 1/16
	58	8 1/4	1/2	3 7/8	1 3/16
	67	8 1/4	9/16	3 7/8	1 5/16
10	15	4	1/4	1 7/8	9/16
	21	5 3/4	1/4	2 3/4	11/16
	25	5 3/4	1/4	2 3/4	13/16
	29	5 3/4	5/16	2 3/4	7/8
	33	8	5/16	3 7/8	15/16
	39	8	5/16	3 7/8	1 1/16
	45	8	3/8	3 7/8	1 1/8
	49	10	3/8	4 7/8	1 1/16
	60	10 1/8	7/16	4 7/8	1 3/16
	72	10 1/8	1/2	4 7/8	1 5/16
	100	10 3/8	11/16	4 7/8	1 5/8
12	27	6 1/2	1/4	3 1/8	13/16
	36	6 5/8	5/16	3 1/8	15/16
	40	8	5/16	3 7/8	1 1/8
	50	8 1/8	3/8	3 7/8	1 1/4
	58	10	3/8	4 7/8	1 1/4
	65	12	3/8	5 3/4	1 3/16
	106	12 1/4	5/8	5 3/4	1 9/16
14	30	6 3/4	5/16	3 1/4	7/8
	48	8	3/8	3 7/8	1 3/16
	68	10	7/16	4 3/4	1 5/16
	84	12	7/16	5 3/4	1 3/8

Table 39-4 Square Tubing

Outside Dimension in Inches	*Wall Thickness*	*Weight per Foot in Lbs*
½ × ½	.049	.301
	.065	.385
⅝ × ⅝	.049	.384
	.065	.500
¾ × ¾	.035	.340
	.049	.467
	.065	.606
	.120	1.03
⅞ × ⅞	.065	.716
1 × 1	.035	.459
	.049	.634
	.065	.827
	.072	.909
	.073	.920
	.083	1.04
	.109	1.32
	.120	1.44
1¼ × 1¼	.049	.800
	.065	1.05
	.083	1.32
	.090	1.42
	.120	1.84
	.134	2.03
	.188	2.62
1½ × 1½	.049	.967
	.065	1.27
	.083	1.60
	.120	2.25
	.140	2.50
	.180	3.23
	.188	3.23
	.250	4.11
1¾ × 1¾	.083	1.88
	.120	2.66

(continued)

Table 39-4 *(continued)*

Outside Dimension in Inches	Wall Thickness	Weight per Foot in Lbs
2 × 2	.065	1.71
	.083	2.16
	.095	2.46
	.120	3.07
	.125	3.07
	.145	3.51
	.180	4.46
	.188	4.46*
	.250	5.41*
2½ × 2½	.083	2.73
	.120	3.88
	.125	3.88
	.141	4.52
	.180	5.68
	.188	5.68*
	.250	7.10*
	.313	8.44
3 × 3	.083	3.29
	.125	4.70
	.180	6.62
	.188	6.86
	.250	8.80
	.313	10.00
3½ × 3½	.125	5.52
	.188	8.14
	.250	10.50
	.313	12.12
4 × 4	.125	6.33
	.180	9.07
	.188	9.31
	.250	12.02
	.313	14.52
	.375	16.84
	.500	20.88
4½ × 4½	.188	10.58
	.250	13.72

Outside Dimension in Inches	*Wall Thickness*	*Weight per Foot in Lbs*
5 × 5	.180	11.40
	.188	11.86
	.250	15.42
	.313	18.77
	.375	21.94
	.500	27.68
6 × 6	.250	14.41
	.313	18.82
	.375	23.02
	.500	27.04
	.625	34.48
7 × 7	.188	16.85
	.250	22.04
	.313	26.99
	.375	31.73
	.500	40.55
8 × 8	.250	25.44
	.313	31.24
	.375	36.83
	.500	47.35
	.625	56.98
10 × 10	.250	32.23
	.313	39.74
	.375	47.03
	.500	60.95
	.625	77.40
12 × 12	.250	39.03
	.375	57.23
	.500	74.54
14 × 14	.375	67.43
	.500	88.14
16 × 16	.375	77.63
	.500	101.7

**Weight is for Electric Resistance Welded tubes—slightly less than butt welded tubes.*

Table 39-5 Rectangular Tubing

Outside Dimension in Inches	*Wall Thickness*	*Weight per Foot in Lbs*
1½ × ¾	.075	1.07
1½ × 1	.074	1.19
	.083	1.32
	.120	1.84
2 × 1	.065	1.27
	.074	1.45
2 × 1¼	.083	1.60
2 × 1½	.065	1.49
	.120	2.66
2½ × 1	.083	1.88
2½ × 1¼	.083	2.02
2½ × 1½	.074	1.94
	.083	2.26
	.145	3.51
	.188	4.43
	.250	5.59
3 × 1	.065	1.71
	.083	2.16
3 × 1½	.083	2.45
	.125	3.45
	.188	5.07
3 × 2	.083	2.73
	.125	3.88
	.180	5.40
	.188	5.59
	.250	7.10
4 × 2	.083	3.294.70
	.125	6.62
	.180	6.86
	.188	8.80
	.250	10.00
	.313	
4 × 2½	.120	5.11

Outside Dimension in Inches	*Wall Thickness*	*Weight per Foot in Lbs*
4 × 3	.120	5.52
	.180	7.86
	.188	8.14
	.250	10.50
	.313	12.69
5 × 2	.188	8.14
	.250	10.50
5 × 2½	.120	5.92
	.180	8.74
	.188	8.88
5 × 3	.180	9.07
	.188	9.31
	.250	12.02
	.313	14.52
	.375	16.84
	.500	20.88
6 × 2	.188	9.31
	.250	12.02
6 × 3	.180	10.20
	.188	10.58
	.250	13.72
	.313	16.65
	.375	19.39
	.500	24.28
6 × 4	.180	11.40
	.188	11.86
	.250	15.42
	.313	18.77
	.375	21.94
	.500	27.68
7 × 5	.188	14.41
	.250	18.82
	.313	23.02
	.375	27.04
	.500	34.48
8 × 2	.188	11.86

(continued)

Table 39-5 (continued)

Outside Dimension in Inches	Wall Thickness	Weight per Foot in Lbs
8 × 3	.188	13.13
	.250	17.12
	.375	24.49
8 × 4	.188	14.41
	.250	18.82
	.313	23.02
	.375	27.04
	.500	34.48
8 × 6	.188	16.85
	.250	22.04
	.313	26.99
	.375	31.73
	.500	40.55
10 × 2	.188	14.41
10 × 3	.250	20.34
10 × 4	.188	16.85
	.250	22.04
10 × 5	.250	23.74
10 × 6	.250	25.44
	.313	31.24
	.375	36.83
	.500	47.35
10 × 8	.250	28.83
	.375	41.93
	.500	54.15
12 × 2	.188	16.85
12 × 4	.250	25.44
	.375	36.83
12 × 6	.250	28.83
	.375	41.93
	.500	54.15

Outside Dimension in Inches	*Wall Thickness*	*Weight per Foot in Lbs*
12 × 8	.375	47.03
16 × 8	.375	57.23
16 × 12	.375	67.43
	.500	88.14

40. WELDING

Because of the development of numerous welding processes, and also because of the development of new steels and other metals that can be welded, welding has become the most important metal-joining process. Following is a concise explanation of the most widely used of these processes, giving basic information that should be of value in understanding the welding processes in general use. This includes information regarding welding equipment, procedures, filler metals, and so on that are involved in the operation called welding.

Welding Safety

Welding involves several potentially hazardous conditions: very high temperatures, use of explosive gases, possible exposure to harmful light, toxic fumes, etc., molten metal spatter, and flying particles. Welding hazards can be controlled to ensure the safety of the welder.

Basic protective actions include these: Protective clothing must be worn by the welder to shield skin from exposure to the brilliant light given off by the arc; a helmet is required to protect the face and eyes from the arc; fire-resistant protective clothing, shoes, leather gloves, jacket, apron, etc., are a necessity; a dark-colored filter glass in the helmet allows the welder to watch the arc while protecting the eyes.

Ventilation must be provided when welding in confined areas. The work area must be kept clean and the equipment properly maintained.

Welding Processes

The commonly used welding processes may be grouped into four general categories: *gas oxyacetylene welding, shielded metal-arc welding, gas metal-arc welding,* and *gas tungsten-arc welding.* The word "shielding" is used to describe the creation of an environment of controlled gas or gases around the weld zone to protect the molten weld metal from contamination by the oxygen and nitrogen in the atmosphere.

OxyAcetylene Welding

Oxyacetylene gas welding is perhaps the oldest of the gas welding processes. It came into use about 125 years ago and is still being used in much the same manner. The process is extremely flexible and one of the most inexpensive as far as equipment is concerned. Today, its most popular application is in maintenance welding, small-pipe welding, auto body repairs, welding of thin materials, and sculpture work. The high temperature generated by the equipment is used for soldering, hard soldering, or brazing. Also, using a special torch, a variation of the process allows for flame cutting. This is accomplished by bringing the metal to a high temperature and then introducing a jet of oxygen that burns the metal apart. It is a primary cutting tool for steel.

In the oxyacetylene gas welding process, coalescence is produced by heating with a gas flame obtained from the combustion of acetylene with oxygen, with or without the use of filler metal. An oxyacetylene flame is one of the hottest of flames—6300°F. This hot flame melts the two edges of the pieces to be welded and the filler metal (added to fill the gaps or grooves) so that the molten metal can mix rapidly and smoothly. The acetylene and the oxygen gases flow from separate cylinders to the welding torch, where they are mixed and burned at the torch tip. Fig. 40-1 shows the oxyacetylene gas welding process.

The proportions of oxygen and acetylene determine the type of flame. The three basic types are *neutral, carburizing,* and *oxidizing*. The neutral flame is generally preferred for welding. It has a clear, well-defined white cone, indicating the best mixture of gases and no gas wasted. The carburizing flame has an excess of acetylene, a white cone with a

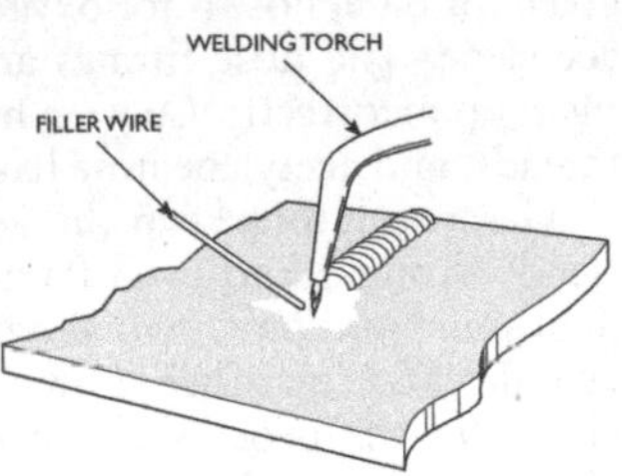

Fig. 40-1 Oxyacetylene gas welding process.

feathery edge, and adds carbon to the weld. The oxidizing flame, with an excess of oxygen, has a shorter envelope and a small, pointed white cone. This flame oxidizes the weld metal and is used only for specific metals. Flame cutting is accomplished by adding an extra oxygen jet to burn the metal being cut. The equipment required for oxyacetylene welding is shown in Fig. 40-2.

The standard torch can be a combination type used for welding, cutting, and brazing. The gases are mixed within the torch. A thumbscrew needle valve controls the quantity of gas flowing into a mixing chamber. A lever type valve controls the oxygen flow for cutting with a cutting torch or attachment. Various types and sizes of tips are used with the torch for specific applications of welding, cutting, brazing, or soldering. The usual welding outfit has three or more tips. Too small a tip will take too long or will be unable to melt the base metal. Too large a tip may result in burning the base metal.

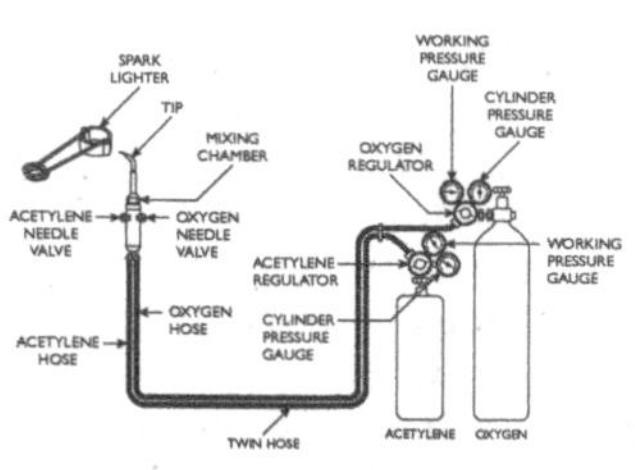

Fig. 40-2 Oxyacetylene welding equipment.

The gas hoses may be separate or molded together. The green (or blue) hose is for oxygen, and the red (or orange) is for acetylene. The hose fittings are different to prevent hooking them up incorrectly. Oxygen hose has fittings with right-hand threads, and acetylene hose has fittings with left-hand threads.

Gas regulators keep the gas pressure constant, ensuring steady volume and even flame quality. Most regulators are dual stage and have two gauges; one tells the pressure in the cylinder and the other shows the pressure entering the hose. Gases for the process are oxygen and, primarily, acetylene. Other gases, including hydrogen, city gas, natural gas, propane, and mapp gas, are used for specific applications.

With its higher burning temperature, acetylene is the preferred gas in most instances.

Gas cylinders for acetylene contain porous material saturated with acetone. Because acetylene cannot safely be compressed over 15 psi, it is dissolved in the acetone that keeps it stable and allows pressure of 250 psi. Because of the acetone in the acetylene cylinders, they should always stand upright. The oxygen cylinder capacities vary from 60 to 300 cu. ft. with pressures up to 2400 psi. The maximum charging pressure is always stamped on the cylinder.

Use of the Cutting Torch

Oxyacetylene cutting, often called flame or oxygen cutting, is the most widely used process for thermal cutting of carbon steel. Just as a saw is to a carpenter, the cutting torch is to the steel fabricator.

The process involves preheating the steel to a temperature of 1500°F to establish an ignition temperature for the steel and then introducing a stream of oxygen under pressure from the tip of the torch. This causes the metal to oxidize or burn rapidly and produces a cut or kerf along the direction of travel.

Use of the cutting torch follows a set of steps. The first step is to select a tip or nozzle that is suitable for the job. This tip is screwed into the torch securely. The torch with a cutting tip installed is shown in Fig. 40-3. While manufacturers' charts are available for cutting data, the values shown in Table 40-1 will prove valuable in matching the tip to the job.

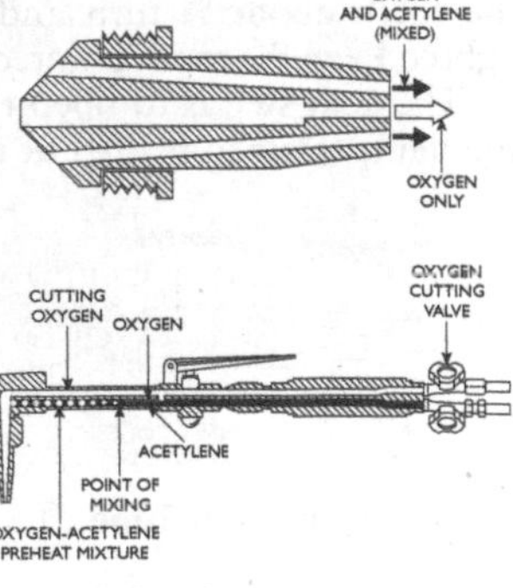

Fig. 40-3 Oxyacetylene torch cutting tip.

Table 40-1 Cutting Torch Tip Selections

Thickness of Steel Plate (Inches)	*Diameter of Cutting Tip Orifices (Inches)*	*Approximate Cutting Speed (Inches per Minute)*
⅛	0.020–0.040	16–32
¼	0.030–0.060	16–26
⅜	0.030–0.060	15–24
½	0.040–0.060	12–23
¾	0.045–0.060	12–21
1	0.045–0.060	9–18
1½	0.060–0.080	6–14
2	0.060–0.080	6–13
3	0.065–0.085	4–11
4	0.080–0.090	4–10
5	0.080–0.095	4–8
6	0.095–0.105	3–7
8	0.095–0.110	3–5
10	0.095–0.110	2–4
12	0.110–0.130	2–4

Make sure that the oxygen and acetylene pressure regulators on the tanks are set to the manufacturer's recommendations. The next step is to partly open the acetylene valve on the torch about ¼ turn and light the torch using a spark lighter. Keep the torch pointed away from people or property.

The next step is to open the acetylene valve further until the flame starts to feather at the end and a condition of "no-smoke" exists. At this point, the oxygen valve on the torch is slowly opened until a neutral flame is established. A neutral flame has approximately equal volumes of both acetylene and oxygen and has the appearance shown in Fig. 40-4.

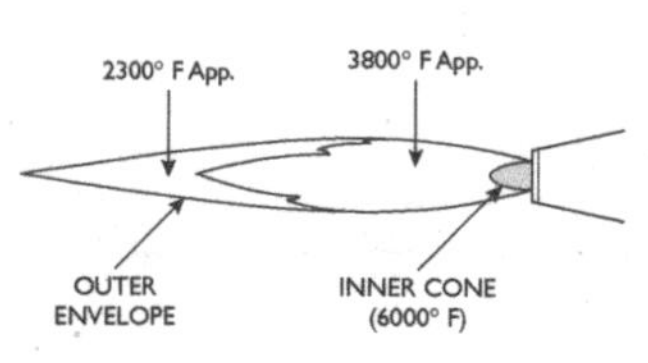

Fig. 40-4 Neutral flame.

After the neutral flame is established, the oxygen lever on the torch is depressed. The individual flames that preheat flames around the oxygen orifices at the tip may change from neutral to slightly carburizing, as shown in Fig. 40-5.

Keeping the oxygen lever in the depressed position, the oxygen valve is further adjusted until these preheat flames again appear neutral, as shown previously in Fig. 40-4. At this point, after releasing the lever, the torch is set for preheating.

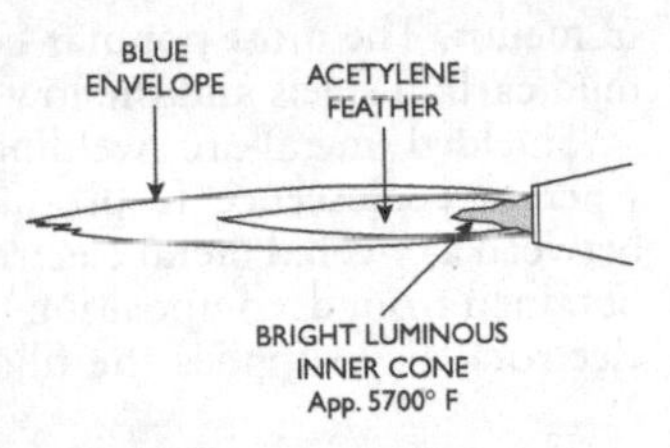

Fig. 40-5 Slightly carburizing flame.

Usually the torch is held at the edge of the carbon steel material and held there until the steel becomes red hot. Then the oxygen lever is depressed, and a jet of oxygen is sent onto the red-hot metal, which ignites or oxidizes the metal, as shown in Fig. 40-6. This is the point where cutting begins. Hold the torch steady until the cut is all the way through the metal. Then move the torch slowly along the line to be cut, forming a kerf through the metal.

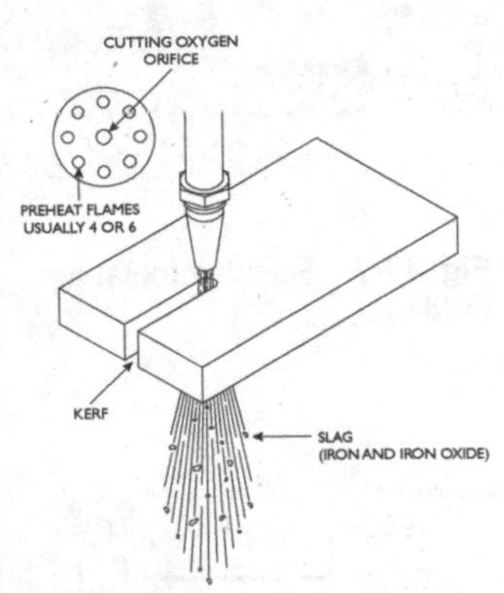

Fig. 40-6 Cutting with the torch.

When the cut is finished, release the oxygen lever, close the acetylene valve, and close the oxygen valve—in that order.

Shielded Metal-Arc Welding—Stick Electrode

This is perhaps the most popular welding process in use today. The high quality of the metal produced by the shielded-arc process, plus the high rate of production, has made it a replacement for other fastening methods. The process can be used in all positions and will weld a wide variety of metals. The most popular use, however, is the welding of mild carbon steels and the low-alloy steels.

Shielded metal-arc welding is an arc-welding process wherein coalescence is produced by heating with an arc between a covered metal electrode and the work. Shielding is obtained from decomposition of the electrode covering. The electrode also supplies the filler metal. Fig. 40-7 shows the covered electrode, the core wire, the arc area, the shielding atmosphere, the weld, and solidified slag.

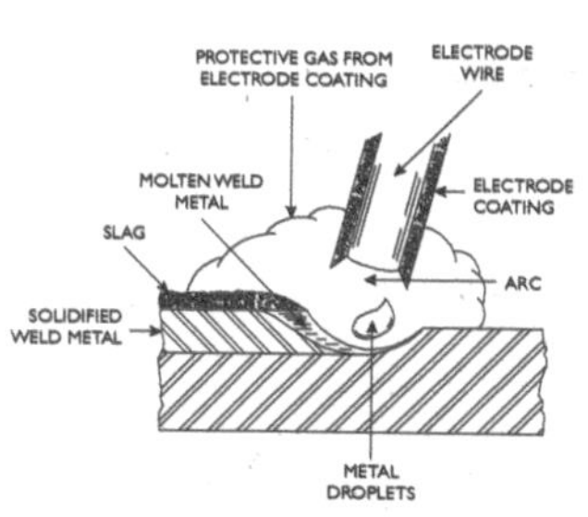

Fig. 40-7 Shielded metal-arc welding.

This manually controlled process welds all nonferrous metals ranging in thickness from 18 gauge to the maximum encountered. For material thicknesses of ¼ in., a beveled-edge preparation is used and the multipass welding technique is employed. The process allows for all position welding. The arc is under the control of, and is visible to, the welder. Slag removal is required. The major components required for shielded metal-arc welding are shown in Fig. 40-8.

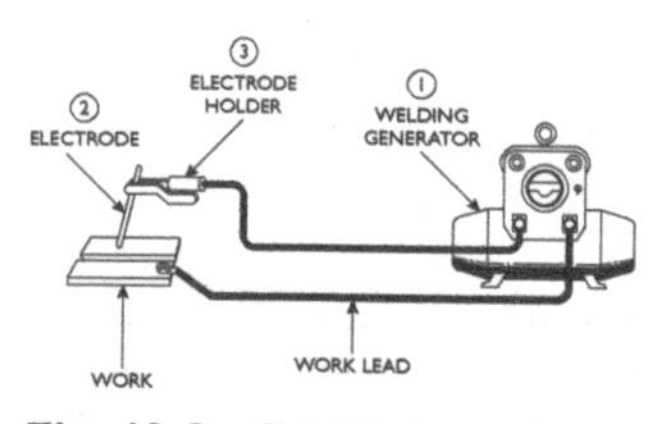

Fig. 40-8 Shielded metal-arc welding equipment.

The welding machine (power source) is a major item of required equipment. Its primary purpose is to provide electric power of the proper current and voltage sufficient to maintain a welding arc. Shielded metal-arc welding can be accomplished by either alternating current (AC) or direct current (DC). Direct current can be employed straight (electrode negative) or reverse (electrode positive). A variety of welding machines are used, each having specific advantages or special features. The AC transformer is simple, inexpensive, and quiet. The transformer-rectifier-type machine converts AC power to DC power and provides direct current at the arc. There is also the AC-DC transformer-rectifier-type machine, which combines the features of both the transformer and the rectifier. Probably the most versatile welding power source is the direct current generator. The conventional dual-control, single-operator generator allows the adjustment of the open-circuit voltage and the welding current. When electric power is available, the generator is driven by an electric motor. Away from power lines, a gasoline internal combustion engine or a diesel engine can drive the generator.

The electrode holder is held by the operator and firmly grips the electrode, carrying the welding current to it. The insulated pincer-type holders are the most popular. Electrode holders come in various sizes and are designated by their current-carrying capacity.

The welding circuit consists of the welding cables and connectors used to provide the electrical circuit for conducting the welding current from the machine to the arc. The electrode cable forms one side of the circuit and runs from the electrode holder to the electrode terminal of the welding machine. Welding cable size is selected based on the maximum welding current used. Sizes range from AWG No. 6 to AWG No. 4/0 with amperage ratings from 75 amps upward. The work lead is the other side of the circuit and runs from the work clamp to the work terminal of the welding machine.

Covered electrodes, which become the deposited weld metal, are available in sizes from 1⁄16 to 5⁄16 in diameter and from 9 to 18 in. in length, with the 14-in. length the most popular. The covering on the electrode dictates the usability of the electrode and provides the following:

- Gas shielding
- Deoxidizers for purifying the deposited weld metal
- Slag formers to protect weld metal from oxidation

- Ionizing elements for smooth operation
- Alloying elements to strengthen deposited weld metal
- Iron powder to improve the productivity of the electrode

The usability characteristics of different types of electrodes are standardized and defined by the American Welding Society (AWS). The AWS identification system indicates the characteristics and usability by classification numbers printed on the electrodes.

Gas Metal-Arc Welding (MIG)

Gas metal-arc (MIG) welding is an arc-welding process wherein coalescence is produced by heating with an arc between a continuous filler-metal (consumable) electrode and the work. The electrode is a wire that is continuously and automatically fed into the arc to maintain a steady arc. This electrode wire, melted into the heat of the arc, is transferred across the arc and becomes the deposited weld metal. Shielding is obtained entirely from an externally supplied gas mixture. Fig. 40-9 shows the electrode wire, the gas shielding envelope, the arc, and the deposition of the weld metal.

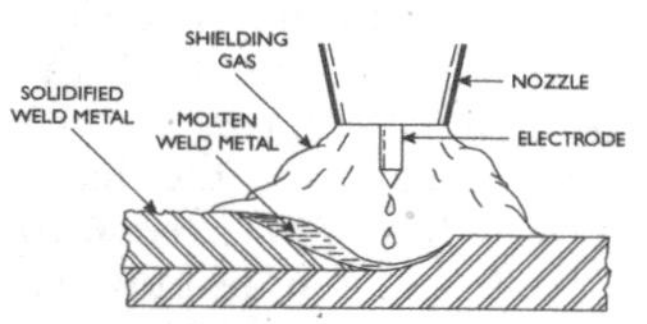

Fig. 40-9 Shielded gas metal-arc welding.

Some of the outstanding features of gas metal-arc welding are top-quality welds in almost all metals and alloys, little after-weld cleaning, relatively high speed, and no slag production. Some of the variations of the process involve microwire for thin gauge materials, CO_2 for low-cost high-speed welding, and argon/oxygen for stainless steels. The major components required for gas metal-arc welding are shown in Fig. 40-10.

The welding machine or power source for consumable-electrode welding is called a constant voltage (CV) type of welder, meaning that its output voltage is essentially the same with different welding-current levels. These CV power sources do not have a welding-current control and cannot be used for welding with electrodes. The welding-current output is determined by the load on the machine, which is dependent on the electrode wire-feed speed. The wire-feeder system must be matched to the constant-voltage power supply. At a given wire-feed speed rate, the welding machine will supply the proper amount of current to maintain a steady arc. Thus the electrode wire-feed rate determines the amount of welding current supplied to the arc.

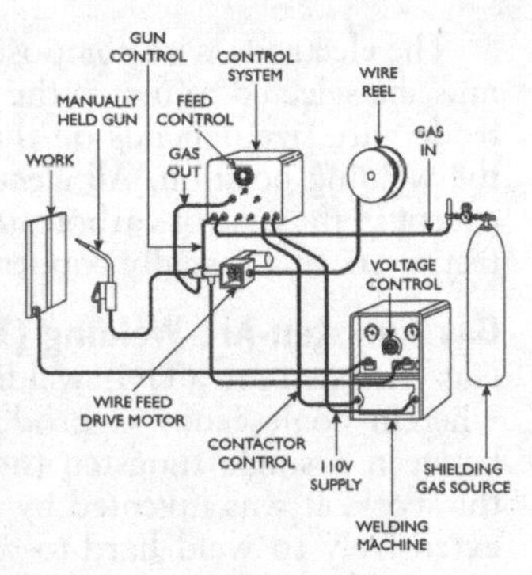

Fig. 40-10 Gas metal-arc welding equipment.

The welding gun and cable assembly are used to carry the electrode wire, the welding current, and the shielding gas to the welding arc. The electrode wire is centered in the nozzle, with the shielding gas supplied concentric to it. The gun is held fairly close to the work to control the arc properly and provide an efficient gas-shielding envelope. Guns for heavy-duty work at high currents, and guns using inert gas and medium currents, must be water cooled.

The shielding gas displaces the air around the arc to prevent contamination by the oxygen or nitrogen in the atmosphere. This gas-shielding envelope must efficiently shield the area in order to obtain high-quality welds. The shielding gases normally used for gas metal-arc welding are argon, helium, or mixtures for nonferrous metals; CO_2 for steels; CO_2 with argon and sometimes helium for steel and stainless steel.

The electrode wire composition for gas metal-arc welding must be selected to match the metal being welded. The electrode wire size depends on the variation of the process and the welding position. All electrode wires are solid and bare except in the case of carbon steel wire, when a very thin protective coating (usually copper) is employed.

Gas Tungsten-Arc Welding (TIG)

Gas tungsten-arc (TIG) welding is an arc-welding process wherein coalescence is produced by heating with an arc between a single tungsten (non-consumable) electrode and the work. It was invented by the aircraft industry and used extensively to weld hard-to-weld metals, primarily magnesium and aluminum, and also stainless steels. Shielding is obtained from an inert gas mixture. Filler metal may or may not be used. Fig. 40-11 shows the arc, the tungsten electrode, and the gas-shield envelope all properly positioned above the work piece. The filler-metal rod is fed manually into the arc and weld pool.

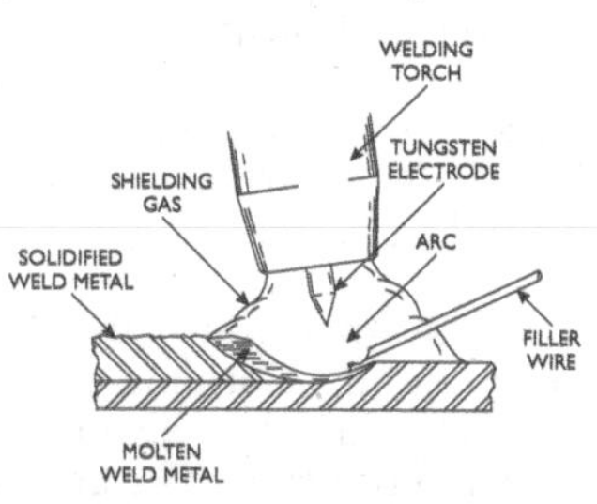

Fig. 40-11 Gas tungsten-arc welding.

Outstanding features of gas tungsten-arc welding include the top quality in hard-to-weld materials and alloys, practically no after-weld cleanup, no weld spatter, and no slag production. The process can be used for welding aluminum, magnesium, stainless steel, cast iron, and mild steels. It will weld a wide range of metal thicknesses. The major components required for gas tungsten-arc welding are shown in Fig. 40-12.

A specially designed welding machine (power source) is used for tungsten-arc welding. Both AC and DC machines are built for the welding of specific materials: AC is usually used for welding aluminum and magnesium; DC is used for stainless steel, cast iron, mild steel, and several alloys. High-frequency current is used in starting the welding arc when using DC current and continuously with AC current. A typical gas tungsten-arc welding machine operates with a range of 3 to 350 amps, with 10 to 35 V at a 60 percent duty cycle.

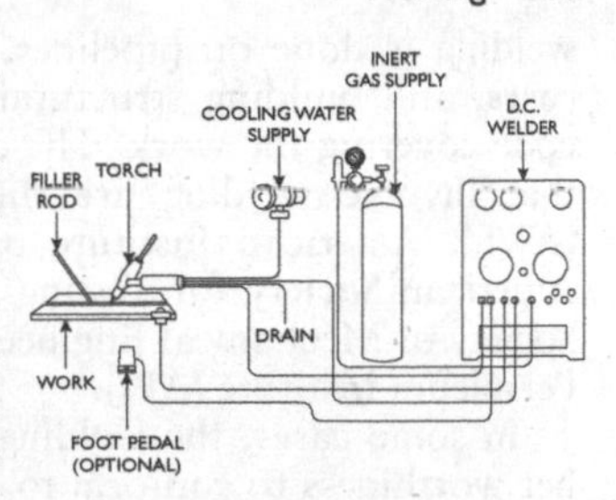

Fig. 40-12 Gas tungsten-arc welding equipment.

The torch holds the tungsten electrode and directs shielding gas and welding power to the arc. Most torches are water cooled; some air-cooled torches are in use. The electrodes are made of tungsten and tungsten alloys. They have a very high melting point (6170°F) and are practically nonconsumable. The electrode does not touch the molten weld puddle; properly positioned, it hangs over the work and the arc keeps the puddle liquid. Electrode tips contaminated by contact with the weld puddle must be cleaned, or they will cause a sputtering arc.

Filler metals are normally used except when very thin metal is welded. The composition of the filler metal should be matched to that of the base metal. The size of the filler-metal rod depends on the thickness of the base metal and the welding current. Filler metal is usually added to the puddle manually, but automatic feed may on occasion be used. An inert gas, either argon, helium, or a mixture of both, shields the arc from the atmosphere. Argon is more commonly used because it is easily obtainable and, being heavier than helium, provides better shielding at lower flow rates.

Code Welding

A code is a set of rules covering procedures and standards to secure uniformity and to protect the public interest. When welding is done on pipelines, pressure vessels, metal staircases, and building structural members, there is usually a code covering the work. The organizations that write codes that involve welding are the American Welding Society (AWS), American Institute of Steel Construction (AISC), American Society for Testing Materials (ASTM), American Society of Mechanical Engineers (ASME), and the American Petroleum Institute (API).

In some cases, the welding mechanic must prove his or her worthiness to conform to a code by welding a test piece checked using X-ray equipment to make sure that the weld is uniform and demonstrates good workmanship. For instance, certain welds and high-pressure gas lines might require complete X-ray checking after each and every weld is accomplished to make certain that no cracks, slag, or porosity might be present.

Plasma Cutting and Welding

Plasma cutting is faster than oxyacetylene cutting but does not produce a cut of similar quality and economy. A plasma-forming gas, such as hydrogen or nitrogen, is heated electrically to such a high temperature that its molecules are transformed into ionized atoms. These atoms possess extremely high energy in their plasma state. When they are changed back into a gas, the energy is released in the form of heat. The heat is of such intensity that it can cut any known metal.

Plasma arc welding is a process in which a column of plasma-producing gas (argon, nitrogen, or hydrogen) is ionized by the heat of an electric arc and passed through a small welding torch orifice. The result is a plasma arc capable of delivering a high concentrated heat to the area being welded. The plasma arc causes deep penetration and produces welds with narrow beads and sharply limited heat-affected zones.

Plasma welding, as shown in Fig. 40-13, uses an appropriate shielding gas and also requires water as a cooling agent. The high temperatures produced in plasma arc welding require a cooling system to prevent damage to the welding torch and other components. Usually a nonconsumable tungsten electrode is used in plasma arc welding and additional metal is added to the weld using a filler rod.

Plasma welding allows higher welding speeds, lowers current requirements, and gives cleaner welds.

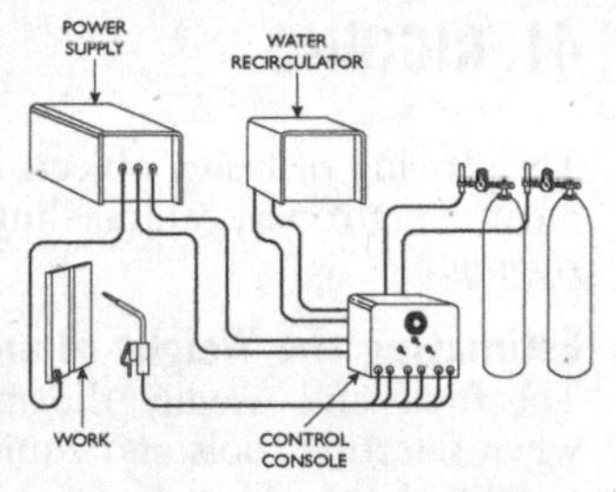

Fig. 40-13 Plasma welding.

41. RIGGING

The moving of heavy objects such as machinery and equipment using ropes, cables, slings, rollers, and hoists is called *rigging*.

Estimating the Weight of an Object

The first, and usually the most important, consideration when selecting tools and equipment for rigging work is the weight of the object to be moved. Reasonable accuracy in the determination of an object's weight is a requirement for safe rigging. When it is not known and dependable information is not available, the weight must be estimated. This should be an approximate calculation, not a guess.

In most cases, a rigging weight estimate is made by roughly calculating the object's volume and multiplying this by the unit weight of the material of which it is made. As only an approximate figure is required, simplify the calculation by using approximate values, which allow many of the calculations to be made mentally.

For example, most heavy objects are made of iron or steel, which ranges in weight from 475 to 490 lb per cu. ft. This value can be rounded off to 500. For cylindrical calculations, the value of Pi ( ) can be rounded off from 3.1416 to 3. The object's dimensions can be rounded off to the closest even numbers, preferable multiples of 10. When rounding off decimals, alternately increasing and decreasing to get even numbers will help to cancel out errors.

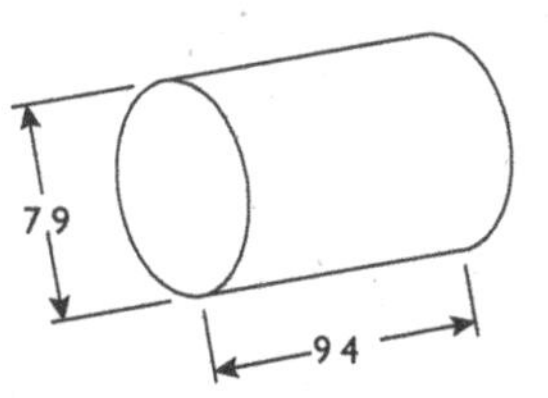

Fig. 41-1 Cylindrical tank.

In Fig. 41-1, values and dimensions are rounded off to allow a quick and accurate weight estimate of the tank. Imagine a cylindrical tank with a circular top and bottom that are ¾-in. thick steel and a ¾-in. thick shell. Of course, the tank is empty and must be rigged from the truck that delivered it to the concrete pad that will hold it. We want to determine a good estimate for the weight.

Known Values

Top and bottom circles are ¾-in. thick steel plate

Shell is ¾-in. thick steel plate

Rounding Off

Round off 7 ft 9 in. to 8-ft diameter or 4-ft radius

Round off 9-ft 4-in. length to 9-ft length

Determine the Area of the Top (and Bottom) of the Tank

$A = \pi r^2$ or A = 3 (approximate value for Pi) × 4 ft × 4ft

A = 48 square feet for the top (and bottom)

Total area of top and bottom = 2 × calculated area = 2 × 48 = 96 square feet

Determine the Area of the Shell

Shell area = Circumference × Length

Circumference = πd

Area = $\pi d \times l$ or 3 × 8 × 9 = 216 square feet for the shell area

Find the total area for all tank components.

A (total) = A (tank top and bottom) + A (shell)

A (total) = 96 square feet + 216 square feet = 312 square feet

Converting Tank Thickness from Inches to Feet

The calculation for volume must be in cubic feet, so...

$3/4$ in. must be converted from inches to feet

$3/4$ in. divided by 12 = $3/48$ or $1/16$ ft

$3/4$ in. = $1/16$ ft

Obtaining the Volume of the Tank Components

Volume = Total square feet of tank components × thickness of tank walls

Volume = 312 square feet × $1/16$ ft = 19.5 rounded off to 20 cu. ft.

Calculate Weight

Weight = volume × pounds per cubic foot (steel)

Weight = 20 cu. ft. × 500 lb per cubic foot = 10,000 lb of total weight

An alternative method is to obtain the weight per square foot of steel plate from a table of weights and multiply this by the total area. The weight per square foot of a $3/4$-in steel plate is 30.6 lb, which can be rounded off to 31 lb. The weight of the tank is as follows:

312 square feet of surface area of components × 31 lb per square foot = 9,672 lb

Rounding up the estimate of 9,672 lb would give 10,000 lb of total weight

Estimating the weight of any regular-shaped object may be done in the same manner as the preceding example. In some cases several calculations may be required, as with the chambered roll in Fig. 41-2.

For calculating purposes, the roll is considered to be made up of three parts, two shafts and a body. The total solid volume is the sum of the solid volumes of the three parts. To determine actual volume, the total chamber volume, which is the sum of the three chamber volumes, is subtracted from the total solid volume. Weight is then determined by multiplying the actual volume by the unit weight of the roll material.

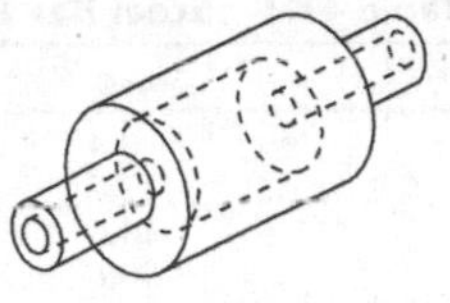

Fig. 41-2 Chambered roll.

The weight of irregularly shaped objects may be estimated with a high degree of accuracy by visualizing the object as a regular shape, or as made up of a group of regular shapes. For example, the irregularly shaped object in Fig. 41-3a may be visualized as a regularly shaped object of lesser dimensions, as shown in Fig. 41-3b.

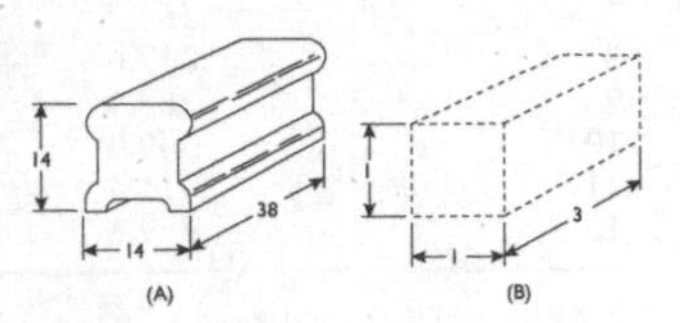

Fig. 41-3 Use regular objects to represent irregular ones.

Machines are usually an assembly of components of varying shapes, sizes, and construction. They may be visualized as a group of regularly shaped solid units when weight estimating. Each unit must be reduced in size to approximate the actual volume of material it contains. The machine shown in Fig. 41-4a could be visualized as shown in Fig. 41-4b.

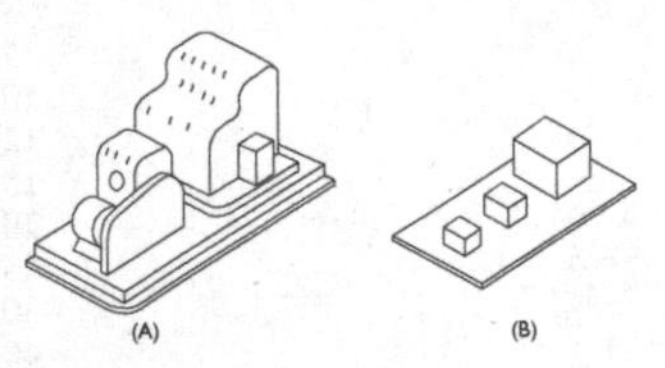

Fig. 41-4 Visualize a machine as a group of regular shaped objects.

Table 41-1 Steel Bar Pounds per Lineal Foot

Size	Square	Round
1	3.4	2.7
1½	7.7	6.0
2	13.6	13.6
3	30.6	24.0
4	54.4	42.7
5	85.0	66.8
6	122.4	96.1
7	166.7	130.8
8	217.6	171.0
9	283.1	222.3
10	340.0	267.0
11	411.4	323.1
12	489.6	384.5

Table 41-2 Steel Plate Pounds per Square Foot

Thickness	Weight
1/16	2.55
1/8	5.1
3/16	7.65
1/4	10.2
5/16	12.75
3/8	15.3
1/2	20.4
5/8	25.5
3/4	30.6
1	40.8
1¼	51.0
1½	61.2
2	81.6

Table 41-3 Weights of Materials

Material	Weight/in.³	Weight/ft³
Aluminum	0.093	160
Brass	0.303	524
Cast Iron	0.260	450
Concrete	0.083	144

Material	*Weight/in.³*	*Weight/ft³*
Sand	0.070	120
Steel	0.281	490
Water	0.036	62½
Wood	0.020	36

Wire Rope and Wire Rope Inspection

The basic element in the construction of wire rope is a single metallic *wire*. Several of these wires are laid helically around a center to form a *strand*. Finally, a number of strands are laid helically around a *core* to form the *wire rope*.

The primary purpose of the core is to serve as a foundation for the rope, to keep it round and to keep the strands correctly spaced and supported.

During construction, the wires that make up the strand may be laid around the center in a clockwise or counterclockwise direction. The same is true of the strands when they are laid around the core. This direction of rotation is called the *lay* of the rope. In *right* lay the strands rotate around the core in a clockwise direction, as the threads do in a right-hand thread. In *left* lay the stands rotate counterclockwise, as do left-hand threads.

The terms *regular* and *lang* are used to designate direction of the wires around the center. Regular lay means that the wires rotate in a direction opposite to the direction of strands around the core. This results in the wires being roughly parallel to the centerline of the rope. Lang lay means the wires rotate in the same direction as the strands, resulting in the wires being at a diagonal to the rope centerline.

A right regular lay rope is shown in Fig. 41-5. The strands rotate clockwise and the wires counterclockwise. This is the most widely used rope lay and is commonly referred to simply as "regular lay."

Wire rope is classified by the number of strands and the approximate number of wires in each strand (Fig. 41-6). For example, the 6 × 7 classification indicates the rope has

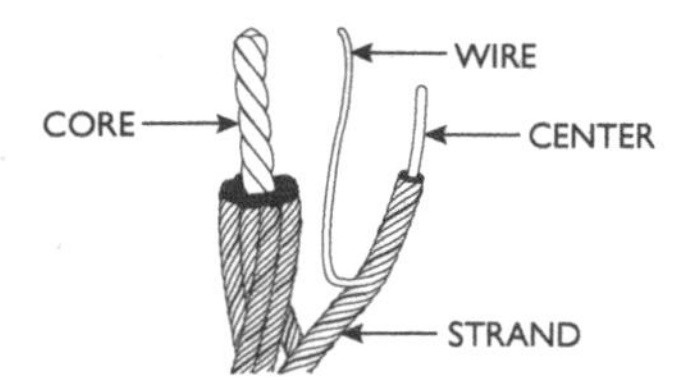

Fig. 41-5 Regular lay wire rope.

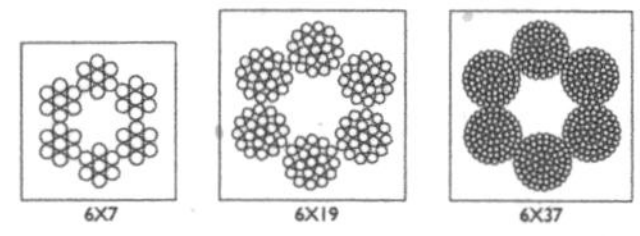

Fig. 41-6 Classifying wire rope.

6 strands and that each strand contains 7 wires. The wires in a strand are placed in layers around a center wire, each layer containing 6 more wires than the preceding one. These arrangements are referred to by the number of wires in each layer. The 7-wire strand is 6-1, the 19-wire strand is 12-6-1, and the 37-wire strand is 18-12-6-1.

The designations for wire rope classifications are only nominal as the actual number of wires in a strand varies with the style of construction. For example, the 6 × 19 classification is made up of wire rope having anywhere from 15 to 26 wires per strand.

Factors of Safety

The safe use of wire rope requires that loads be limited to a portion of the rope's ultimate or breaking strength. The safe load for a wire rope is determined by dividing its breaking strength by a *factor of safety*. Factors of safety for wire rope range from 5 for steady loads to 8 or more for uneven and shock loads.

For example, the breaking strength for a ½-in. diameter improved plow-steel rope is listed in Table 41-4 at 10.5 tons. If this rope were to be used with hoisting tackle at a factor of safety of 5, its maximum safe load would be one-fifth the breaking strength, or 2.1 tons. If, however, it were to be used in a sling at a factor of safety of 8, its maximum safe load would be one-eighth of the breaking strength, or 1.4 tons.

Table 41-4 Breaking Strength

	Breaking Strength, in tons, of 2000 lb		
	Improved Plow Steel		*Extra Improved Plow Steel*
Diameter, in.	*Fiber Core*	*IWRC*	*IWRC*
3/16	1.46	—	—
1/4	2.59	2.78	3.20
5/16	4.03	4.33	4.98
3/8	5.77	6.20	7.14
7/16	7.82	8.41	9.67
1/2	10.5	11.0	12.6
9/16	12.9	13.9	15.9
5/8	15.8	17.0	19.6
3/4	22.6	24.3	27.9
7/8	30.6	32.9	37.8
1	39.8	42.8	49.1

Table 41-5 End Attachment Efficiencies

Fitting	*Nominal Efficiency, percent of catalog rated rope strength*
Wire Rope Sockets	100
Spelter (Zinced) Attachments	100
Fittings (Swaged or Pressed)	100
Torpedo Collar (with or without thimble)	100
Open Wedge Sockets	80–90
Clips (U-bolt Type)	80
Clips (Twin-base Type)	80
Spliced-in Thimbles:	
1/4 and smaller	90
5/16	89
3/8	88
1/2	86
5/8	84
3/4	82
7/8 to 2 1/2, incl.	80

Table 41-6 Safe Load in Tons

Nominal Size, in.	*Single*	*Choker*	*U-sling*	*Basket*	*60°*	*45°*	*30°*
1/4	.5	.3	.7	.6	.57	.5	.3
5/16	.8	.6	1.1	1.0	.9	.7	.6
3/8	1.1	.8	1.5	1.4	1.3	1.1	.8
1/2	2.0	1.4	2.7	2.4	2.3	1.9	1.3
5/8	2.9	2.1	4.2	3.8	3.7	3.0	2.1
3/4	4.1	3.0	6.0	5.4	5.2	4.2	3.0
7/8	5.6	3.8	7.7	6.8	6.7	5.4	3.8
1	7.2	5.0	10.0	9.3	8.7	7.1	5.0
1 1/8	9.0	5.6	11.2	10.5	9.7	7.9	5.6

Wire Rope Attachment

The U-bolt or Crosby (Fig. 41-7a) wire rope clip is probably the most common method of attaching a wire rope to equipment. All U-bolt clips must be placed on the rope with U bolts bearing on the short or "dead" end of the rope. Illustrations of the correct and incorrect application of U-bolt clips are shown in Fig. 41-8.

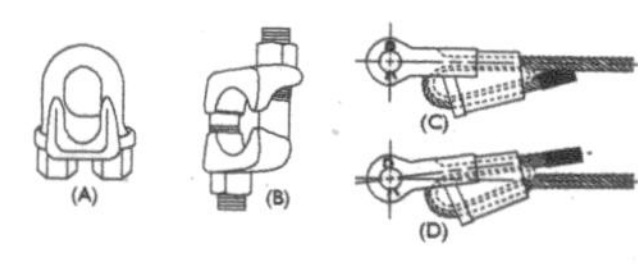

Fig. 41-7

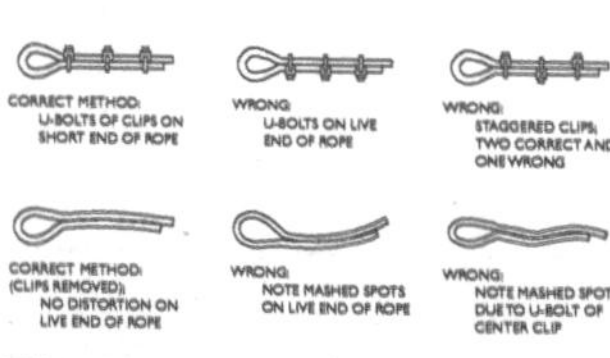

Fig. 41-8

An improved type wire rope clip is called the double-base safety or fist grip, shown in Fig. 41-7b. It has corrugated jaws to fit both parts of the rope, allowing it to be installed without regard to the live or dead part.

When making an eye attachment with clips, a thimble should be used, plus the correct number of clips (Table 41-7). All clips should be spaced not less than six rope diameters apart. Apply the clip farthest from the thimble first, at about four inches from the end of the rope, and screw up tightly. Next, put on the clip nearest the thimble and apply the nuts handtight. Then put on one or more intermediate clips handtight. Take a strain on the rope, and while the rope is under this strain, tighten all the clips previously left loose. Tighten alternately on the two nuts so as to keep the clip square. After the rope has been in use for a short time, retighten all clips.

Table 41-7 Clips

Rope Size	*U-Bolt*	*Safety*
1/4 to 3/8	2 to 4	2
7/16 to 5/8	3 or 4	2
3/4	4 or 5	3
7/8 to 1 1/8	4 or 5	4
1 1/4 to 1 1/2	5 to 8	5

Wedge socket attachments are used on equipment where frequent changes are required. Care must be exercised to install the rope so that the pulling part is directly in line with the clevis pin, as shown in Fig. 41-7c. If incorrectly installed, as shown in Fig. 41-7d, a sharp bend will be produced in the rope as it enters the socket.

Multiple Reeving

A single rope supporting a load is referred to as single part line, and the tension in the rope is equal to the suspended weight. When a load is supported by a multipart wire rope tackle, as shown in Fig. 41-9, and the rope is not moving, the load on each line, including the lead line, is equal to the weight of the load divided by the number of parts of rope supporting the load. When this load is

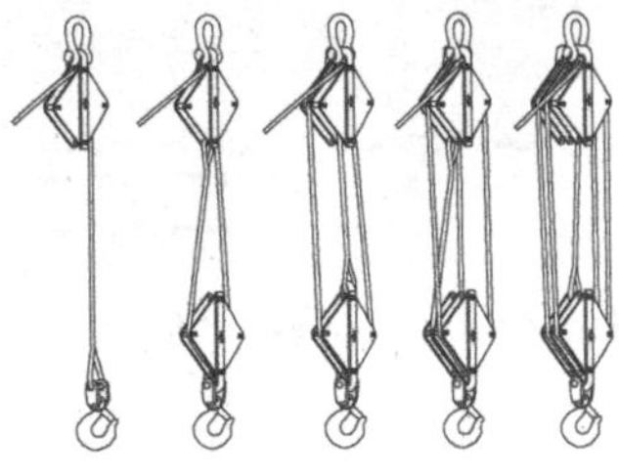

Fig. 41-9 Multiple reeving.

raised, however, the loads on the individual supporting ropes change, increasing from the dead end to the lead line.

How to Measure Wire Rope

Measurements should be made carefully with calipers. Fig. 41-10 shows the correct and incorrect method of measuring the diameter of wire rope.

Wire Rope Slings and Chokers

The single sling with loop ends, the most widely used of all sling types, lends itself readily for use in a basket hitch, in a choker hitch, or as a straight rope. Blocking should always be used to protect the sling from sharp corners.

Sling-end fittings of the more popular styles are shown in Fig. 41-11. Any combination of these, to suit the job requirements, is available.

The lengths of slings having loop- or ring-style end fittings are measured from the weight-bearing surface. Those with pin-style end fittings are measured from the center of the pin.

Eyebolts and Shackles

The strength of an eyebolt is influenced greatly by the direction of pull to which it is subjected. For loads involving angular forces, the shoulder type has several times the strength of the conventional eyebolt. Tables 41-8 and 41-9 list the safe loads by sizes and direction of load.

The shackle is the recommended fastener for attaching slings to eyebolts, chain, or wire rope. Table 41-10 lists the safe loads by nominal size and gives the dimensions of standard shackles.

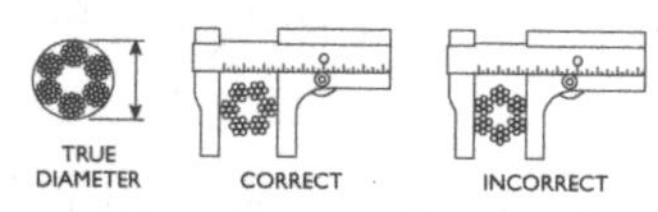

Fig. 41-10 Measuring wire rope.

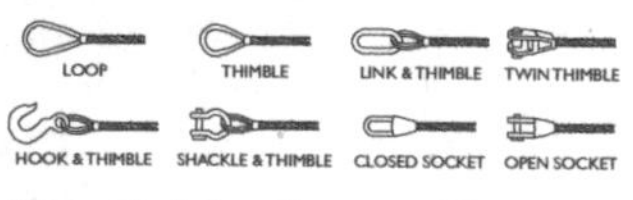

Fig. 41-11 Sling-end fittings.

Table 41-8 Safe Loads for Ordinary Drop-Forged Steel Eyebolts, lb

Size, in.	PULL	PULL	PULL
½	1,100	50	40
⅝	1,800	90	65
¾	2,800	135	100
⅞	3,900	210	150
1	5,100	280	210
1¼	8,400	500	370
1½	12,200	770	575
1¾	16,500	1,080	800
2	21,800	1,440	1,140

Table 41-9 Safe Loads for Shoulder-Type Drop-Forged Steel Eyebolts, lb

Size, in.	PULL	PULL	PULL
¼	300	30	40
½	1,300	140	150
¾	3,000	250	300
1	6,000	500	600
1¼	9,000	800	900
1½	13,000	1,200	1,300
2	23,000	2,100	2,300
2½	37,000	3,800	4,300

Table 41-10 Safe Loads for Anchor Shackles

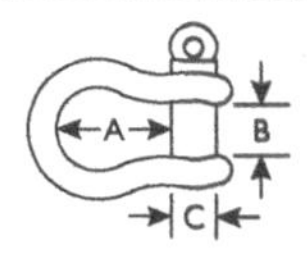

Nominal Size	Tons Safe Load	Dimensions A	Dimensions B	Dimensions C
3/8	.8	1 1/2	1 1/16	7/16
1/2	1.4	2	7/8	5/8
5/8	2.2	2 3/8	1 1/16	3/4
3/4	3.2	2 7/8	1 1/4	7/8
7/8	4.3	3 1/4	1 3/8	1
1	5.6	3 5/8	1 11/16	1 1/8
1 1/8	6.7	4 1/4	1 7/8	1 1/4
1 1/4	8.2	4 3/4	2	1 3/8
1 1/2	11.8	5 1/2	2 1/4	1 5/8
2	21.1	7 3/4	3 1/4	2 1/4

Fiber Rope

For many years the principal materials from which fiber rope was manufactured were *manila* (hemp) and *cotton.* Synthetic materials, such as nylon, polypropylene, and polyester, have replaced manila and cotton. While the reasons for this are numerous, probably the most important are increased resistance to deterioration, greater strength, and greater pliability. Most rope is manufactured in three or more strands. Three-strand construction has been the most widely used style for many years, but since the eight-strand plaited construction has been introduced, it is widely used, particularly in synthetic materials. Increasing the number of strands reduces the size of each strand for a given rope size and increases the rope's pliability.

Because of the variety of rope materials and constructions now in use, as well as the wide range of factors affecting rope behavior, it is impossible to cover the multitude of possible rope applications. Perhaps the single most important consideration in rope use is that of safety. While all aspects cannot be specifically detailed, several general safety considerations are important if the mechanic is to use rope properly and avoid possible rope failure. Some of the safety considerations are as follows:

1. Practically all rope failure accidents are caused by improper care and use rather than poor engineering or original product defect.
2. Rope must be adequate for the job. Choosing a rope of correct size, material, and strength must not be done haphazardly. Consult the dealer, distributor, or manufacturer for information and assistance if needed.
3. Do not overload rope. Sudden strains or shock loading can cause failure. Working load specifications may not be applicable when rope is subject to significant dynamic loading. Loads must be handled slowly and smoothly to minimize dynamic effects.
4. Avoid using rope that shows signs of aging and wear. If in doubt, destroy the used rope. If the fibers show wear in a given area, the rope should be respliced, downgraded, or replaced.
5. Avoid chemical exposure. Rope can be severely damaged by contact with some chemicals. Special attention must be given for applications where exposure to either fumes or actual contact may occur.
6. Avoid overheating. Heat can seriously affect the strength of rope. The frictional heat from slippage on capstan or winch may cause localized heating, which can melt synthetic fibers or burn natural fibers.
7. Never stand in line with rope under strain. If a rope or attachment fails it can recoil with sufficient force to cause physical injury. The snap-back action can propel fittings and rope with possibly disastrous results.

Rope Characteristics

Manila. Made from fine Abaca (hemp) fiber. Excellent resistance to sunlight, low stretch, and easy to knot.

Nylon. The strongest fiber rope manufactured. High elasticity allows absorption of shock loads. Resistant to rot, oils, gasoline, grease, marine growth, and most chemicals. High abrasion resistance.

Polypropylene. A lightweight fiber with good strength. It floats, is resistant to rot, gasoline, and most chemicals, and is waterproof. Some manufacturers' products contain additives to reduce sunlight resistance.

Polyester. Less strength than nylon, but better resistance to sunlight deterioration. Low stretch and excellent surface abrasion resistance. Other characteristics similar to nylon.

Table 41-11 New Rope Working Load, lb

Diameter, in.	*Safety Factor*	*Manila*	*Nylon*	*Polypropylene*	*Polyester*
1/4	10	54	124	113	120
5/16	10	90	192	171	180
3/8	10	122	278	244	270
1/2	9	264	525	420	520
5/8	9	496	935	700	925
3/4	7	695	1420	1090	1400
1	7	1160	2520	1800	2490
1 1/8	7	1540	3320	2360	3280
1 1/4	7	1740	3760	2700	3700
1 1/2	7	2380	5320	3820	5260

Working loads are for rope in good condition with appropriate splices, in noncritical applications, and under normal service conditions. Loads must be reduced for exceptional service conditions such as shock loads, sustained loads, etc.

Knots

Bight. Formed by simply bending the rope and keeping the sides parallel.

Fig. 41-12

Loop or turn. Formed by crossing the sides of a bight.

Fig. 41-13

Round turn. Further bending of one side of a loop.

Fig. 41-14

Standing part. That part of the rope this is not used in tying a knot, the long part that is not worked upon.

End. As the name implies, the very end of the rope.

Whipping. Common, plain, or ordinary whipping is tied by laying a loop along the rope and then making a series of turns over it. The working end is finally stuck through this loop, and the end is hauled back out of sight. Both ends are then trimmed short. A whipping should be, in width, about equal to the diameter of the rope.

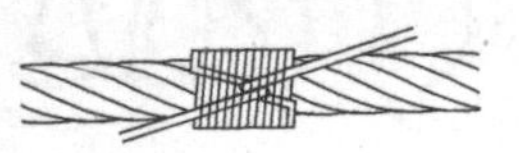

Fig. 41-15

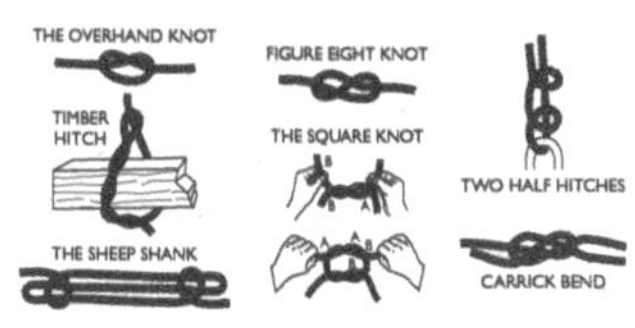

Fig. 41-16

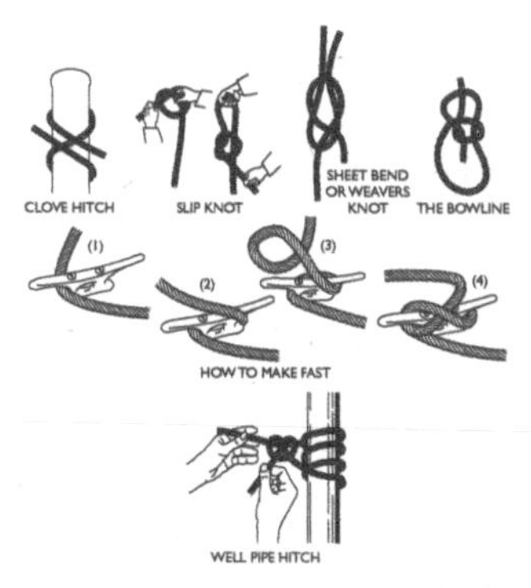

Fig. 41-17

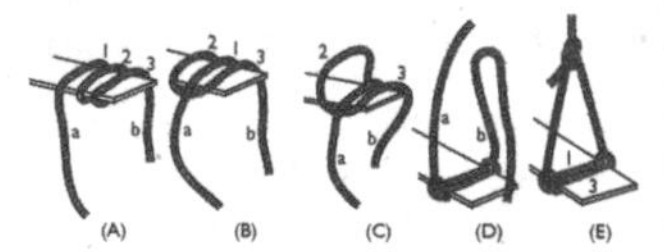

Fig. 41-18 Scaffold hitch.

Scaffold hitch. Lay the short end (a) of the rope over the top of the plank (Fig. 41-18a), leaving enough hanging down to the left to tie to the long rope (Fig. 41-18e). Wrap the long end (b) loosely around the plank twice, letting it hang down to the right (Fig. 41-18a). Now, carry rope 1 over rope 2 and place it next to rope 3 (Fig. 41-18b). Pick up rope 2 (Fig. 41-18c) and carry it over 1 and 3 and over the end of the plank. Take up the slack by pulling rope (a) to the left and rope (b) to the right. Draw ropes (a) and (b) above the plank (Fig. 41-18d), and join the short end (a) to the long rope (b) by an overhand bowline (Fig. 41-18e). Pull the bowline tight, at the same time adjusting the lengths of the two ropes so that they hold the plank level. Attach a second rope to the other end of the plank in the same way, and the scaffold is now ready for use.

Using a Cable Winch (Come-a-Long)

A come-a-long is a portable cable winch that uses wire rope and a ratcheting handle to allow one-man operation. They are available in ½- to 3-ton capacity and usually weigh from 7 to 35 lb. Their rated factor of safety is from 1.5 to 3. They should be inspected before each and every use. Never extend the handle with a cheater bar or pipe; the rating of the hoist will be compromised. These devices are typically 2 to 3 ft long and cannot be used in tight spaces. Make sure to take care in rewinding the cable — it is easy to foul.

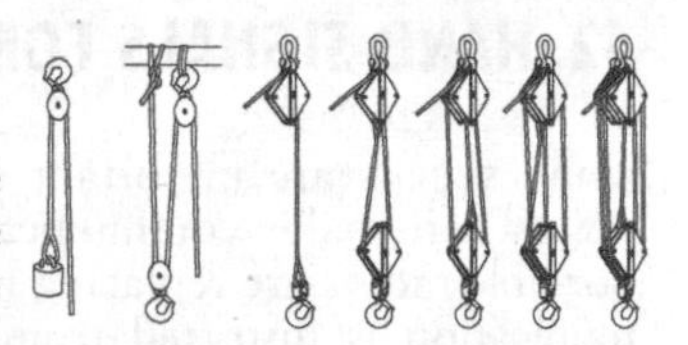

Fig. 41-19

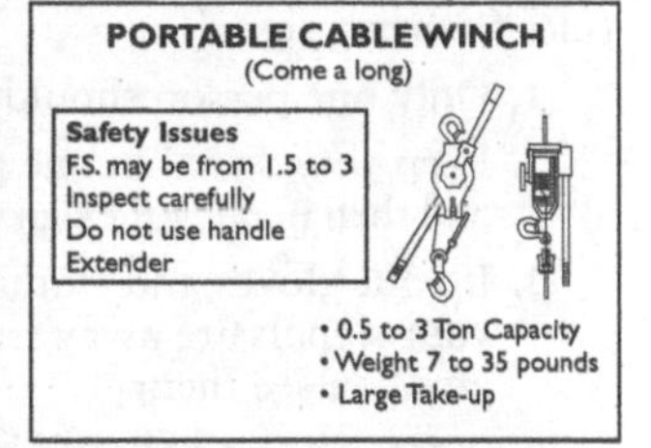

Fig. 41-20 Come-a-long.

Using a Chain Hoist

Chain hoist can handle up to 6 tons with a 100-lb pull. A chain hoist is not meant to be pulled by more than one person. A chain hoist has a large take-up — usually 10 ft — and requires only a small amount of clearance for operation.

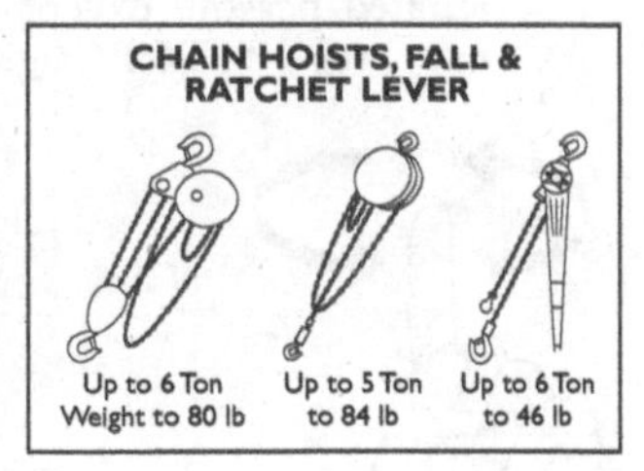

Fig. 41-21 Chain hoist.

42. HAND SIGNALS FOR CRANES AND HOISTS

Hand signals are important to learn because background noises can make communication difficult. Two-way FM portable radios are replacing hand signals, but if radio communication is distorted from static or other interference, hand signals are a reliable fallback. If the designated signal person is not in line-of-sight with the operator, then a third person needs to relay the signals. Following are some simple rules for using signals:

1. Only one person should be designated to give signals.
2. Keep your signal in one place. Try to make eye contact, and then make your signals in front of your face or body.
3. If your gloves and clothing are of similar colors, make your signals are away from your body where the operator can see them.
4. A set of signals should be *agreed upon* and adopted at each operation where hoisting equipment is used. Only the agreed-upon signals should be used by a designated person, except that in an emergency, *anyone* may give a "STOP" signal.

Common Hand Signals

With forearm vertical and index finger pointing up, make a small horizontal circular motion with hand and forearm.

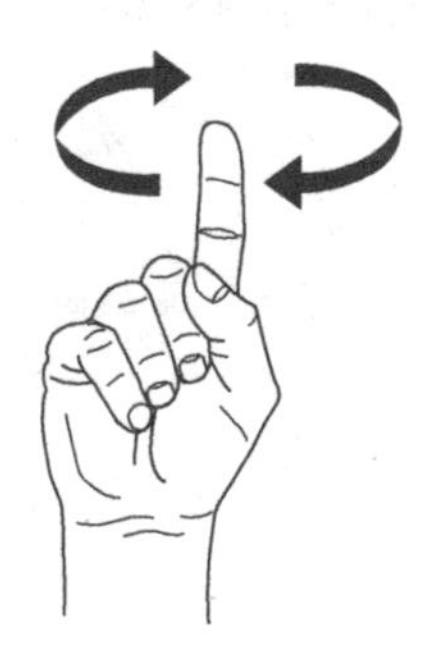

Fig. 42-1 Raise load.

With whole arm extended down and palm downward, point index finger down and move hand and arm in small horizontal circles.

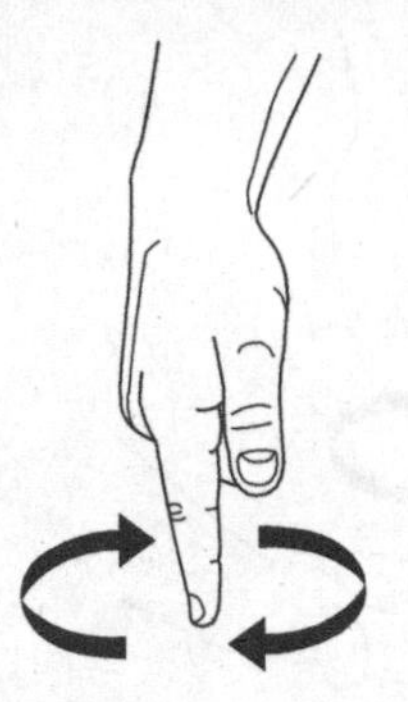

Fig. 42-2 Lower load.

Using two arms, one hand with index finger pointing up into downturned palm of other hand, make circular motion with index finger.

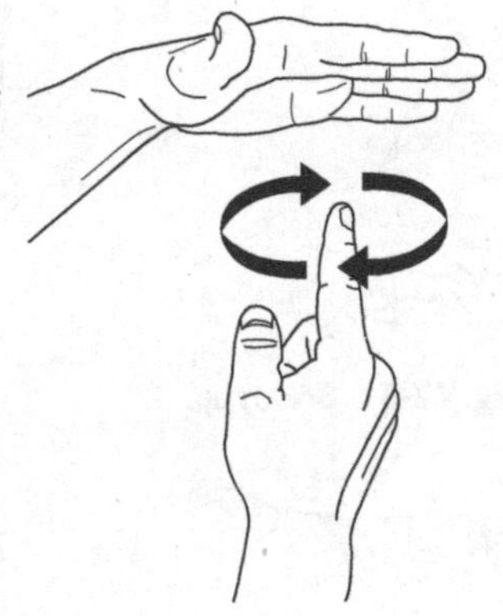

Fig. 42-3 Raise load slowly.

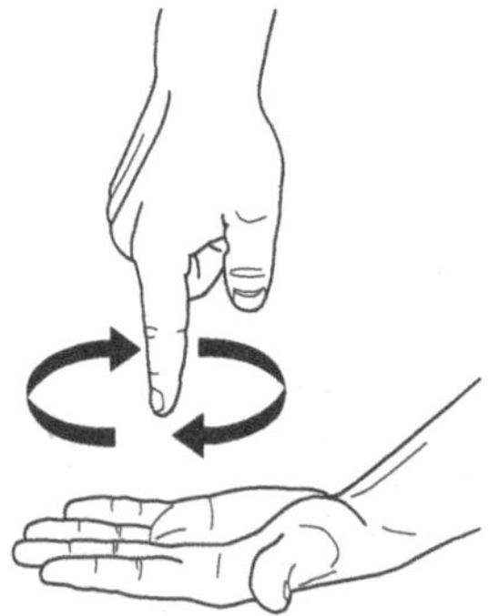

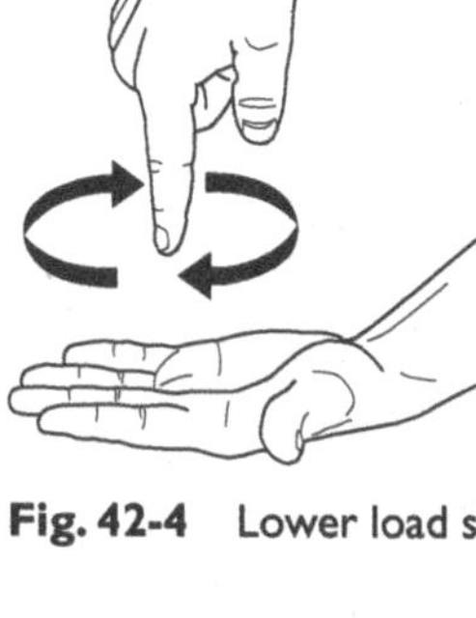

Fig. 42-4 Lower load slowly.

Using two arms, one hand with index finger pointing down into upturned palm of other hand, make circular motion with index finger.

Fig. 42-5 Boom up.

With arm extended and finger clenched, point thumb up.

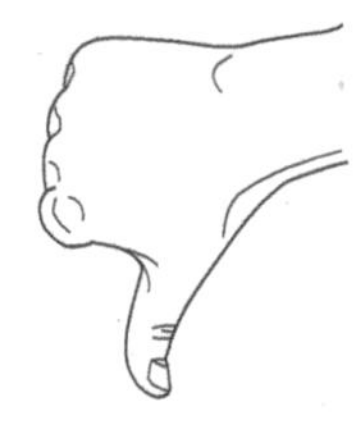

Fig. 42.6 Boom down.

With arm extended and finger clenched, point thumb down.

With one arm partially extended and fingers clenched, point thumb up into palm of other hand.

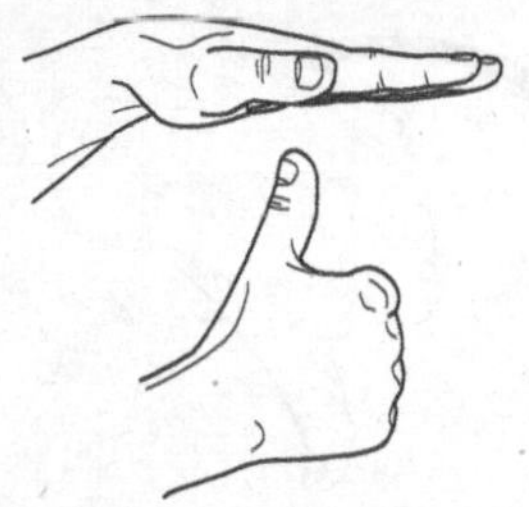

Fig. 42-7 Raise boom slowly.

With one arm partially extended and fingers clenched, point thumb down into palm of other hand.

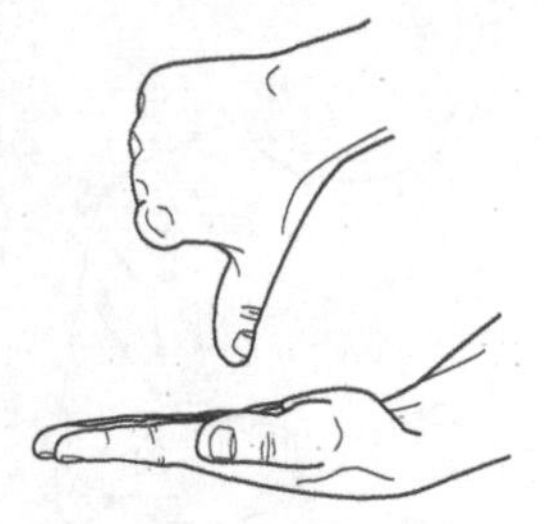

Fig. 42-8 Lower boom slowly.

With arm extended and thumb pointing upward, flex the fingers in and out.

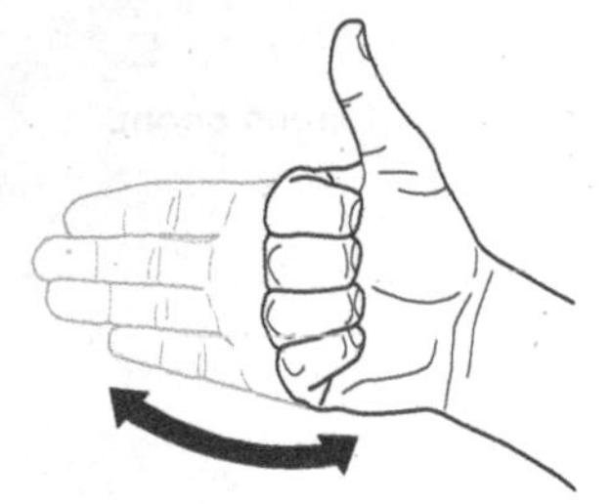

Fig. 42-9 Raise boom and lower load.

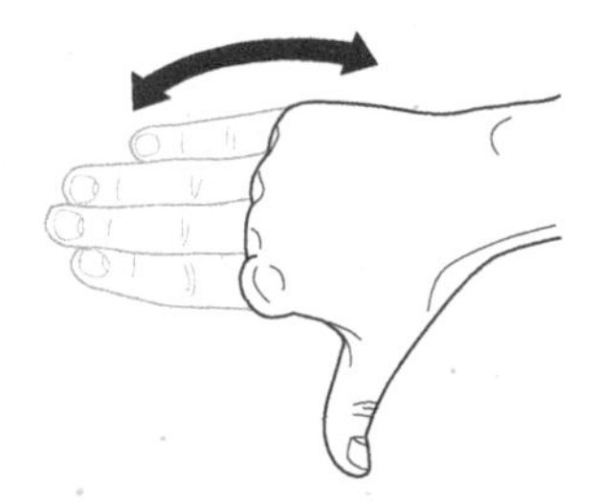

Fig. 42-10 Lower boom and raise load.

With arm extended and thumb pointing downward, flex the fingers in and out.

Fig. 42-11 Extend boom.

With both fists clenched in front of body, point the thumbs outward away from each other.

With both fists clenched in front of body, point the thumbs inward toward each other.

Fig. 42-12 Retract boom.

With one arm raised, forearm vertical, and palm pointing toward crane operator, use the index finger of the other hand to trace a small circle in the palm.

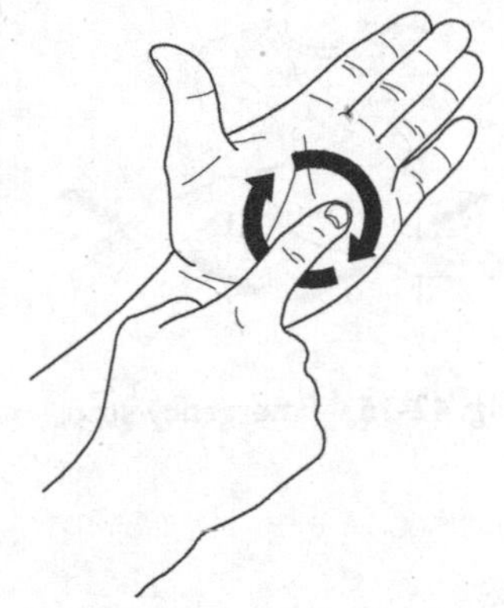

Fig. 42-13 Go-Slow all movements.

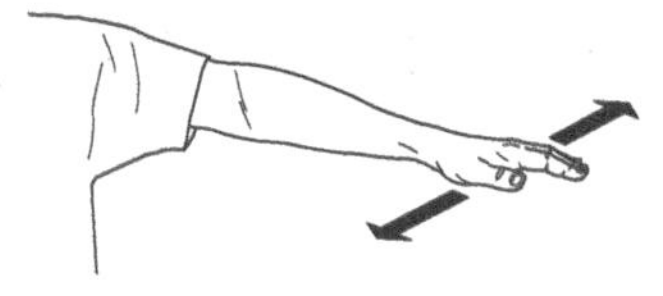

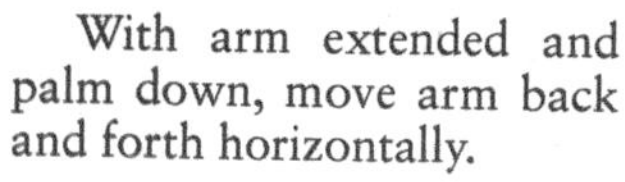

With arm extended and palm down, move arm back and forth horizontally.

Fig. 42-14 Stop.

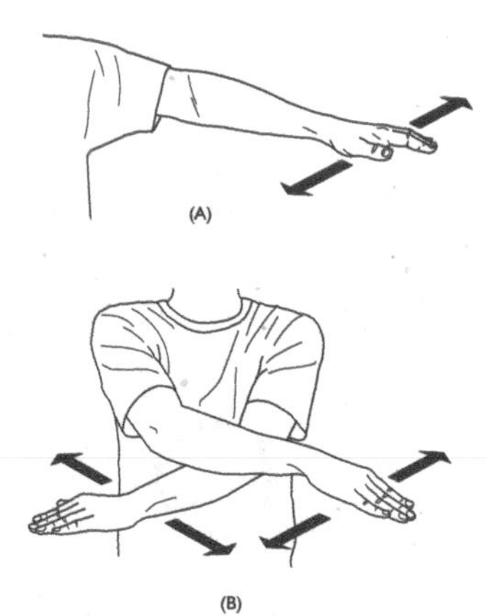

With *both* arms extended and palms down, move both arms back and forth horizontally. Also one arm can be used with arm extended and palm down, moving hard back and forth *very* rapidly.

Fig. 42-15 Emergency stop.

43. PIPING AND PIPEFITTING

Thread Designations

American Standard

American Standard Pipe Threads are designated by specifying in sequence the nominal size, number of threads per inch, and the thread series symbols:

Nominal Size	*No. of Threads*	*Symbols*
$3/8$	18	NPT

Where:

N—American (Nat.) Standard
P—Pipe
T—Taper
C—Coupling
S—Straight
L—Locknut
R—Railing Fittings
M—Mechanical

Examples

$3/8$—18 NPT, American Standard Taper Pipe Thread

$3/8$—18 NPSC, American Standard Straight Coupling Pipe Thread

$1/8$—27 NPTR, American Standard Taper Railing Pipe Thread

$1/2$—14 NPSM, American Standard Straight Mechanical Pipe Thread

1 — 11½ NPSL, American Standard Straight Locknut Pipe Thread

Left-hand threads are designated by adding LH.

American Standard Taper Pipe Threads (NPT)

Basic Dimensions

Taper pipe threads are engaged or made up in two phases, *hand engagement* and *wrench makeup* (Fig. 43-2). Table 43-1 lists the basic hand engagement and wrench makeup for American Standard Taper Pipe Threads. Dimensions are rounded off to the closest 1/32 in.

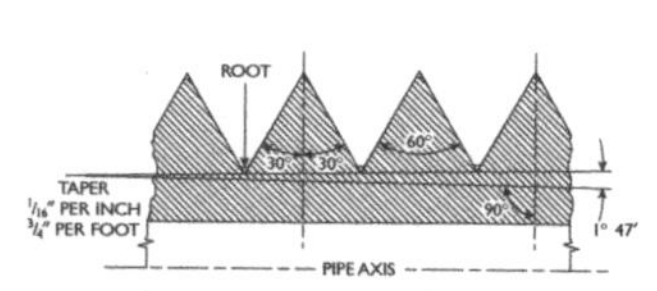

Fig. 43-1 American Standard taper pipe thread form.

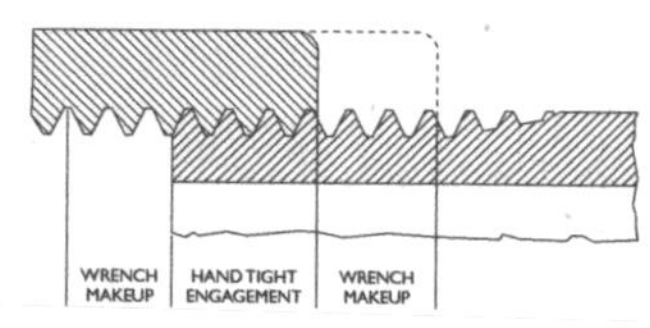

Fig. 43-2 Pipe thread engagement.

Pipe Measurement and Identification

Dimensions on pipe drawings specify the location of centerlines and/or points on centerlines; they do not specify pipe lengths. This system of distance dimensioning and measurement is also followed in the fabrication and installation of pipe assemblies (Fig. 43-3).

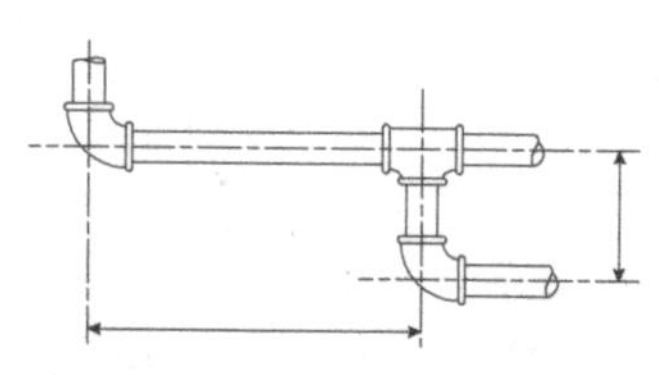

Fig. 43-3

Takeout Allowances for Screwed Pipe

To determine actual pipe lengths, allowances must be made for the length of the fittings and the distance threaded pipe is made up into the fittings. To do this, subtract an amount

Table 43-1 Taper Pipe Makeup Dimensions

		Hand Tight		Wrench Makeup		Dimension	Turns
Pipe Size	Threads per Inch	Dimension	Turns	Dimension	Turns	Total Makeup	
1/8	27	3/16	4 1/2	3/32	2 1/2	9/32	7
1/4	18	7/32	4	3/16	3	13/16	7
3/8	18	1/4	4 1/2	3/16	3	7/16	7 1/2
1/2	14	5/16	4 1/2	7/32	3	17/32	7 1/2
3/4	14	5/16	4 1/2	7/32	3	17/32	7 1/2
1	11 1/2	3/8	4 1/2	1/4	3 1/4	11/16	8
1 1/4	11 1/2	13/32	4 1/2	9/32	3 1/4	11/16	8
1 1/2	11 1/2	7/16	5	1/4	3	11/16	8
2	11 1/2	7/16	5	1/4	3	11/16	8 1/2
2 1/2	8	11/16	5 1/2	3/8	3	1 1/8	9
3	8	3/4	6	3/8	3	1 1/8	9

This table includes basic makeup dimensions. Commercial products may vary as much as one turn larger or smaller and still be within standard tolerance. In practice, pipe threads are usually cut to give a connection that makes up less than the basic standard. Common practice is about 3 turns by hand and 3 to 4 turns by wrench.

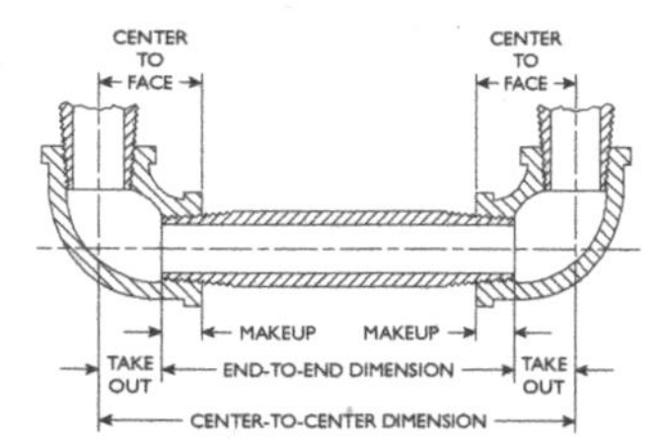

Fig. 43-4

called *takeout* from the *center-to-center* dimension. The relationships of takeout to other threaded pipe connection distances, termed *makeup, center-to-center,* and *end-to-end,* are illustrated in Fig. 43-4.

To determine end-to-end pipe length, the takeout is subtracted from the center-to-center dimension. Standard tables may be used for this purpose (Table 43-2). These tables should be used with judgment, however, because commercial product tolerance is one turn plus or minus. On critical connections, materials should be checked and compensation made for variances.

Table 43-2 Takeout Allowances

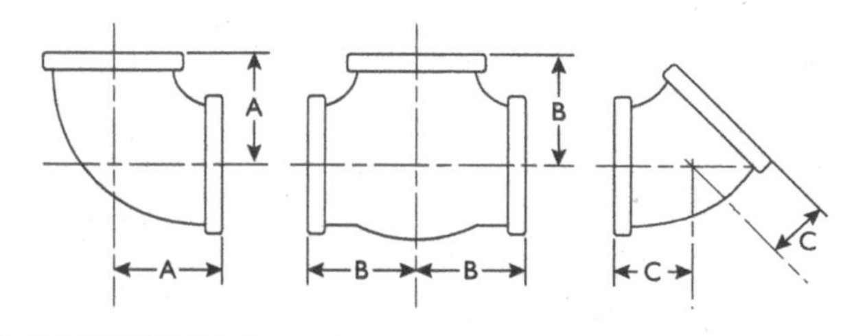

	90° Elbow		*Tee*		*45° Elbow*	
Pipe Size	***A***	***Takeout***	***B***	***Takeout***	***C***	***Takeout***
1/8	11/16	7/16	11/16	7/16	9/16	1/4
1/4	13/16	7/16	13/16	7/16	3/4	3/8
3/8	15/16	9/16	15/16	9/16	13/16	7/16
1/2	1 1/8	5/8	1 1/8	5/8	7/8	3/8
3/4	1 5/16	3/4	1 15/16	3/4	1	7/16
1	1 1/2	7/8	1 1/2	7/8	1 1/3	9/16

	90° Elbow		Tee		45° Elbow	
Pipe Size	A	Takeout	B	Takeout	C	Takeout
1¼	1¾	1⅛	1¾	1⅛	1 15/16	11/16
1½	1 15/16	1¼	1 15/16	1¼	1 7/16	¾
2	2¼	1⅝	2¼	1⅝	1 11/16	1

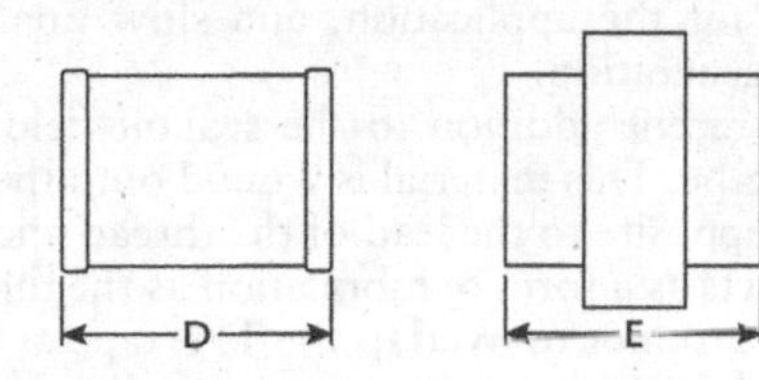

		Coupling		Union	
Pipe Size	Thread Makeup	D	Takeout	E	Takeout
⅛	¼	1	¼	1½	¾
¼	⅜	1⅛	⅜	1⅝	⅞
⅜	⅜	1¼	⅜	1⅝	⅞
½	½	1⅜	⅜	1⅞	1
¾	9/16	1½	⅜	1⅞	1
1	9/16	1¾	½	2⅜	1¼
1¼	⅝	2	¾	2⅝	1⅜
1½	⅝	2½	⅞	3	1½
2	11/16	2½	1¼	3¼	1¾

Threading Pipe

As the threads are made up in the fitting, high forces and pressures are developed by the wedging action of the taper. Also, frictional heat develops as the surfaces are deformed to match the variations in thc thread form. It is important that the threads be clean and well lubricated and that the connection is not screwed up fast enough to generate excessive heat.

Using Pipe Sealant

Using a lubricant called "dope". allows thread surfaces to deform and mate without galling and seizing. The dope also helps to plug openings resulting from improper threads, and it acts as a cement.

The requirements for tight makeup of threaded pipe connections are good quality threads, clean threads, proper dope for the application, and slow final makeup to avoid heat generation.

A recent addition to the sealant field is the use of TFE pipe tape. This material is wound onto the threads in a direction opposite to the lead of the thread and forms a good seal and acts as a form of lubrication as the threads are pulled up. Take care not to overlap the TFE tape at the end of the pipe threads, which might cause parts of the tape to enter the final fluid stream after the line is commissioned. This can cause problems if the tape becomes lodged in a critical sensing or flow-control device used in the line.

Pipe Schedules

Pipe schedule numbers shown in Table 43-3 indicate pipe strength. The higher the number in a given size, the greater the strength. The schedule number indicates the approximate values of the following expression:

$$\text{Schedule number} = \frac{1000 \times \text{internal pressure}}{\text{allowable stress in pipe}}$$

Copper Water Tube

Seamless copper water tube, popularly called *copper tubing,* is widely used for plumbing, water lines, heater coils, fuel oil lines, gas lines, and so on. Standard copper tubing is commercially available in three types, designated as type "K," "L," and "M." The type "K" has the thickest wall and is generally used for underground installations. The type "L" has a thinner wall and is most widely used for general plumbing, heating, and so on. The type "M" is the thinwall tubing used for low-pressure and drainage applications.

Table 43-3 Commercial Pipe Sizes and Wall Thickness

Nominal Pipe Size	Outside Diameter	Nominal Wall Thickness					
		Sched. 5	Sched. 10	Sched. 40 Std.	Sched. 80 Ex. Std.	Sched. 160	Ex. Ex. Strong
1/8	.405	-	.049	.068	.095	-	-
1/4	.540	-	.065	.088	.119	-	-
3/8	.675	-	.065	.091	.126	-	-
1/2	.840	-	.083	.109	.147	.187	.294
3/4	1.050	.065	.083	.113	.154	.218	.308
1	1.315	.065	.109	.133	.179	.250	.358
1 1/4	1.660	.065	.109	.140	.191	.250	.382
1 1/2	1.900	.065	.109	.145	.200	.281	.400
2	2.375	.065	.109	.154	.218	.343	.436
2 1/2	2.875	.083	.120	.203	.276	.375	.552
3	3.500	.083	.120	.216	.300	.438	.600
3 1/2	4.000	.083	.120	.226	.318	-	-
4	4.500	.083	.120	.237	.337	.531	.674
5	5.563	.109	.134	.258	.375	.625	.750
6	6.625	.109	.134	.280	.432	.718	.864
8	8.625	.109	.148	.322	.500	.906	.875

The outside diameter of copper tubing is uniformly ⅛ in. larger than the nominal size; for example, the outside diameter of 1-in tubing is 1⅛ in. All three types of copper tubing in a given size have the same outside diameter. The inside diameter of each type will vary as the wall thickness varies (Table 43.4).

The three types of tubing are made in either "hard" or "soft" grade. Hard tubing is used for applications where lines must be straight without kinks or pockets. Soft tubing is used for bending around obstructions and inaccessible places.

Table 43-4 Standard Copper Tubing Dimensions

Nominal Size	Outside Diameter	Inside Diameter		
		Type "K"	Type "L"	Type "M"
⅜	½	.402	.430	.450
½	⅝	.528	.545	.569
⅝	¾	.652	.668	.690
¾	⅞	.745	.785	.811
1	1⅛	.995	1.025	1.055
1¼	1⅜	1.245	1.265	1.291
1½	1⅝	1.481	1.505	1.527
2	2⅛	1.959	1.985	2.009
2½	2⅝	2.435	2.465	2.495
3	3⅛	2.907	2.945	2.981
3½	3⅝	3.385	3.425	3.459
4	4⅛	3.857	3.905	3.935

Plastic Pipe

Plastic pipe, available in a great variety of types and sizes, is used in a multiplicity of applications. This is in part because it can be readily cut, fitted, and assembled before joining. Also because the joining is accomplished with suitable cement that is quick setting, it does not require heating, thus resulting in an efficient and relatively low-cost pipe system. Another reason, and in many instances the principal reason, for the use of plastic pipe is that it is the most effective and economical piping system available to handle corrosive materials. While cement jointing of plastic pipe is commonly practiced, the heavier wall thicknesses, schedules 80 and 120, may be threaded and joined in the same manner as regular threaded pipe. The outside diameter of plastic pipe is the same as that of commercial metal pipe, and while the inside diameter is slightly smaller, the difference is negligible. Table 43-5 lists the outside and inside diameters of plastic pipe in the schedule 40, 80, and 120 strength groups. Pipe made from thermoplastic (becoming soft when heated) is limited to applications where exposure temperatures are well below its softening temperature. One of the principal uses for plastic pipe is that of waste-water handling. Pipe for this purpose is available, with suitable cleaners and cement, at pipe and plumbing supply outlets. Because this is a relatively low-temperature, low-hazard application, there is little reason to be concerned about material formulation. Other applications, particularly if hazardous materials are involved, should be referred to suppliers or others qualified to make formulation selections to match application exposures.

Plastic pipe systems employ a variety of fittings, valves, pumps, and more, to complement the pipe. Cement-joined pipe system measurements are made in the same manner as copper water type—that is, the pipe bottoms in the fitting socket and takeout dimensions are calculated by subtracting the socket depth from center-to-face dimensions. Threaded plastic pipe measurements are handled in the same manner as those of regular metal threaded pipe.

Table 43-5 Plastic Pipe Dimensions

Nominal Size	*Outside Diameter*	*Inside Diameter*		
		Schedule 40	*Schedule 80*	*Schedule 120*
⅛	.405	.261	.203	-
¼	.540	.354	.288	-
⅜	.675	.483	.408	-
½	.840	.608	.528	.480
¾	1.050	.810	.725	.690
1	1.315	1.033	.935	.891
1¼	1.660	1.364	1.256	1.204

(continued)

Table 43-5 *(continued)*

Nominal Size	Outside Diameter	Inside Diameter		
		Schedule 40	Schedule 80	Schedule 120
1½	1.900	1.592	1.476	1.423
2	2.375	2.049	1.913	1.845
2½	2.875	2.445	2.289	2.239
3	3.500	3.042	2.864	2.758
4	4.500	3.998	3.786	3.572
6	6.625	6.031	5.709	5.434

45-Degree Offset Calculations

The *travel* distance of a 45° offset is calculated in the same manner as the diagonal of a square. Multiply the distance across flats by 1.414. The *run* and *offset* represent the two equal sides of a square, and the *travel*, the diagonal, is shown in Fig. 43-5.

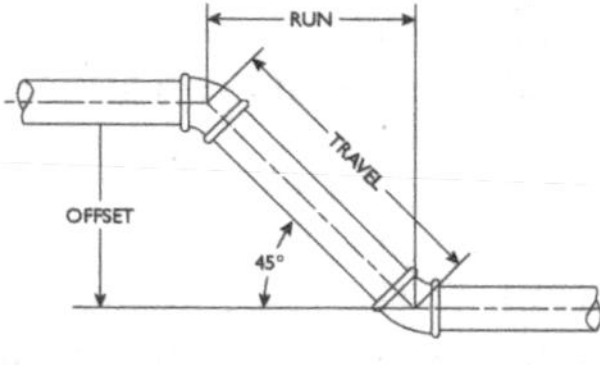

Fig. 43-5

There are also occasions when the travel is known and the offset and run dimensions are wanted. This may be done in the same manner as finding the distance across the flats of a square when the across-corners dimension is known. Multiply the travel dimension by 0.707.

Examples

What is the travel of a 16-in. 45° offset?

$$16 \times 1.414 = 22.625 \text{ or } 22\tfrac{5}{8} \text{ inches}$$

What are the offset and run of a 45° offset having a travel of 26 in.?

$$26 \times 0.707 = 18.382 \text{ or } 18\tfrac{3}{8} \text{ inches}$$

Other Offsets

The dimensions of piping offsets of several other common angles may be calculated by multiplying the known values by the appropriate constants listed in Table 43-6.

Table 43-6 Common Angle Pipe Offsets—Multipliers

Angle	To Find Travel Offset Known	To Find Travel Run Known	To Find Run Travel Known	To Find Offset Travel Known
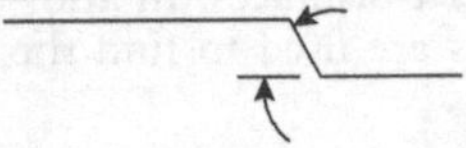 60°	1.155	2.000	0.500	0.866
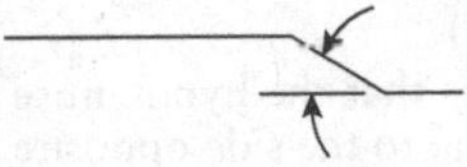 30°	2.000	1.155	0.866	0.500
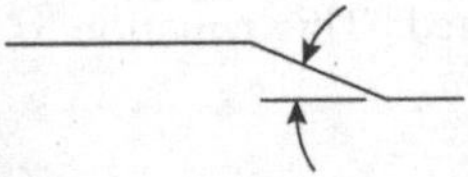 22½°	2.613	1.082	0.924	0.383
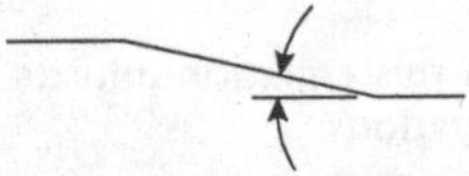 11¼°	5.126	1.000	0.980	0.195

45-Degree Rolling Offset Calculations

The 45° offset is often used to offset a pipeline in a plane other than the horizontal or vertical. This is done by rotating the offset out of the horizontal or vertical plane, and it is known as a *rolling offset*. The rolling offset can best be visualized as contained in an imaginary isometric box, as shown in Fig. 43-6.

The run and offset distances are equal, as they are in the plain 45° offset; however, there are two additional dimensions, *roll* and *height*. Two right-angle triangles must now be

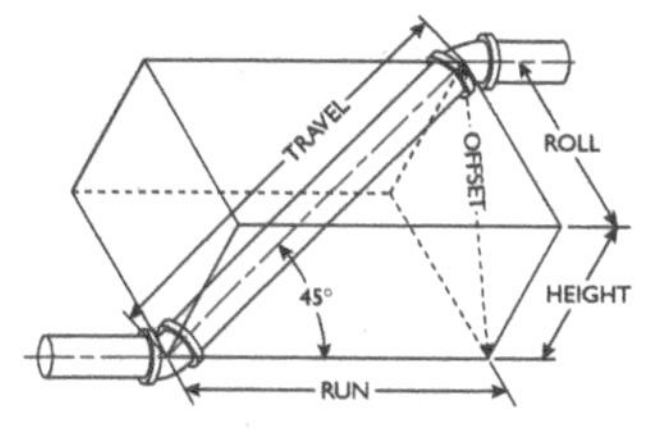

Fig. 43-6

considered. The original one remains the same, with the offset and run as equal sides and the travel as the hypotenuse. The new triangle has the roll and height as sides and the offset as the hypotenuse.

The method of finding distances for plain 45° offsets is also used for calculating rolling offset distances. In addition, the sum-of-the-squares equations are used to find the values of the second triangle.

Sum of the Squares

The sum-of-the-squares equation states that the hypotenuse of a right-angle triangle squared is equal to the side opposite squared plus the side adjacent squared. This equation is commonly written as follows:

$$c^2 = a^2 + b^2$$

Substitution of pipe-offset terms in this equation and its rearrangements gives the following equations:

$$\text{offset}^2 = \text{roll}^2 + \text{height}^2$$
$$\text{run}^2 = \text{roll}^2 + \text{height}^2$$
$$\text{roll}^2 = \text{offset}^2 - \text{height}^2$$
$$\text{roll}^2 = \text{run}^2 - \text{height}^2$$
$$\text{height}^2 = \text{offset}^2 - \text{roll}^2$$
$$\text{height}^2 = \text{run}^2 - \text{roll}^2$$

Depending on what the known values are, it may sometimes be necessary to solve two equations to find the distance wanted.

Examples

What is the *travel* for a 6-in. *roll* with a 7-in. *height?*

Using the formula:

$$\text{offset}^2 = \text{roll}^2 + \text{height}^2$$
$$\text{offset}^2 = (6 \times 6) + (7 \times 7) = 36 + 49 = 85$$
$$\text{offset} = \sqrt{85} = 9.22 = 9\tfrac{7}{32} \text{ inches}$$

Followed by:

$$\text{travel} = \text{offset} \times 1.414$$
$$\text{travel} = 9\tfrac{7}{32} \times 1.414 = 13.035 = 13\tfrac{1}{32} \text{ inches}$$

What is the *roll* for an 11-in. *offset* with an 8-in. *height?*

Using the formula:

$$\text{roll}^2 = \text{offset}^2 - \text{height}^2$$
$$\text{roll}^2 = (11 \times 11) - (8 \times 8) = 121 - 64 = 57$$
$$\text{roll} = \sqrt{57} = 7.55 = 7\tfrac{9}{16} \text{ inches}$$

What is the *height* for a 16-in. *offset* with a 12-in. *roll?*

Using the formula:

$$\text{height}^2 = \text{offset}^2 - \text{roll}^2$$
$$\text{height}^2 = (16 \times 16) - (12 \times 12) = 256 - 144 = 112$$
$$\text{height} = \sqrt{112} = 10.583 = 10\tfrac{9}{32} \text{ inches}$$

Flanged Pipe Connections

Flanged pipe connections are widely used, particularly on larger pipes, as they provide a practical and economical piping connection system. Flanges are commonly connected to the pipe by screw threads or by welding. Several types of flange facings are in use, the simplest of which are the plain *flat face* and the *raised face.*

The plain flat-faced flange is usually used for cast-iron flanges where pressures are under 125 lb. Higher-pressure cast-iron flanges and steel flanges are made with a raised face. Generally, full-face gaskets are used with flat-face flanges and ring gaskets with raised-face flanges. The function of the gasket is to provide a loose, compressible substance between the faces with sufficient body resiliency and strength to make the flange connection leak-proof.

Flange Bolting

The assembly and tightening of a pipe-flange connection is a relatively simple operation; however, certain practices must be followed to obtain a leak-proof connection. The gasket must line up evenly with the inside bore of the flange face, with no portion of it extending into the bore. When tightening the bolts, the flange faces must be kept parallel and the bolts tightened uniformly.

The tightening sequence for round flanges is shown in Fig. 43-7a. The sequence is to lightly tighten the first bolt, then move directly across the circle for the second bolt, then move a quarter way around the circle for the third, and directly across for the fourth, continuing the sequence until all are tightened.

When tightening an oval flange, the bolts are tightened across the short centerline, as shown in Fig. 43-7b.

A four-bolt flange, either round or square, is tightened with a simple criss-cross sequence, as shown in Fig. 43-7c.

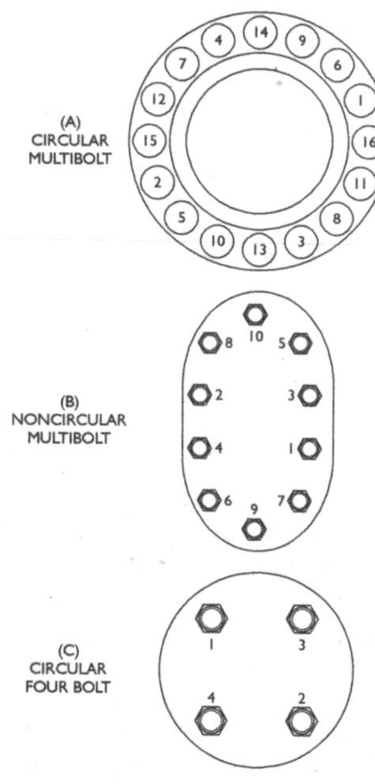

Fig. 43-7

Do not snug up the bolts on the first go-around. This can tilt the flanges out of parallel. If using an impact wrench, set the wrench at about one-half final torque for the first go-around. Pay particular attention to the hard-to-reach bolts.

Pipe Flange Bolt-Hole Layout

Mating pipe flanges to other flanges or circular parts requires correct layout of boltholes. In addition, the holes must be located around the circle to line up when the flanges are mated. The usual practice is to specify the location of the holes as either "On" or "Off" the vertical centerline. The shop term commonly applied to "On" the centerline layout is "One Hole Up" and to "Off" the centerline is "Two Holes Up." An "Off" the centerline or "Two Holes Up" layout is illustrated in Fig. 43-8.

While boltholes may be laid out with a protractor using angular measurements to obtain uniform spacing, this method is most satisfactory when there are six or fewer holes. Also, layout by stepping off spacing around the circle with dividers, by trial and error, is a time-consuming operation. To eliminate the trials and errors, a system of multipliers or constants may be used to calculate the chordal distance between bolthole centers. Simply multiply the constant for the appropriate number of boltholes, as shown in Table 43-7, by the bolt circle diameter to determine the chordal distance between holes.

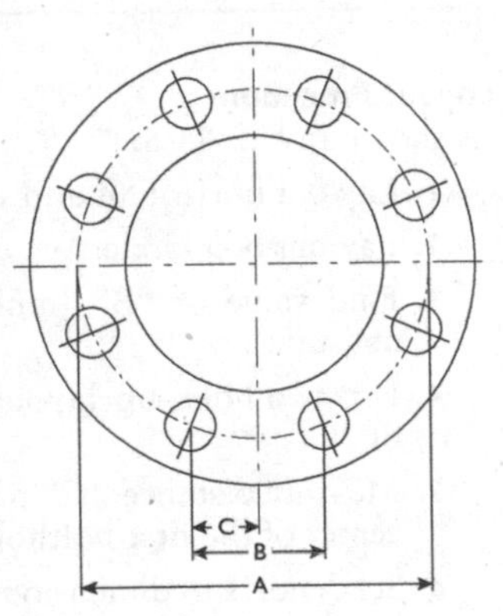

Fig. 43-8

Table 43-7 Flange Hole Constants

No. of Boltholes	Constant
4	.707
6	.500
8	.383
10	.309
12	.259
16	.195
20	.156
24	.131
28	.112
32	.098
36	.087
40	.079

Layout Procedure

As shown in Fig. 43-8:

1. Lay out horizontal and vertical centerlines.
2. Lay out bolt circle.
3. Find value of "B" (multiply bolt circle diameter by constant).
4. For two-holes-up layout, divide "B" by 2 for value of "C."
5. Measure distance "C" off the centerline and locate the center of the first bolthole.
6. Set dividers to dimension "B" and lay out center points by swinging arcs, starting from first center point.

Types of Valves

The principal function of two-way pipe valves is to open and close a line to flow.

Gate

Fluid flows through a gate valve in a straight line. Its construction offers little resistance to flow and causes a minimum of pressure drop. A gate-like disc—actuated by a stem screw and handwheel—moves up and down at right angles to the path of flow, and it seats against two seat faces to shut off flow.

Gate valves are best for services that require infrequent valve operation and where the disc is kept either fully opened or closed.

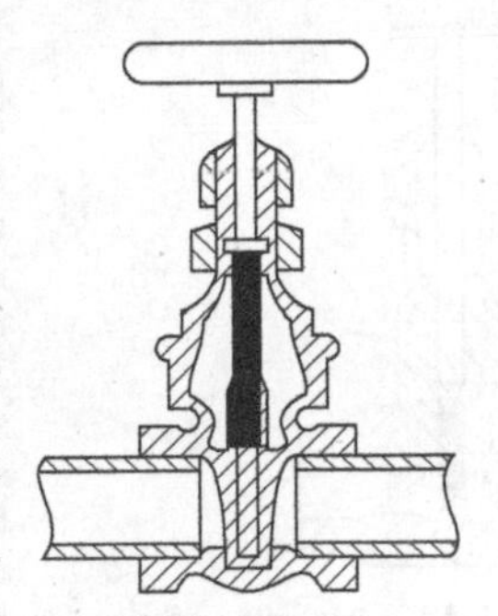

Fig. 43-9 Gate valve.

Globe

Fluid changes direction when flowing through a globe valve. The construction increases resistance to—and permits close regulation of—fluid flow. Disc and seat can be quickly and conveniently replaced or reseated. Angle valves are similar in design to globe valves. They are used when making a 90° turn in a line, as they reduce the number of joints and give less resistance to flow than the elbow and globe valve.

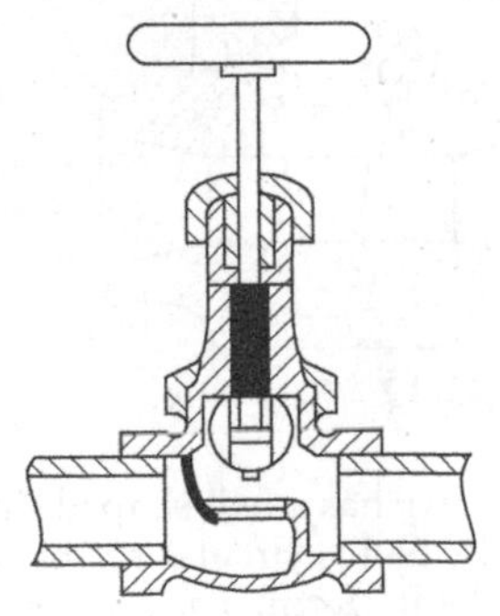

Fig. 43-10 Globe valve.

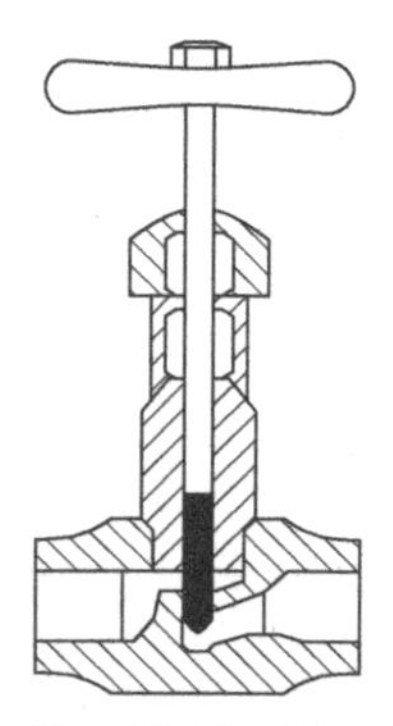

Fig. 43-11 Needle valve.

Needle

The needle valve can be used to shut off flow; however, it is designed primarily as a throttling valve. The pointed disc can be adjusted in the mating seat to give small increments of flow change. As the port diameter is smaller than the connection size, resistance to flow is high, making the needle valve unsuitable for high-volume flow.

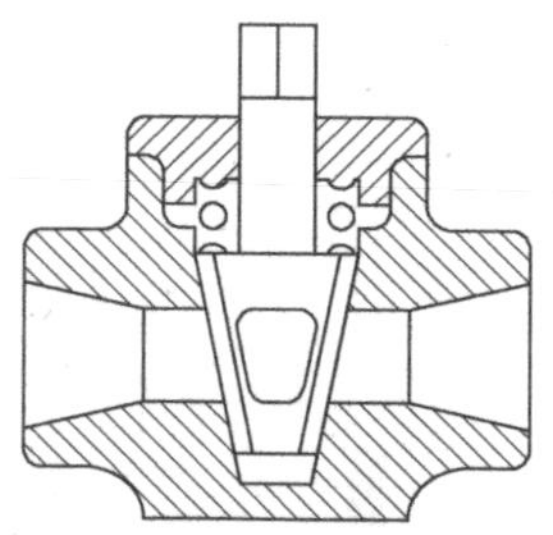

Fig. 43-12 Plug valve.

Plug

Fluid flows through a plug valve in a straight line. Its advantages are low cost, small pressure loss because of its straight-through construction, and fast operation. Only a quarter turn is needed to fully open or close it.

Ball

Most like the plug valve, fluid flows through a ball valve in a straight line. The ball valve has a ball-shaped closure member. The seat matching the ball is circular so that the seating stress is circumferentially uniform. The sealing characteristic of a ball valve is excellent because most ball valves have soft seats that conform readily to the surface of the ball. Again, much like a plug valve, only a quarter turn is needed to open or close it fully.

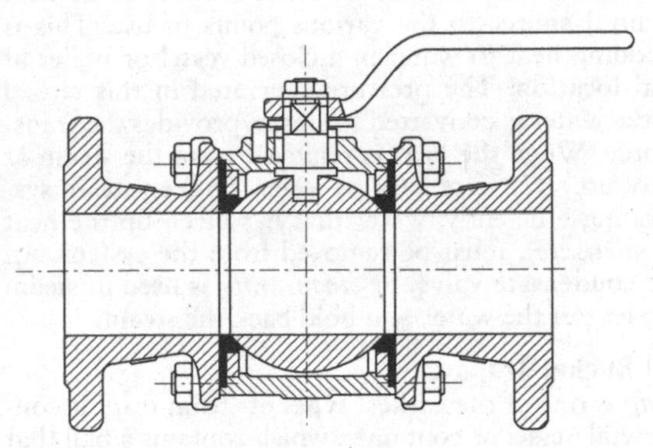

Fig. 43-13 Ball valve.

44. STEAM TRAPS

Steam distribution systems make up a major share of the piping in many industrial plants. These systems carry heat from a central source to the various points of use. This is done by adding heat to water in a closed vessel or boiler at the central location. The pressure generated in this closed vessel, as the water is converted to steam, provides the transporting force. When the heat is removed from the steam at the point of use, it reverts back to water. To keep steam systems operating efficiently, water that has given up the heat (called *condensate*), must be removed from the system. An automatic condensate valve, or *steam trap*, is used in steam systems to let out the water and hold back the steam.

Float and Bucket Traps

A *float trap* is one of the earliest types of steam trap. It consists of a metal vessel or container, which contains a ball that floats on the condensate and controls the valve. Fig. 44-1 shows the key elements of this trap. In this trap the float (A) is lifted by the condensate. The lever (B) attached to the float moves with it, causing the valve (C) to open gradually. As the condensate flow increases and raises the float, the opening of the valve is increased.

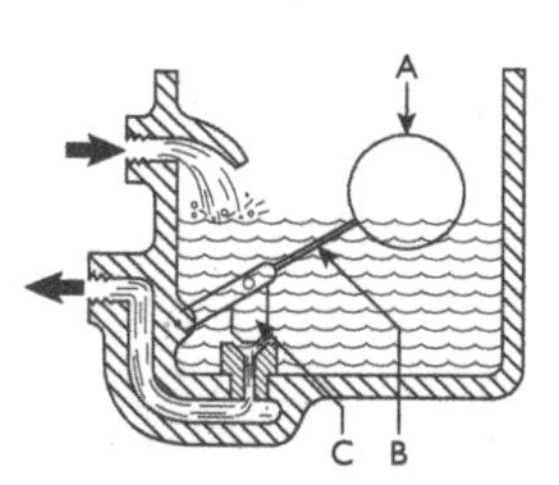

Fig. 44-1 Float trap.

An *open-bucket* trap is shown in Fig. 44-2. In operation, the condensate floats the bucket (A), which pushes up on rod (B), closing the valve (C). As the condensate continues to build up and it spills over the top of the bucket, sinking it, thus opening the valve.

Thermostatic Traps

A *thermostatic* trap responds to temperature changes. Hot steam causes it to open, and the cooler condensate causes it to close. Fig. 44-3 shows a typical thermostatic trap. The operating element (A) is a corrugated bellows filled with a liquid, such as alcohol and water, having a boiling point below that of water. Attached to the bottom of the bellows is a valve (B), which closes when the bellows expands. When cool condensate is flowing, the bellows is contracted and the valve is open, allowing the condensate to discharge. As the condensate temperature approaches steam temperature, the liquid inside the bellows is vaporized, causing a pressure increase inside the bellows that expands it and closes the valve.

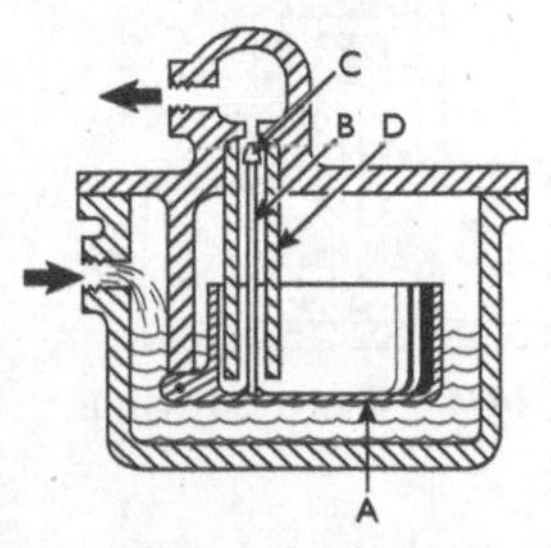

Fig. 44-2 Open-bucket trap.

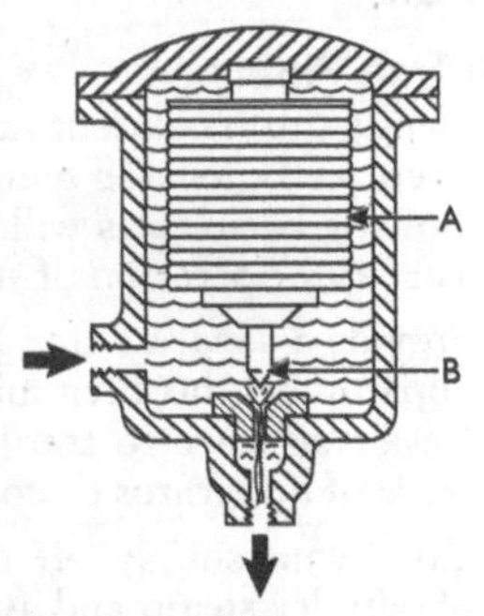

Fig. 44-3 Thermostatic trap.

Thermodynamic Traps

The *thermodynamic* trap uses the heat energy in hot condensate and steam to control its operation. Fig. 44-4 shows a thermodynamic trap in operation. This trap operates by the flashing action of hot condensate discharging into a chamber that is at a lower pressure than the discharging condensate. As cool condensate flows into the trap, pressure is exerted on the bottom side of the piston disc (A), thus lifting the piston and holding open the valve (B). When the condensate

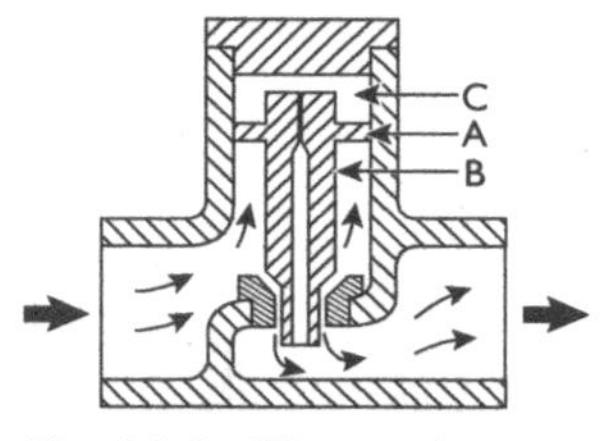

Fig. 44-4 Thermodynamic trap.

flowing through the trap nears steam temperature, it flashes into vapor in the chamber (C) above the piston disc. This increase in pressure above the piston disc causes the valve to close, preventing the discharge of live steam. When the condensate cools to the point where flashing no longer occurs, the pressure above the piston disc is reduced and the piston is lifted, thus opening the valve to allow discharge of the cool condensate.

Steam Trap Piping

To efficiently drain condensate from steam lines, steam traps must be correctly located and the piping properly arranged. The following basic rules will go a long way toward providing satisfactory operation of steam traps (Fig. 44-5):

1. Provide a separate trap for each piece of equipment or apparatus. Short circuiting (steam follows path of least resistance to trap) may occur if more than one piece of apparatus or coil is connected to a single trap.
2. Tap steam supply off the top of the steam main to obtain dry steam and avoid steam line condensate.
3. Install a supply valve close to the steam main to allow maintenance and/or revisions without steam main shutdown.
4. Install a steam supply valve close to the equipment entrance to allow equipment maintenance work without supply line shutdown.
5. Connect the condensate discharge line to the lowest point in the equipment to avoid water pockets and water hammer.

6. Install a shutoff valve upstream of condensate-removal piping to cut off discharge of condensate from equipment and allow service work to be performed.

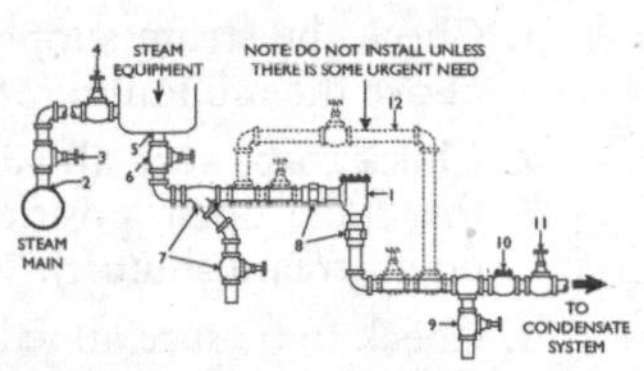

Fig. 44-5

7. Install a strainer and strainer flush valve ahead of the trap to keep rust, dirt, and scale out of working parts and to allow blow-down removal of foreign material from the strainer basket.
8. Provide unions on both sides of the trap for its removal and/or replacement.
9. Install a test valve downstream of the trap to allow observation of discharge when testing.
10. Install a check valve downstream of the trap to prevent condensate flowback during shutdown or in the event of unusual conditions.
11. Install a downstream shutoff valve to cut off equipment condensate piping from the main condensate system for maintenance or service work.
12. Do not install a bypass unless there is some urgent need for it. Bypasses are an additional expense to install and are frequently left open, resulting in loss of steam and inefficient operation of equipment.

Steam Trap Troubleshooting

In .cases of improper functioning of steam equipment, a few simple checks of the steam system should be made before looking for trap malfunction. The following preliminary checks should precede checking the operation of a stream trap:

1. Check the steam supply pressure—it should be at or above the minimum required.
2. Check to be sure all valves required to be open are in the fully open position (supply, upstream shutoff, downstream shutoff).
3. Check to be sure all valves required to be closed are in the tightly closed position (bypass, strainer, test).

The initial step in checking the operation of a steam trap is to check its temperature. Because a properly functioning steam trap is an automatic valve that allows condensate to be discharged but closes to prevent the escape of steam, it should operate very close to the steam temperature. For exact checks, a surface pyrometer should be used. A convenient and dependable operating test is simply to sprinkle water on the trap. If the water spatters, rapidly boils, and vaporizes, the trap is hot and probably is very close to steam temperature.

If the trap is very close to steam temperature, the next step is to determine if condensate or steam is being discharged. When a test valve is provided for this purpose, the check is made by closing the downstream shutoff valve and opening the test valve. The discharge from the test valve should be carefully observed to determine whether condensate or live steam is escaping. If the trap being tested is the type that has an opening and closing cycle, condensate should flow from the test valve and then it should stop as the trap shuts off. The flow should resume when the trap opens again. Steam should not discharge from the test valve if the trap is operating properly. If the trap is a continuous-discharge type, there should be a continuous discharge of condensate but no steam.

In the event that the installation does not have a test valve, the trap may be checked by listening to its operation. The ideal instrument to do this is an industrial stethoscope. If one is not available, a suitable device for this purpose is a screwdriver or metal rod. By holding one end against the trap and the other end against the ear, you can hear the sound of the trap's operation. If the trap is operating properly, you should hear the flow of condensate for a few seconds, a click as the valve closes, and then silence, indicating that the valve has closed tightly. This cycle of sounds should repeat in a regular pattern. The listening procedure, however, is not suitable for checking continuous-discharge traps, because these automatically regulate to an open position in balance with the condensate flow.

Another check on trap operation is to open the strainer valve and observe the discharge at this point. There should be an initial gush of condensate and steam as the valve is opened, followed by a continuous flow of live steam. If condensate flows for a prolonged period before steam is observed, the condensate is not being properly discharged from the system.

Unsatisfactory performance of a steam unit may not be due to improper steam trap operation. When testing steam traps, conditions other than trap malfunction must also be considered. Some of the common faults that cause troubles are these:

1. Inadequate steam supply
2. Incorrectly sized trap
3. Improperly connected piping
4. Improper pitch of condensate lines
5. Inadequate condensate lines

45. STEAM PIPING

Opening the Steam Supply Valve

The opening of valves controlling steam flow in steam supply lines, called steam *mains,* requires care and following the correct procedure. The expansion or growth of the piping system, as the temperature increases when steam is introduced, must be carefully controlled. Also the air in the lines, and the large volume of condensate formed as the line heats up, must be removed. To facilitate removal of condensate during normal operation, as well as at startup, steam lines incorporate *drip pockets, drip legs,* and *drip valves,* as shown in Fig. 45-1.

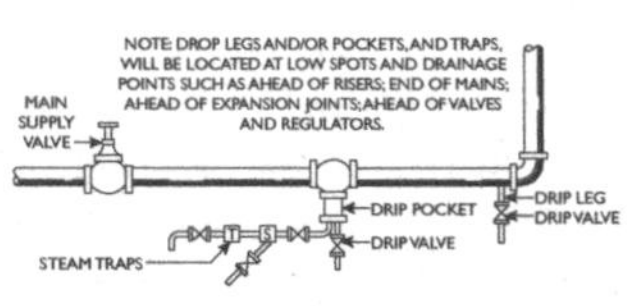

Fig. 45-1

Supply Line Safety

The following procedure should be followed when opening a steam main supply valve:

1. Open all drip valves full open to act as air vents and condensate discharge openings. Check setting of distribution valves to be sure steam goes only to those branch lines ready to receive it.
2. Open main supply valve slowly and in stages to control steam flow volume and provide gradual heatup of the line.
3. Watch discharge at drip valves. Do not close drip valves until warmup condensate has been discharged (except for the next item).

4. Condensate should not be drained from drip pockets. An accumulation of condensate is necessary in drip pockets as they are in the line to do the following:
   a. Let condensate escape by gravity from the fast-moving steam
   b. Store condensate until the pressure differential is great enough for the steam trap to discharge
   c. Provide storage of condensate until there is positive pressure in the line
   d. Provide static head, enabling the trap to discharge before a positive pressure exists in the line
5. Check to see that line pressure comes up to the required operating pressure.
6. Check operation of all steam traps draining condensate from the line to be sure they are operating properly (check temperature, discharge, etc.).

Steam Systems Troubleshooting Chart

The following chart (Table 45-1) will allow troubleshooting and corrective action for many problems noted in malfunctioning steam systems.

Table 45-1 Steam Troubleshooting Chart

Condition	Reason	Corrective Action
Trap blows live steam	1. No prime (bucket traps):	1. No prime (bucket traps):
	a. Trap not primed when originally installed	a. Prime the trap
	b. Trap not primed after cleanout	b. Prime the trap
	c. Open or leaking bypass valve	c. Remove or repair bypass valve
	d. Sudden pressure drops	d. Install check valve ahead of trap
	2. Valve mechanism does not close:	2. Valve mechanism does not close:
	a. Scale or dirt lodged in orifice	a. Clean out the trap
	b. Worn or defective valve or disk mechanism	b. Repair or replace defective parts
	3. Ruptured bellows (thermostatic traps)	3. Replace bellows
	4. Back pressure too high (thermodynamic trap):	4. Back pressure too high (thermodynamic trap):
	a. Worn or defective parts	a. Repair or replace defective parts
	b. Trap stuck open	b. Clean out the trap
	c. Condensate return line or pig tank undersized	c. Increase line or pig tank size
	5. Blowing flash steam:	5. Normal condition
	a. Forms when condensate released to lower or atmospheric pressure	a. No corrective action necessary

Condition	Reason	Corrective Action
Trap does not discharge	1. Pressure too high:	1. Pressure too high:
	a. Trap pressure rating too low	a. Install correct trap
	b. Orifice enlarged by normal wear	b. Replace worn orifice
	c. Pressure-reducing valve set too high or broken	c. Readjust or replace pressure-reducing valve
	d. System pressure raised	d. Install correct pressure change assembly
	2. Condensate not reaching trap:	2. Condensate not reaching trap:
	a. Strainer clogged	a. Blow out screen or replace
	b. Obstruction in line to trap inlet	b. Remove obstruction
	c. Bypass opening or leaking	c. Remove or repair bypass valve
	d. Steam supply shut off	d. Open steam supply valve
	3. Trap clogged with foreign matter	3. Clean out and install strainer
	4. Trap held closed by defective mechanism	4. Repair or replace mechanism
	5. High vacuum in condensate return line	5. Install correct pressure change assembly
	6. No pressure differential across trap:	6. No pressure differential across trap:
	a. Blocked or restricted condensate return line	a. Remove restriction
	b. Incorrect pressure change assembly	b. Install correct pressure change assembly

(continued)

Table 45-1 (continued)

Condition	*Reason*	*Corrective Action*
Continuous discharge from trap	1. Trap too small: a. Capacity undersized b. Pressure rating of trap too high	1. Trap too small: a. Install properly sized larger trap b. Install correct pressure change assembly
	2. Trap clogged with foreign matter: a. Dirt or foreign matter in trap internals b. Strainer plugged	2. Trap clogged with foreign matter: a. Clean out and install strainer b. Clean out strainer
	3. Bellows overstressed (thermostatic traps)	3. Replace bellows
	4. Loss of prime	4. Install check valve on inlet side
	5. Failure of valve to seat: a. Worn valve and seat b. Scale or dirt under valve and in orifice c. Worn guide pins and lever	5. Failure of valve to seat: a. Replace worn parts b. Clean out the trap c. Replace worn parts
Sluggish or uneven heating	1. Trap has no capacity margin for heavy starting loads	1. Install properly sized larger trap (bucket traps)
	2. Insufficient air-handling capacity	2. Use thermic buckets or increase vent size
	3. Short circuiting (group traps)	3. Trap each unit individually

Condition	Reason	Corrective Action
	4. Inadequate steam supply: a. Steam supply pressure valve has changed b. Pressure-reducing valve setting off	4. Inadequate steam supply: a. Restore normal steam pressure b. Readjust or replace reducing valve
Back pressure troubles	1. Condensate return line too small 2. Other traps blowing steam into header 3. Pig tank vent line plugged 4. Obstruction in condensate return line 5. Excess vacuum in condensate return line	1. Install larger condensate return line 2. Locate and repair other faulty traps 3. Clean out pig tank vent line 4. Remove obstruction 5. Install correct pressure change assembly

46. AUTOMATIC SPRINKLER SYSTEMS

Wet Sprinkler Systems

A *wet sprinkler system* (Fig. 46-1) is described by the National Board of Fire Underwriters as "a system employing automatic sprinklers attached to a piping system containing water and connected to a water supply so that water discharges immediately from sprinklers opened by a fire." A vital component of a wet sprinkler system is the *alarm check valve,* also called the *wet sprinkler valve.* Its function is to direct water to alarm devices and sound the sprinkler alarm. *It does not control the flow of water into the system.*

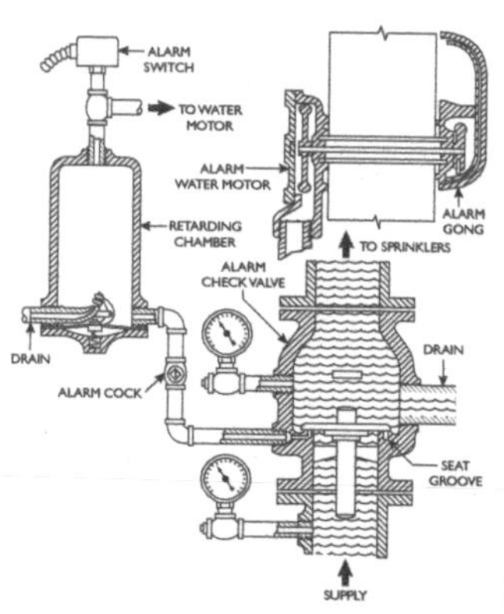

Fig. 46-1

The alarm check valve is located in the pipe riser at the point the water line enters the building. An underground valve with an indicator post is usually located a safe distance outside the building. In design and operation, the alarm check valve is a globe-type check valve. A groove cut in the seat is connected by passage to a threaded outlet on the side of the valve body. When the valve lifts, the water can flow to the outlet. There is also a large drain port above the seat, which is connected to the *drain valve.* Two additional ports, one above and one below the seat, allow the attachment of pressure gauges.

A plug or stop-cock type of valve called the *alarm control cock* is connected to the outlet from the seat groove. This cock controls the water flow to the alarm devices, allowing their silencing. This cock *must* be in the alarm position when the system is "set," or the alarm check valve will be unable to perform its function of sounding an alarm.

Operation of Wet Sprinkler Alarm Systems

When a sprinkler head opens, or for any reason water escapes from a wet sprinkler system, the flow through the alarm check valve causes the check valve disc to lift. Water entering the seat groove flows through the alarm cock to the retarding chamber. The function of the retarding chamber is to avoid unnecessary alarms caused by slight leakage. It will allow a small volume of water flow without actuating the alarm. When there is a large flow, as occurs when a sprinkler opens, the chamber is quickly filled and pressure closes the diaphragm-actuated drain valve in the bottom of the retarding chamber. The pressure of the water then actuates the electrical alarm. The water also flows to the water motor, causing it to be rotated and whirl hammers inside the alarm gong, thus mechanically sounding the bell alarm.

Placing a Wet Sprinkler System in Service

1. Check the system to be sure it is ready to be filled with water. If the system has been shut down because a head has opened, be sure the head has been replaced with one of the proper rating.
2. Open "vent valves"—located at high points.
3. Place "alarm cock" in CLOSED position. This will prevent sounding of the alarm while flowing water fills the systems.
4. Place "drain valve" in nearly closed position. A trickle of water should flow from the drain valve during filling.
5. Open "indicator post valve" slowly. When the system has filled, there will be a quieting of the sound of rushing water. Open the valve full open, then back off a quarter turn.

6. Observe the water flow at the vent valves. When a steady flow of water occurs (no air), close the vent valves.
7. Check the water flow by quickly opening the "drain valve" and closing it. The water pressure should drop about 10 lb when the valve is opened and immediately return to full pressure when valve is closed. The excessive pressure drop indicates insufficient water flow.
8. Open "alarm cock"—system is now in SET condition.
9. Test—Open "drain valve" several turns. Water should flow, sounding the mechanical gong alarm and the electrical alarm.
10. Close "drain valve."—If the alarms have functioned properly, the system is operational.

Dry-Pipe Sprinkler Systems

The National Board of Fire Underwriters describes a dry-pipe sprinkler system as "a system employing automatic sprinklers attached to a piping system containing air under pressure, the release of which as from the opening of a sprinkler permits the water pressure to open a valve known as a *dry-pipe valve*. The water then flows into the piping system and out the open sprinkler."

The dry-pipe valve is located in the pipe riser at the point the water line enters the building. An underground valve with indicator post is usually located a safe distance outside the building.

The dry-pipe valve is a dual style valve, as both an air valve and a water valve are contained inside its body. These two internal valves may be separate units, one positioned above the other, or they may be combined in a single unit, one within the other. The function of the air valve is to retain the air in the piping system and to hold the water valve closed, thus restraining the flow of water. As long as sufficient air pressure is maintained in the system, the air valve can do this because it has a larger surface than the water valve. The air in the system acting on this large surface provides enough force to hold the water valve closed against the pressure of the water (Fig. 46-2).

Operation of Dry-Pipe Sprinkler Systems

When a sprinkler head opens, or when for any reason air escapes from a dry-pipe system, the air pressure above the internal air valve is reduced. When the air pressure falls to the point where its force is exceeded by the force of the water below the internal valve, both valves are thrown open. This allows an unobstructed flow of water through the dry-pipe valve into the system piping and to the open sprinklers. As the valves are thrown open, water fills the intermediate chamber. To avoid unnecessary alarms the chamber is equipped with a "velocity drip valve." This drip valve is normally open to the atmosphere and allows drainage of any slight water leakage past the internal water valve seat.

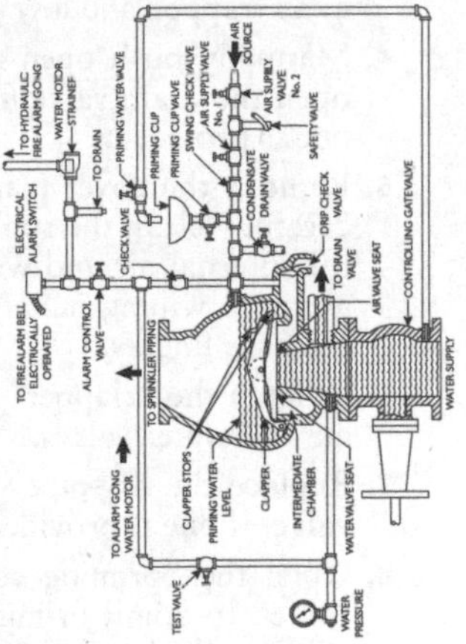

Fig. 46-2

When a sprinkler head opens and falling air pressure allows the internal valves to be thrown open, the intermediate chamber instantly fills with water and the velocity drip valve is forced closed. The water now under pressure in the chamber flows to the electrical alarm switch and to the alarm gong.

Placing a Dry-Pipe Sprinkler System in Service

1. Close the valve controlling water flow to the system. This may be located in the riser, or it may be an underground valve with an indicator post. If a fire has occurred and water is flowing from opened sprinklers, the approval of the person in authority must be obtained before closing the valve.
2. Open the "drain valve" and allow the water to drain from the sprinkler system piping.
3. Open all "vent" and "drain" valves throughout the system. Vents will be located at the high points and drains at all trapped and low portions of the piping system.
4. Manually push open the "velocity drip valve." Also open the drain valve for the "dry-pipe valve body," if one is provided.
5. Remove the cover plate from the dry-pipe valve and carefully clean the rubber facings and seat surfaces of the internal air and water valves. *Do not* use rags or abrasive wiping materials. Wipe the seats clean with the bare fingers.
6. Unlatch the "clapper" and carefully close the internal air and water valves.
7. Replace the dry-pipe valve cover and close the drain valve, if one is provided.
8. Open the "priming cup valve" and "priming water valve" to admit priming seal water into the dry-pipe valve to the level of the pipe connection. The priming water provides a more positive seal to prevent air from escaping past the air valve seat into the intermediate chamber.
9. Drain excess water by opening the "condensate drain valve." Close tightly when water no longer drains from the valve.
10. Open "air supply valve" and admit air to build up a few pounds of pressure in the system.

11. Check all open vents and drains throughout the system to be sure all water has been forced from the low points. As soon as dry air exhausts at the various open points, the openings should be closed. Close air supply valve.
12. Replace any open sprinklers with new sprinkler heads of the proper rating.
13. Open air supply valve and allow system air pressure to build up to the required pressure. The air pressure required to keep the internal valves closed varies directly with water supply pressure. Consult pressure-setting tables.
14. Open the system water supply valve slightly to obtain a small flow of water to the dry-pipe valve.
15. When water is flowing clear at the drain valve, slowly close it, allowing water pressure to build up gradually below the internal water valve as observed on the water pressure gauge.
16. When water pressure has reached the maximum below the internal water valve, open the supply valve to the full open position. Back off the valve about a quarter turn from full open.
17. Test the alarms—open the "test valve," or if system has a three-position test cock, place the cock in the TEST position. Water should flow to the electrical alarm switch and also to the alarm gong water motor.
18. If alarms have functioned properly, close the test valve or place the three-position test cock in the ALARM position.

The alarm test is a test of the functioning of the alarm system only and does not indicate the condition of the dry-pipe valve. Test the dry-pipe valve operation by opening a vent valve to allow air in the piping system to escape, causing the dry-pipe valve to trip. It must then be reset, going through the procedure listed previously.

47. CARPENTRY

Commercial Lumber Sizes

Two words, *timber* and *lumber*, are commonly used to describe the principal material used by carpenters. In the early stages of lumber production the word "timber" is usually applied to wood in its natural state. Wood cut and sawed into standard commercial pieces is called *lumber*.

Stock lumber may be *green*, meaning that the wood contains a large percentage of its natural moisture, or it may be *seasoned*. Seasoning is the process, either naturally or by exposure to heat, of removing about 85 percent of the moisture contained in freshly cut timber. Lumber is usually classified according to the three types into which it is rough sawed. These are *dimension stock*, which is 2 in. thick and from 4 to 12 in. wide; *timbers*, which are 4 to 8 in. thick and 6 to 10 in. wide; and *common boards*, which are 1 in. thick and 4 to 12 in. wide.

Rough lumber is *dressed* or *surfaced* by removing about ½ in. from each side and about ⅜ in. from the edges. This planing operation is called *dressing* or *surfacing*, and the letter "D" for dressed or "S" for surface is used to indicate how many sides or edges are planed. For example, D1S or S1S indicates one side has been planed. Lumber planed on both sides and edges is designated S4S.

Traditional sawmill operations called for logs sawed to nominal dimensions—that is, dimension stock was sawed to 2 in. thickness, boards to 1 in., and so on. Seasoning or drying further reduced the thickness roughly ⅛ in. and the width a lesser amount. Table 47-1 lists the dimensions of lumber produced as standard during the many years this practice was followed.

The practice of sawing to nominal dimensions is no longer followed. Instead, finished standard dimensions have been established and are used to determine sawmill dimensions. This allows more efficient use of logs, as lumber can be rough sawed only to lesser dimensions, providing only enough extra material to surface plane to the new, reduced standard finished dimensions. The present standard lumber sizes are listed in Table 47-2.

Table 47-1 Old Standard Lumber Sizes

	Nominal Size, in.		*Actual S4S Size, in.*	
Lumber Classification	***Thickness***	***Width***	***Thickness***	***Width***
Dimension	2	4	1⅝	3⅝
	2	6	1⅝	5⅝
	2	8	1⅝	7½
	2	10	1⅝	9½
	2	12	1⅝	11½
Timbers	4	6	3⅝	5½
	4	8	3⅝	7½
	4	10	3⅝	9½
	6	6	5½	5½
	6	8	5½	7½
	6	10	5½	9½
	8	8	7½	7½
	10	10	9½	9½
Common Boards	1	4	25/32	3⅝
	1	6	25/32	5⅝
	1	8	25/32	7½
	1	10	25/32	9½
	1	12	25/32	11½

Table 47-2 Standard Lumber Sizes

	Nominal Size, in.		*Actual S4S Size, in.*	
Lumber Classification	***Thickness***	***Width***	***Thickness***	***Width***
Dimension	2	4	1½	3½
	2	6	1½	5½
	2	8	1½	7¼
	2	10	1½	9¼
	2	12	1½	11¼

(continued)

Table 47-2 (continued)

	Nominal Size, in.		Actual S4S Size, in.	
Lumber Classification	Thickness	Width	Thickness	Width
Timbers	4	6	3½	5½
	4	8	3½	7½
	4	10	3½	9½
	6	6	5½	5½
	6	8	5½	7½
	6	10	5½	9½
	8	8	7½	7½
	10	10	9½	9½
Common Boards	1	4	¾	3½
	1	6	¾	5½
	1	8	¾	7¼
	1	10	¾	9¼
	1	12	¾	11¼

Building Layout

The first step in building layout is to establish the wall line for one side of the building by measurement from boundary markers or other reliable positions, and one building corner location point on the line. A stake is driven into the ground at this point and a nail driven into the top of the stake to mark the corner point accurately. The other corner point stakes are then located by measurement and by squaring the corners.

The corners are squared using the 3-4-5 triangle measurement system. Measurements are made along the sides in multiples of 3 and 4, and along the diagonal in multiples of 5. The 6-8-10 and 9-12-15 combination measurements are often used. To further ensure square corners, the diagonals are measured to see if they are the same length.

After the corner stakes are accurately located, *batter board* stakes are driven in at each corner about 4 ft beyond the building lines, and the batter boards are attached. Batter boards are the usual method used to retain the outline of a building layout. The height of the boards may also be positioned to conform to the height of above-grade foundation walls. A line is held across the top of the opposite boards at the corners and adjusted using a plumb bob so that it is exactly over the nails in the stakes. Saw kerfs are cut where the lines touch the boards so that accurate line locations are ensured after the corner stakes are removed.

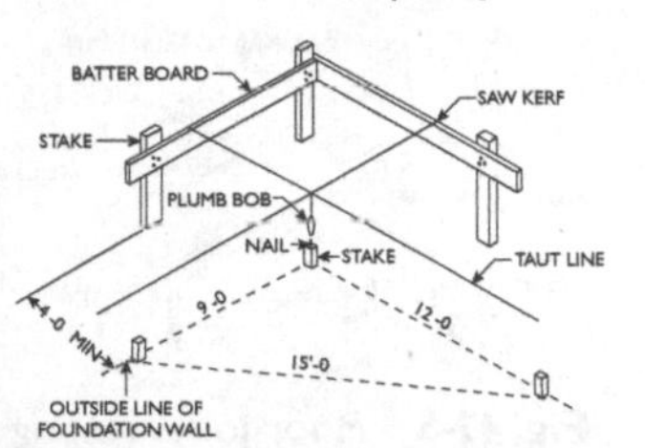

Fig. 47-1 Layout stakes and batter boards.

Wood-Frame Building Construction

While the details of wood-frame building construction may vary in different localities, the fundamental principles are the same. Figures 47-2 through 47-12 show established methods of construction and accepted practices used in wood-frame building construction.

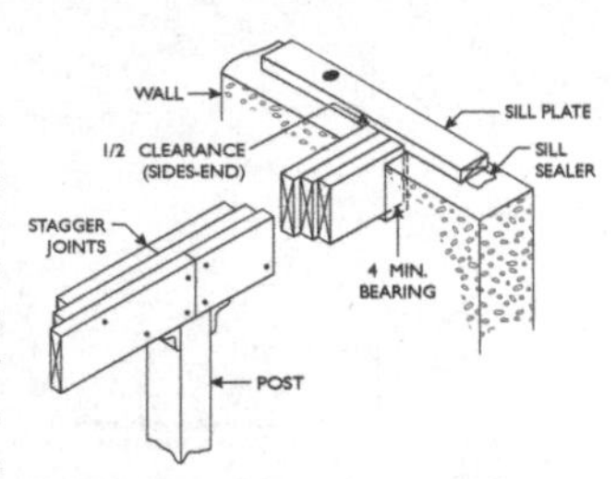

Fig. 47-2 Built-up wooden girder.

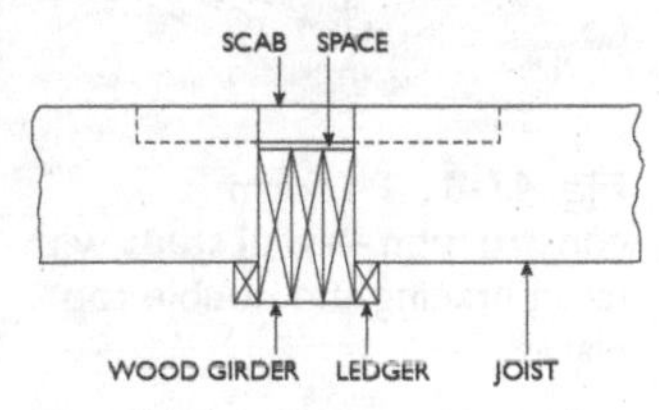

Fig. 47-3 Connecting scab to tie joists together.

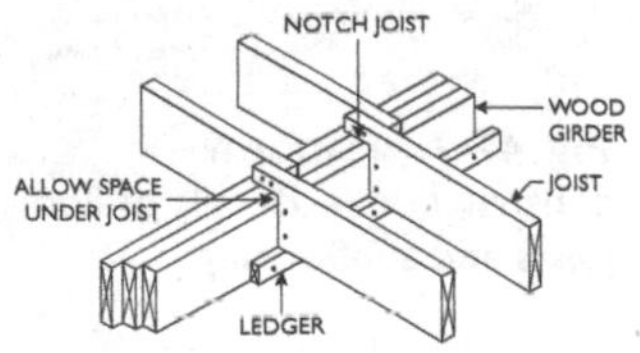

Fig. 47-4 Floor joists notched to fit over girder.

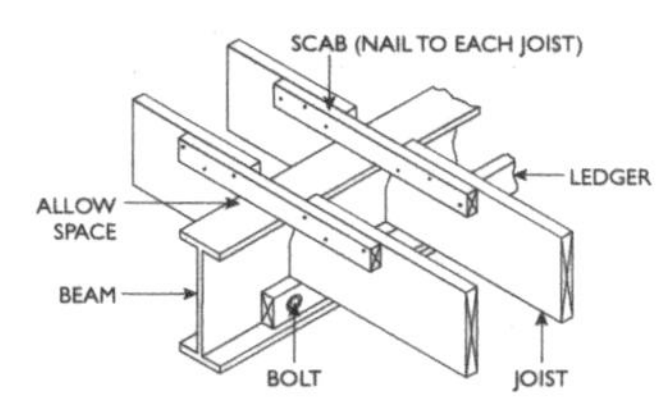

Fig. 47-5 Floor joists resting on wooded ledger fastened to I-beam girder.

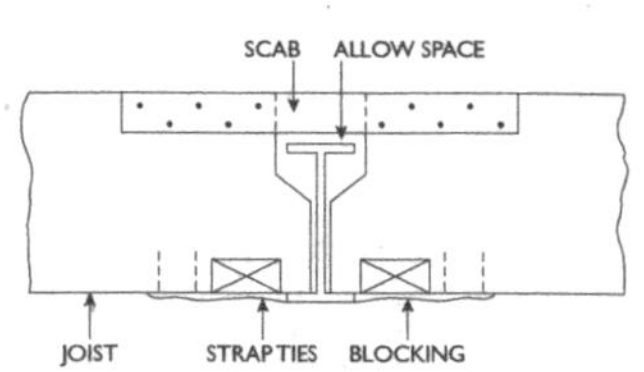

Fig. 47-6 Floor joists resting directly on I-beam girder and connected at top with scab board.

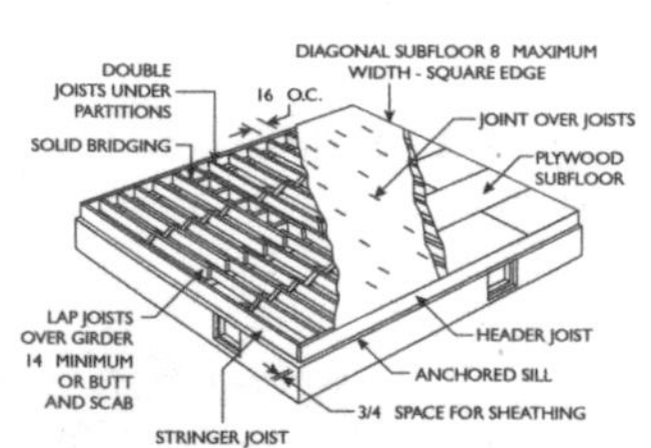

Fig. 47-7 Platform construction — details of floor joists and subflooring.

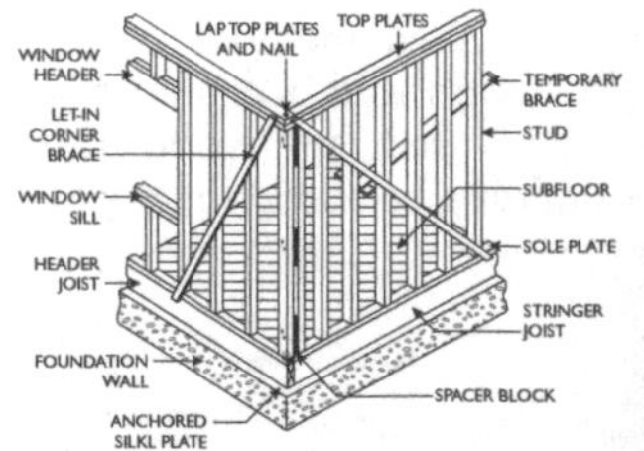

Fig. 47-8 Platform construction — wall studs with let-in bracing and double top plates.

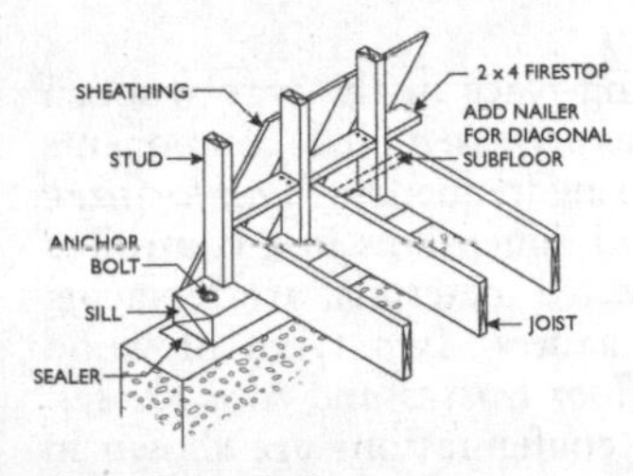

Fig. 47-9 Balloon-frame construction — wall studs and floor joists rest on anchored walls.

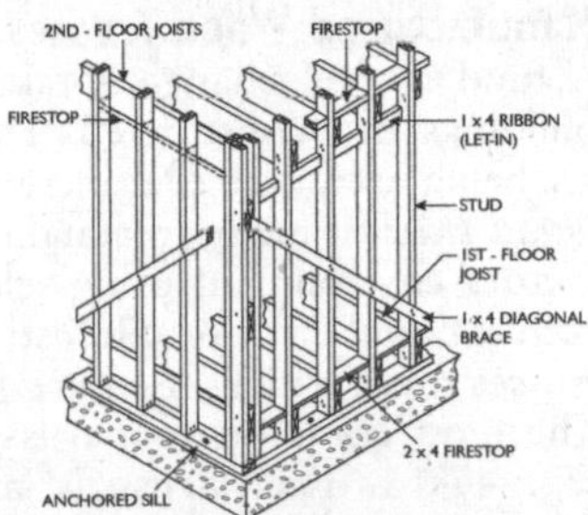

Fig. 47-10 Balloon-frame construction — second-floor joists rest on 1-in. × 4-in. ribbons that have been let into the wall studs. Fire stops prevent the spread of fire through open wall passages.

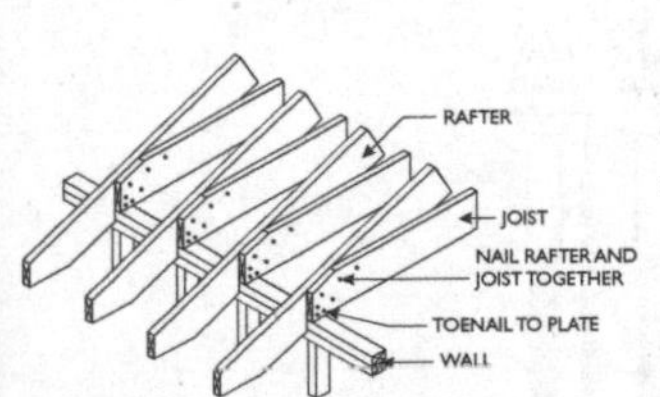

Fig. 47-11 Ceiling joist and rafter construction.

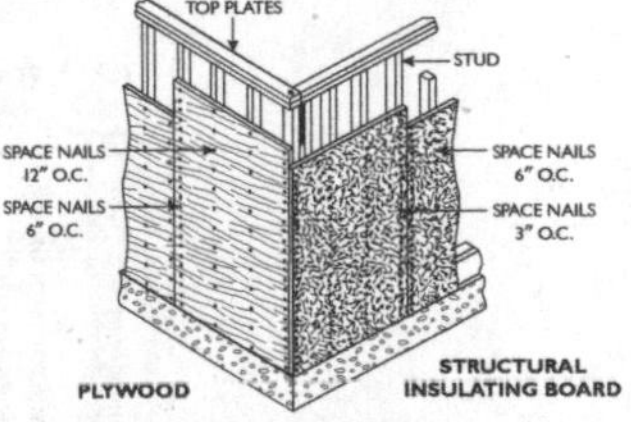

Fig. 47-12 Plywood or insulating board wall sheathing.

Manufactured Wood Trusses

A fundamental change is taking place in the way wooden buildings are constructed. Factory-assembled components are being used with ever-increasing frequency. *Manufactured wood trusses,* made to standard dimensions in a controlled factory environment using selected materials, are replacing conventional "stick-built" structures. Two types of wood trusses most widely used are *floor trusses* and *roof trusses.* The most common roof truss configurations are shown in Fig. 47-13. Each truss is an assembly of precision-cut selected wood members securely fastened together with *metal connector plates.* In addition to reduced cost and improved quality, manufactured trusses make possible longer clear spans, reducing the requirement for interior bearing partitions.

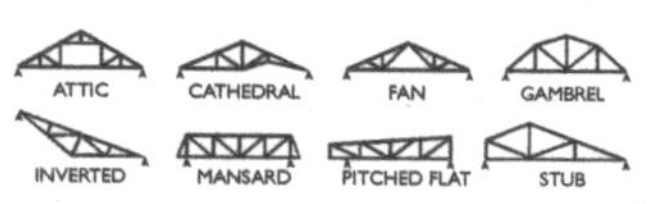

Fig. 47-13 Common roof truss configurations.

Table 47-3 Standard Wood Screw Dimensions

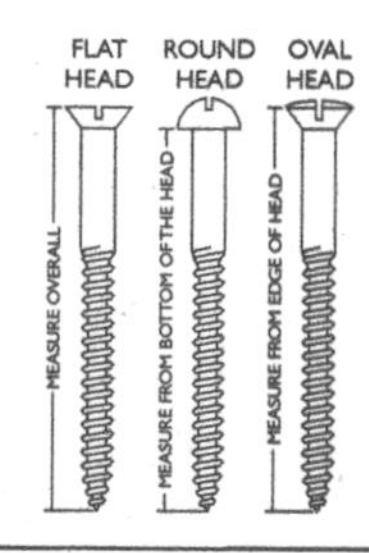

		Head Diameter		
Screw Gauge	***Screw Diameter***	***Flat***	***Round***	***Oval***
0	0.060	0.112	0.106	0.112
1	0.073	0.138	0.130	0.138

Screw Gauge	Screw Diameter	Head Diameter		
		Flat	Round	Oval
2	0.086	0.164	0.154	0.164
3	0.099	0.190	0.178	0.190
4	0.112	0.216	0.202	0.216
5	0.125	0.242	0.228	0.242
6	0.138	0.268	0.250	0.268
7	0.151	0.294	0.274	0.294
8	0.164	0.320	0.298	0.320
9	0.177	0.346	0.322	0.346
10	0.190	0.371	0.346	0.371
11	0.203	0.398	0.370	0.398
12	0.216	0.424	0.395	0.424
13	0.229	0.450	0.414	0.450
14	0.242	0.476	0.443	0.476
15	0.255	0.502	0.467	0.502
16	0.268	0.528	0.491	0.528
17	0.282	0.554	0.515	0.554
18	0.294	0.580	0.524	0.580
20	0.321	0.636	0.569	0.636
22	0.347	0.689	0.611	0.689
24	0.374	0.742	0.652	0.742
26	0.400	0.795	0.694	0.795
28	0.426	0.847	0.735	0.847
30	0.453	0.900	0.777	0.900

Power Nailers and Staplers

Power nailers and *staplers,* first used in production operations, are now commonly used for carpentry projects to promote efficiency and provide dependable fastenings. The nail and staple have certain features that determine which is best used for specific applications. For example, a staple has more holding power in wood than a nail because it has two legs holding it. When conditions of lateral stress exist, or in an application where longer fasteners are needed, however, nails are the correct fastener to use. The types of nails available for use in power nailers vary from the small brad nail for fine finish work to the 5-in. long spike for extra heavy fastening. There are three principal shank types: *smooth* for

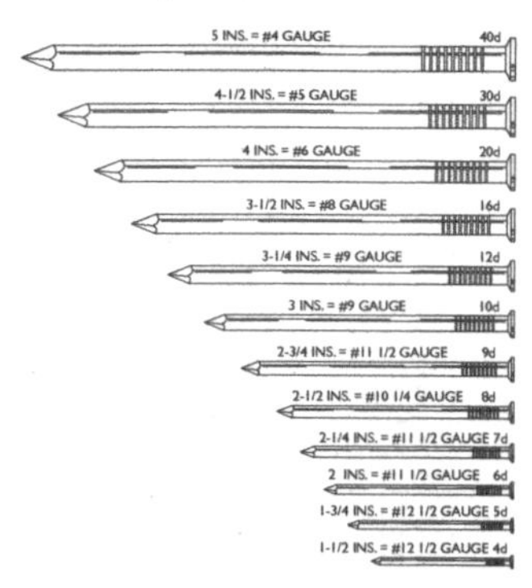

Fig. 47-14 Common wire nails.

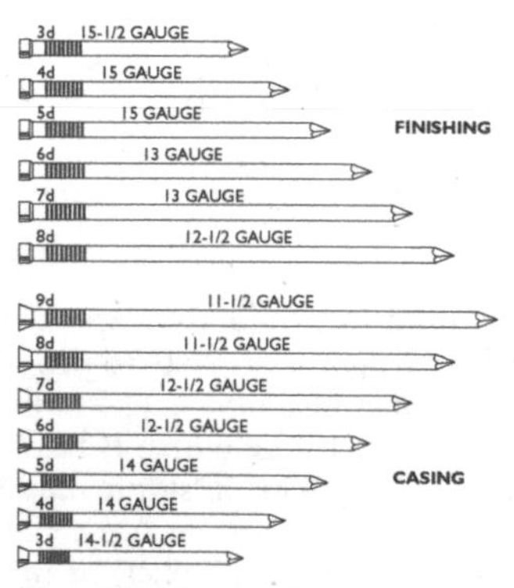

Fig. 47-15 Finishing and casing nails.

general applications, *ring* for superior holding in softwood applications, and *screw* for hardwood applications. The diamond point nail is the most commonly used for general softwood applications.

Most compressed air-powered nailers incorporate a magazine that holds a clip of nails that feed automatically for rapid controlled nailing. The nails or brads are held together with a light glue or plastic that allows them to be loaded conveniently into the tool. When the clip runs out, reloading it is easy.

The *power stapler* is similar in appearance and operation to the power nailer. The fasteners it drives (staples) are available in a great variety of wire gauges, crown widths, and leg lengths. Staplers are in common use for carpentry operations involving the application of sheathing, decking, roofing, and more, as well as in many finishing operations.

Table 47-4 Nominal Dimensions of Standard Machine Bolts

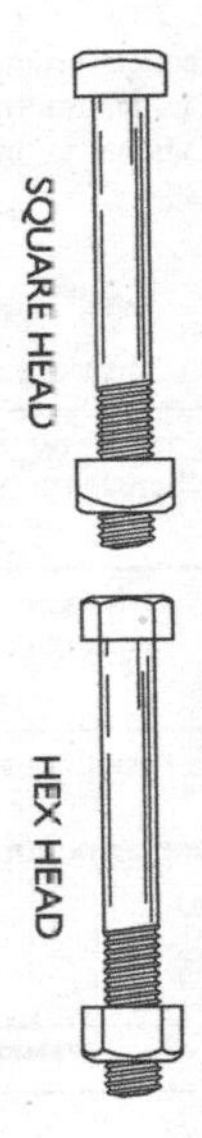

Diameter	UNC Threads per inch	Square Head			Hex Head		
		Width Across Flats	Width Across Corners	Head Height	Width Across Flats	Width Across Corners	Head Height
1/4	20	3/8	17/32	11/64	7/16	1/2	11/64
5/16	18	1/2	45/64	13/64	1/2	9/16	7/32
3/8	16	9/16	51/64	1/4	9/16	2 1/32	1/4
7/16	14	5/8	57/64	19/64	5/8	4 7/64	19/64
1/2	13	3/4	1 1/16	21/64	3/4	5 5/64	11/32
5/8	11	15/16	1 21/64	27/64	1 5/16	1 3/32	27/64
3/4	10	1 1/8	1 19/64	1/2	1 1/8	1 19/64	1/2
7/8	9	1 5/16	1 55/64	19/32	1 5/16	1 33/64	37/64
1	8	1 1/2	2 1/8	21/32	1 1/2	1 47/64	43/64
1 1/8	7	1 11/16	2 25/64	3/4	1 11/15	1 61/64	3/4
1 1/4	7	1 7/8	2 21/32	27/32	1 7/8	2 11/64	27/32
1 3/8	6	2 1/16	2 59/64	29/32	2 1/16	2 3/8	29/64
1 1/2	6	2 1/4	3 3/16	1	2 1/4	2 19/32	1

Table 47-5 Schedule of Common Wire Nail Use in Wood Frame Building Construction

Joining	Nailing method	Nails		
		Number	Size	Placement
Header to joint	End-nail	3	16d	
Joist to sill or girder	Toenail	2 3	10d or 8d	
Header and stringer joist to sill	Toenail		10d	16 in. on center
Bridging to joist	Toenail each end	2	8d	
Ledger strip to beam, 2 in. thick		3	16d	At each joist
Subfloor, boards:				
1 by 6 in. and smaller		2	8d	To each joist
1 by 8 in.		3	8d	To each joist
Subfloor, plywood:				
At edges			8d	6 in. on center
At intermediate joists			8d	8 in. on center
Subfloor (2 by 6 in., T&G) to joist or girder	Blind-nail (casing) and face-nail	2	16d	At each stud
Soleplate to stud, horizontal assembly	End-nail	2	16d	
Top plate to stud	End-nail	2	16d	
Stud to soleplate	Toenail	4	8d	
Soleplate to joist or blocking	Face-nail		16d	16 in. on center
Doubled studs	Face-nail, stagger		10d	16 in. on center
End stud of intersecting wall to exterior wall stud	Face-nail		16d	16 in. on center

Joining	Nailing method	Nails		
		Number	Size	Placement
Upper top plate to lower top plate	Face-nail		16d	16 in. on center
Upper top plate, laps and intersections	Face-nail	2	16d	
Continuous header, two pieces, each edge			12d	12 in. on center
Ceiling joist to top wall plates	Toenail	3	8d	
Ceiling joist laps at partition	Face-nail	4	16d	
Rafter to top plate	Toenail	2	8d	
Rafter to ceiling joist	Face-nail	5	10d	
Rafter to valley or hip rafter	Toenail	3	10d	
Ridge board to rafter	End-nail	3	10d	
Rafter to rafter through ridge board	Toenail	4	8d	
	Edge-nail	1	10d	
Collar beam to rafter:				
2 in. member	Face-nail	2	12d	
1 in. member	Face-nail	3	8d	
1-in. diagonal let-in brace to each stud and plate (4 nails at top)		2	8d	
Built-up corner studs:				
Studs to blocking	Face-nail	2	10d	Each side
Intersecting stud to corner studs	Face-nail		16d	12 in. on center

(continued)

Table 47-5 (continued)

Joining	Nailing method	Nails		
		Number	Size	Placement
Wall sheathing:	Face-nail		20d	32 in. on center, each side
1 by 8 in. or less, horizontal				
1 by 6 in. or greater, diagonal	Face-nail	2	8d	At each stud
Wall sheathing, vertically applied plywood:	Face-nail	3	8d	At each stud
3/8 in. and less thick				
1/2 in. and over thick	Face-nail		6d	6-in. edge
Wall sheathing, vertically applied fiberboard:	Face-nail		8d	12-in. intermediate
1/2 in. thick				
25/32 in. thick	Face-nail			
Roof sheathing, boards, 4-, 6-, 8-in. width	Face-nail			1½-in. roofing nail 3-in. edge
Roof sheathing, plywood:	Face-nail	2	8d	1¾-in. roofing nail 6-in. intermediate
				At each rafter
3/8 in. and less thick	Face-nail		6d	
1/2 in. and over thick	Face-nail		8d	6-in. edge and 12-in. intermediate

Nail Selection Guide for Power Nailer

Shank Types

Smooth. Smooth-shank nails are available in a wide range of lengths and shank diameters for light, standard, and heavy-duty applications. Smooth-shank nails provide excellent holding power in most wood applications and are available with round heads, modified D-heads, or duplex heads for applications that require easy removal of the nail.

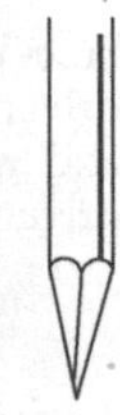

Fig. 47-16

Ring. Ring-shank nails offer superior holding power in softwood applications and are excellent for crating and container construction where fastened joints are subject to lateral stress.

Fig. 47-17

Screw. Screw-shank nails are excellent for hardwood applications, such as pallets. Screw-shank nails provide greater maximum withdrawal loads than smooth-shank nails.

Fig. 47-18

Staple Selection Guide

Crown Widths/Wire Sizes

Fine-Wire Wide Crown. Recommended for applications where the staple crown should not cut into the material being fastened.

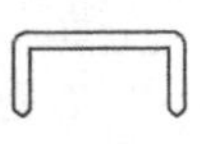

Fig. 47-19

Heavy-Wire Wide Crown. Used for construction-related applications such as asphalt roofing and sheathing where broad holding power is required.

Fine-Wire Intermediate Crown. Holds well when driven into many types of soft material, such as cloth and vinyl. Maintains a neat, clean appearance when driven.

Fig. 47-20

Heavy-Wire Intermediate Crown. Recommended for heavier wood applications, such as decking and wall sheathing, where superior holding power is required.

Fine-Wire Narrow Crown. Used for applications where appearance is important and the staple must be nearly invisible.

Fig. 47-21

Heavy-Wire Narrow Crown. Recommended for furniture, cabinet, and construction-related applications.

Stair Layout

Knowledge and understanding of certain terms and practices help considerably in the layout and installation of simple stairs, as well as a full floor level of stairs, both straight and platform type.

Terms

Total Rise—The vertical distance from floor surface to floor surface.

Total Run—The straight length from first rise to final rise.

Tread Rise—Vertical height of one rise.

Tread Run—Distance from one rise to next rise.

Conditions

A stair layout usually must comply with certain fixed conditions and/or specific dimensions, such as height, size of opening, available space, and direction of run. In addition, there are certain general conditions that the layout must meet, such as the tread-rise dimension and the tread-run dimension.

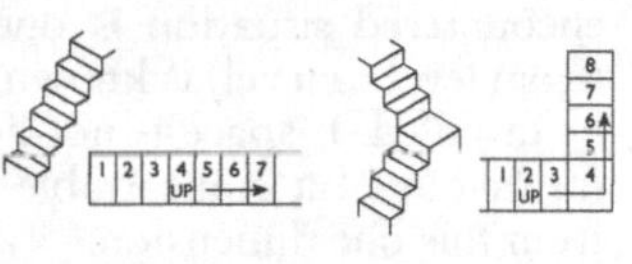

Fig. 47-22

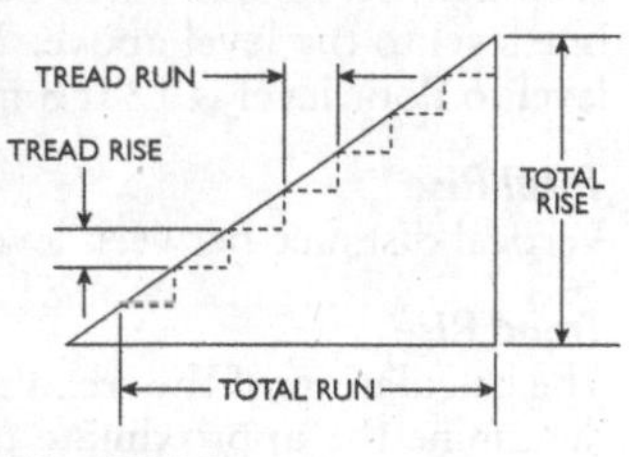

Fig. 47-23

While there is no standard tread-rise dimension as such, experience has shown that it should be in the area of 7 in. Too large a rise results in a steep stairway, sometimes referred to as a "hard" stairway, because it is difficult both to ascend and to descend. Also, the tread rise must be proportional to the tread run; that is, the tread run narrows as the tread rise increases and vice versa. There are numerous methods or rules to determine the dimensions of the tread rise and run. A very simple one, easy to remember and quite satisfactory for general use, is that the sum of the tread rise and run dimensions should total 17 in. The tread-rise dimension should be held as close as possible to the optimum 7 in. Another rule in common use is that the run dimension plus twice the rise dimension should equal 24.

Dimensions

Specific layout dimensions for stairs are determined from the given or known dimensions and/or conditions, such as the size of the opening, the height from floor to floor, the available space, the direction of the run, and so on. The manner or method of layout, therefore, is dictated by the given or known information. For instance, the most frequently encountered situation is one where the total rise distance (from level to level) is known, and a set of straight stairs is to be installed. If space is not limited, all necessary dimensions for layout of a comfortable set of stairs may be calculated from this one dimension.

Example

A straight set of stairs is to be constructed and installed from one level to the level above. The vertical distance from floor level to floor level is 12 ft 8 in.

Total Rise

Vertical distance between levels 12 ft 8 in., or 104 in.

Tread Rise

The calculation of the tread rise is done in three steps. First, determine the approximate number of rises by dividing the total rise in inches by 7. Second, select a whole number that is close to the number calculated. This will be the number of rises or steps in the stairs. Third, calculate the tread rise dimension by dividing the total rise measurement by the number of rises or steps selected.

Using the previous example, 7 will divide into 104 almost 15 times; therefore the selection is between 14 and 15. If 15 rises are selected, the tread rise will be $6\frac{15}{16}$ in. If 14 rises are selected, the tread rise will be $7\frac{7}{16}$ in.

Tread Run

The tread run is calculated by subtracting the tread-rise dimension from 17. As this is an approximate figure, the actual tread-run dimension is rounded off to the closest convenient fraction; for instance, for a $6\frac{15}{16}$-in. tread rise a 10-in. run may be used, while for a $7\frac{7}{16}$-in. rise a $9\frac{1}{2}$-in. run would be proper.

Total Run

The total run is calculated by multiplying the tread-run dimension by one less than the number of rises. Using the figures from the previous example, the total run for 14 rises would be 123½ in. (13 × 9½), and the total run for 15 rises would be 140 in. (14 × 10).

The two examples stated here are illustrated in Fig. 47-24.

Stairs are usually supported by either *stair horses* or *stair stringers*. Stair horses are an underneath style of support member, with the stair tread resting on the steps of the horses. Stair stringers are side support members; the treads are located between the stringers and are supported by end attachment.

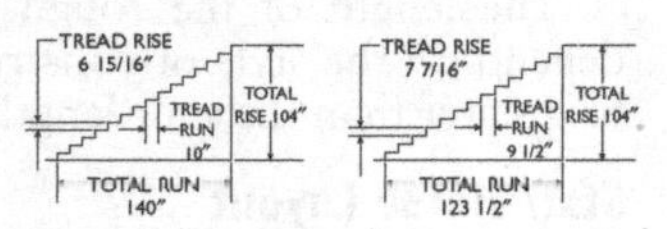

Fig. 47-24

When constructing simple stairs, the stair horse style of construction is usually used with wood materials and the stair stringer style when structural steel materials are used (Fig. 47-25).

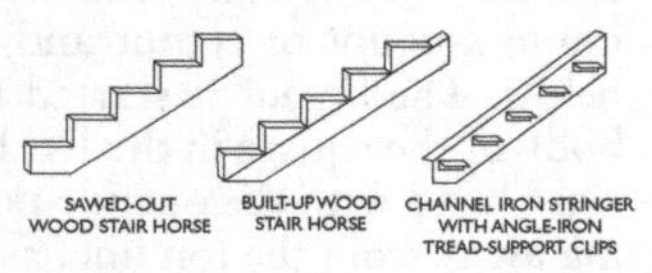

Fig. 47-25

Stair Length

The length of stair horses or stringers may be greater or less than the stair length, depending on the construction that is used. The actual length of the horse or stringer is determined during the layout. A few construction styles are shown in Fig. 47-26.

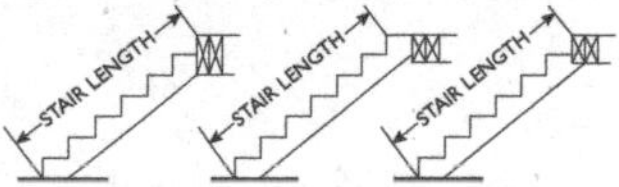

STAIR LENGTH STAIR LENGTH STAIR LENGTH

Fig. 47-26

The usual practice for determining stair length is to lay out to scale the total rise and run on a steel square, as shown in Fig. 47-27. Measuring between the two points on

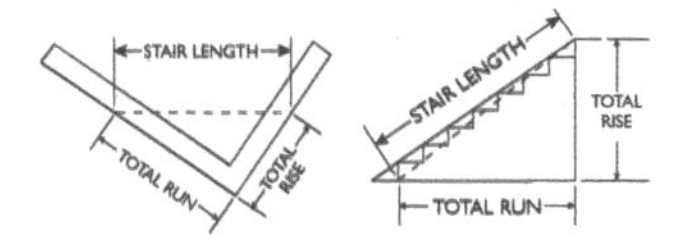

Fig. 47-27

the legs of the square will give the stair length. Even though the stair length is not precisely the hypotenuse of a right triangle whose sides are the total rise and the total run, in practice it is considered to be. The error is slight. Stair length may also be calculated using either trigonometry or the sum-of-the-squares equation.

The length of the rough stock needed for layout will depend on the style of construction. In many cases, it must be greater than the stair length.

Stair Horse Layout

The stair horse should be laid out with the top edge of the stock as the layout *top line*. The edge of the steps is laid out touching this top line or layout line. This will result in a minimum amount of cutout and will leave supporting material below. The layout is started from the left end, holding the body of the square in the left hand. The tongue is held in the right hand with the outside point of the square down or facing away from the top line, as shown in Fig. 47-28.

As shown in Fig. 47-27, the tread-rise measurement point on the tongue of the square and the tread-run measurement point on the body of the square are positioned on the top line. The mark for the tread-run cut is made against the body, and the tread-rise cut mark is made against the tongue of the square.

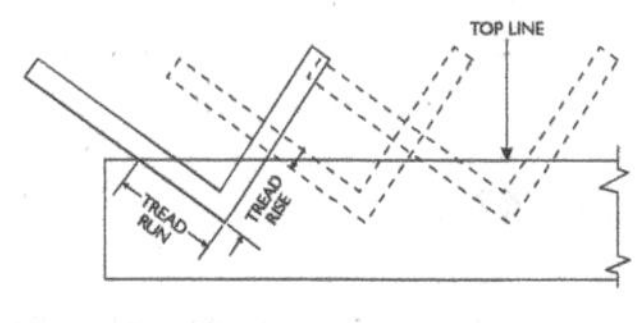

Fig. 47-28

The square is then moved to the next position, again with the run and rise measurement points on the body and tongue matching the top line. The intersecting points of adjacent layouts become the points or edges of the steps. The square is moved along in this manner and layouts made until the required number of steps is laid out.

Stair Stringer Layout

Because stair stringers usually act as enclosures on the sides, it is desirable to have stringer material extend above the treads. To provide this material above the treads, a stair stringer is usually laid out from the bottom or *lower line*. This layout is also made starting from the left, but the square is reversed, as shown in Fig. 47-29.

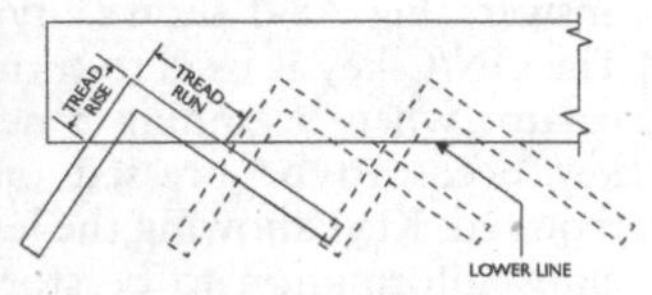

Fig. 47-29

Stair Drop

Layout procedure up to this point has made no provision for the thickness of the stair tread. To compensate for the tread thickness, the stair horse or stringer must be "dropped." If this is not done, the first step will be the thickness of the tread greater than the proper tread rise, and the top step will be the same amount less than the proper rise. Removing the correct amount from the bottom of the horse or stringer automatically corrects the height of both the bottom and top rises (Fig. 47-30).

If finish flooring is to be laid after stairs are installed, the amount that should be removed to allow for "drop" will be the difference between the tread thickness and the finish flooring.

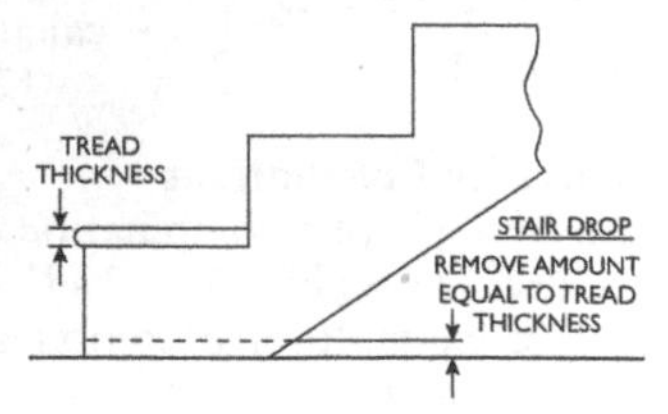

Fig. 47-30

48. SHOP MATHEMATICS

Using a Calculator

Much time can be saved and many mistakes avoided with the use of an electronic pocket calculator for quick and easy answers. Fig. 48-1 shows a typical keyboard for a calculator. The ON/C key is used to turn the unit on or to clear the calculator when beginning a new calculation. The Equals (=) key needs to be pressed only when the final answer is required. Keys showing the letter (M) are memory keys; they allow information to be stored in memory. The MRC key recalls the value stored in memory.

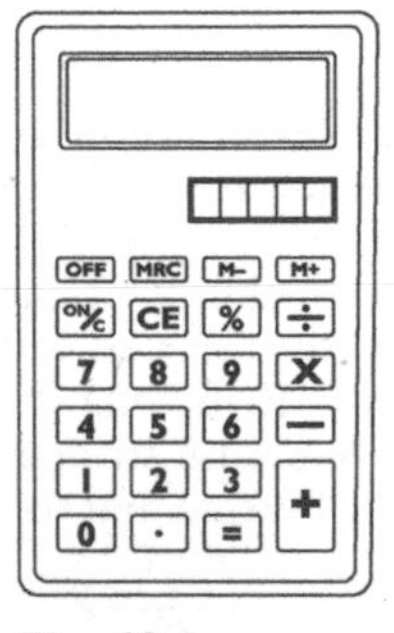

Fig. 48-1

Simple Calculations

An example of addition using the calculator:

$$3 + 6 + 9 + 12 = 30$$

An example of subtraction using the calculator:

$$23 - 2 - 4 - 3.5 = 13.5$$

An example of multiplication using the calculator:

$$14 \times 6.5 \times 2 \times 3 = 546$$

An example of division using the calculator:

$$125 \div 5 \div 2 = 12.5$$

Complex Calculations

An example of a combination of addition and multiplication:

$$(12 + 6) \times (15 + 2) = 289$$

An example of a combination of addition, subtraction, multiplication, and division:

$$\frac{(15 - 6) \times (17 + 4)}{4.5 \times 2} = 21$$

Fractions

A fraction is any part of a whole number. These are examples of fractions:

$1/4$, $3/4$, $5/8$, $2\,3/16$

The top of a fraction is called the numerator, and the bottom is the denominator.

$\frac{3}{4}$ ← NUMERATOR → $\frac{12}{4}$
← DENOMINATOR →

Fig. 48-2

Fractions are one of three types.

Proper. Denominator is larger that numerator.
$3/4$, $1/2$, $5/8$, $15/16$

Improper. Numerator is larger than denominator.
$10/3$, $12/2$, $16/5$, $9/3$

Mixed. Uses whole number and fraction.
$1\,1/2$, $2\,3/4$, $15\,31/32$

Fractional Arithmetic

Addition and Subtraction

To *add* or *subtract* fractions you must have a common denominator. Converting all fractions to one common denominator allows fractional arithmetic to work properly.

Example 1

$1/2 + 2/3 + 5/6 = ?$

$1/2 = 3/6$, $2/3 = 4/6$, $5/6 = 5/6$

$3/6 + 4/6 + 5/6 = 12/6 = 2$

Example 2

$15/16 - 1/8$

$15/16 = 15/16$, $1/8 = 2/16$

$15/16 - 2/16 = 13/16$

Multiplication and Division

To *multiply* or *divide* fractions you must multiply the numerators together and then multiply the denominators together and reduce to the lowest possible fraction.

Example

$\frac{1}{2} \times \frac{1}{4} = \frac{1}{8}$

$\frac{3}{4} \times \frac{5}{8} = \frac{15}{32}$

$1\frac{1}{4} \times \frac{5}{8} = \frac{5}{4} \times \frac{4}{1} = 2$

$\frac{1}{2} \div \frac{1}{4} = \frac{3}{4} \times 8\frac{1}{3} = \frac{24}{12} = 2$

$\frac{3}{4} \div \frac{3}{8} = \frac{3}{4} \times 8\frac{1}{3} = \frac{24}{12} = 2$

Decimals

A decimal is a number expressed to the base 10. There are two types of decimal numbers:

> **Proper.** A decimal that has no whole numbers (.250, .500, .333, .125).
>
> **Mixed.** A decimal that has a whole number and a decimal number separated by a decimal point (1.625, 1.875, 1.913, 2.375).

Converting Fractions to Decimals

Divide the denominator into the numerator, and the result is a decimal number:

$\frac{1}{3} = .333$

$\frac{1}{2} = .500$

$\frac{5}{8} = .625$

$1\frac{1}{4} = 1.250$

$\frac{7}{8} = .875$

Calculations Using Decimals

Using decimals for calculation is easier than using fractions. The basic rules of addition, subtraction, multiplication, and division are used:

$\frac{1}{2} + \frac{1}{4} + \frac{5}{8} = .500 + .250 + .625 = 1.375$

$\frac{1}{2} \times \frac{3}{4} \times \frac{5}{8} = .500 \times .750 \times .625 = .234375$

$\frac{7}{8} \div \frac{3}{4} = .875 \div .750 = 1.166$

$\frac{1}{2} + 1\frac{1}{4} + 3\frac{1}{2} + 5\frac{3}{16} = .500 + 1.250 + 3.500 + 5.1875 = 10.4375$

Ratio

Ratio is the relation of one quantity to another of the same kind, in respect to magnitude.

This relation is the quotient obtained by dividing the first quantity by the second. The first quantity is called the *antecedent*, the second is the *consequent*, and the two form the terms of the ratio. For example, if a gearbox uses 10-tooth input gear and 50-tooth output gear:

Ratio = $^{50}/_{10}$ = 5 to 1 ratio

Calculating Percentage

A *percentage* expresses a part of a whole number in terms of hundredths. A *percent* is one part of 100 parts. All fractions and decimals have equivalent percentages. Conversion to percentage involves dividing the fraction to obtain a decimal and multiplying by 100 to get the percentage.

Examples

$^1/_3$ = .33 × 100 = 33%
$^7/_8$ = .875 × 100 = 87.5%
$1^1/_4$ = $^5/_4$ = 1.250 = 125%

49. SHOP GEOMETRY

Lines

When lines touch at one point they are said to be *tangent* to one another. Two lines that cross are said to *intersect*. Two lines that are always the same distance apart are said to be *parallel*.

Angles

There are four common types of angles: *straight, right, acute,* and *obtuse*.

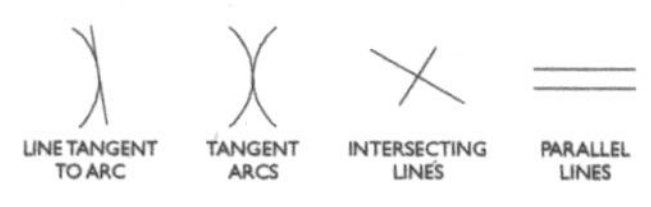

Fig. 49-1

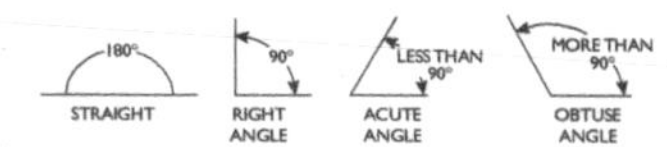

Fig. 49-2

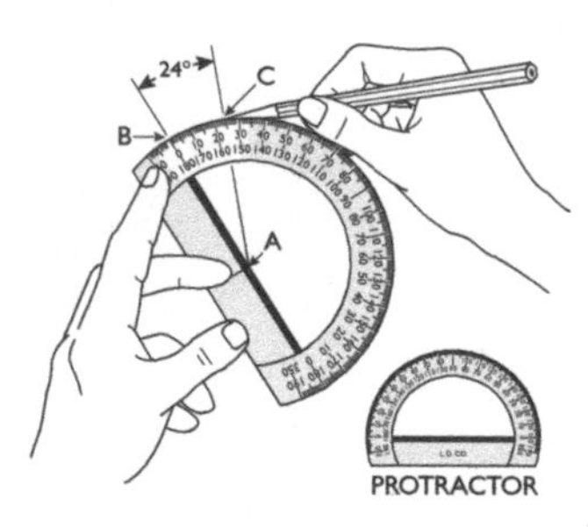

Fig. 49-3 Making an angle with a protractor.

Using a Protractor

A protractor can be used to measure or make angles. Suppose you want to draw a line making an angle of 24° with a given line. Place the vertex indicator at the spot on the line where the vertex of the required angle is to be located (Fig. 49-3a). Then align the 0° mark with the line (Fig. 49-3b). Follow the circular scale up the angular rim of the protractor and make a mark at 24°. Mark this point very lightly; then draw a line through the point and the vertex. The angle formed equals 24°. To measure an angle with the protractor (Fig. 49-4), place the vertex indicator on the vertex of the angle, align one of the lines of the angle with the zero mark on the rim, and read the angle where the other line appears to cross the rim.

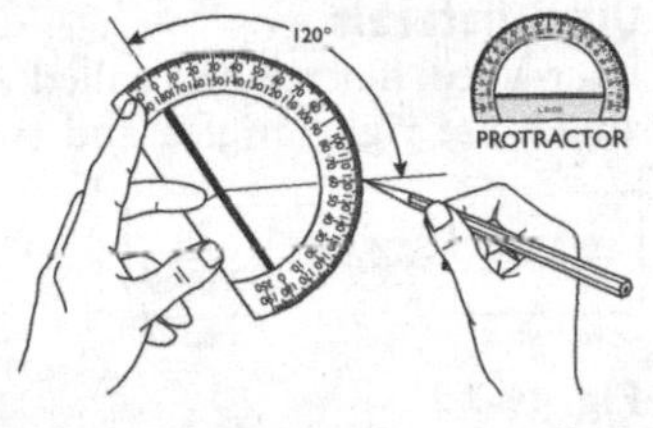

Fig. 49-4 Measuring an angle with a protractor.

Circles

A *circle* is a closed curve on which all points are the same distance from a point (center) inside the circle (Fig. 49-5). A line extending from the center to the closed curve line is the *radius*. The straight line that passes through the center of the circle and ends at the opposite side of the closed curved line is the *diameter.* An *arc* is a portion of the curved line. A *chord* refers to the straight-line segment between the end points of an arc.

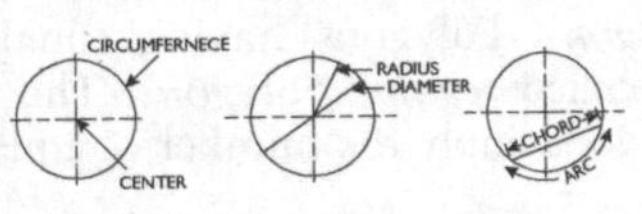

Fig. 49-5

The distance around the circle is called the *circumference.* The circumference is measured in standard linear units or in angular measure. The total angular distance around the circumference of a circle is 360°.

Triangles

A *triangle* consists of three straight lines joined at the end points to form a closed flat shape. The lines are called *sides,* and the angles formed are called *inside angles*. Triangles are described by their included angles (Fig. 49-6). The most commonly used is the *right triangle* in which one angle is a right (90°) angle. An *acute triangle* has three acute angles, and an *obtuse triangle* has one angle greater than 90°.

The sum of the three inside angles of any triangle is always 180°.

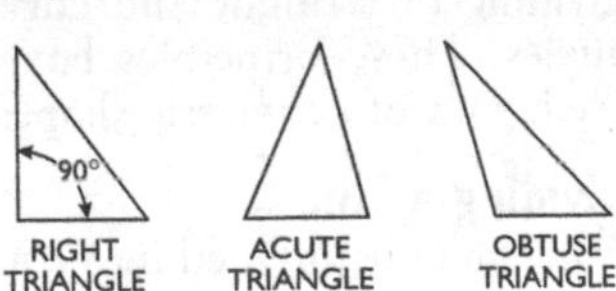

Fig. 49-6

Quadrilaterals

Four-sided figures are called *quadrilaterals*. When all four angles are right angles and two pairs of sides are of equal length, the shape is a *rectangle*. When all four angles are right angles and the four sides are equal, it is called a *square* (Fig. 49-7).

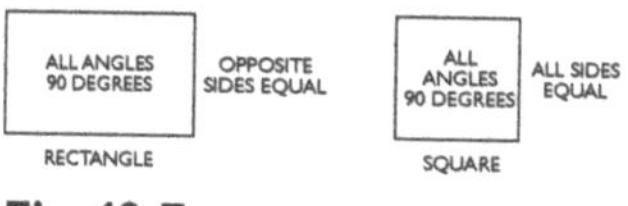

Fig. 49-7

Regular Polygons

All plane figures with three or more sides are called *polygons*. Polygons having equal sides and equal angles are called *regular polygons*. The names given to regular polygons imply the number of equal sides.

Table 49-1 Regular Polygon Identification Table

No. of Sides	*Name*
3	Triangle
4	Square
5	Pentagon
6	Hexagon
7	Heptagon
8	Octagon
9	Nonagon
10	Decagon

Geometric Construction

Practical applications of geometric principles are in the construction of parallel, perpendicular, and tangent lines, the dividing of straight and curved lines, and the bisecting of angles. These principles have further broad application in the layout of geometric shapes.

Dividing a Line

A line may be divided into equal parts by first drawing a line at any angle and length to the given line from one end point; second, stepping off on the angular line the same number of equal spaces as the line is to be divided into; third, connecting the last point with the other end point on the given line; and fourth, drawing lines parallel to this connecting line. Dividing a line into three equal parts is illustrated in Fig. 49-8.

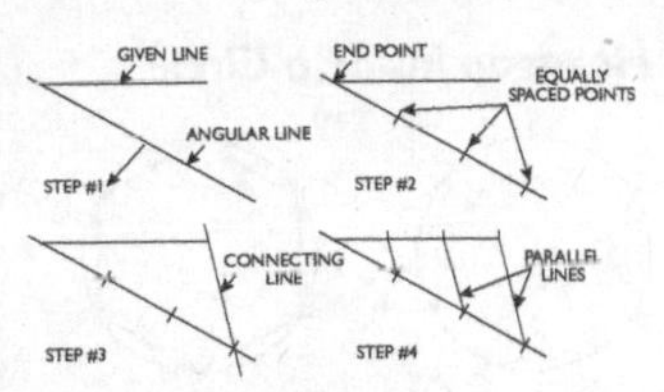

Fig. 49-8 Dividing a line into three equal parts.

Erecting a Perpendicular Line

To erect a perpendicular or right-angle line at a given point on a line, use the following three steps (Fig. 49-9). First, swing equal arcs from the given point to intersect the given line at points (a) and (b). Second, increase the radius by about one-half and swing arcs from points (a) and (b) to locate points (c) and (d). Third, draw a line through point (c) and (d) and the given point.

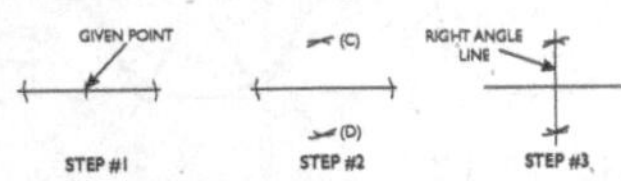

Fig. 49-9 Erecting a perpendicular line.

Polygon Construction

Square Inside a Circle

1. Draw a diameter line across the circle.
2. Construct a perpendicular through the center.
3. Connect the points of the diameter lines.

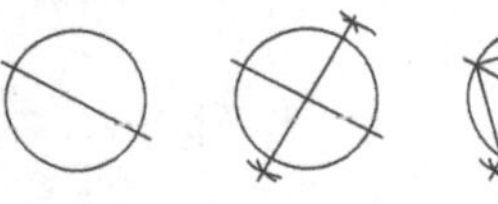

Fig. 49-10

Square Outside a Circle

1. Draw a diameter line across the circle.
2. Construct a perpendicular.
3. Construct tangent lines at the points of the diameter lines.

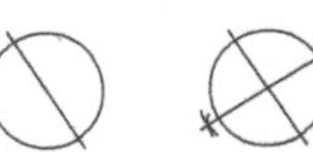
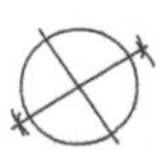
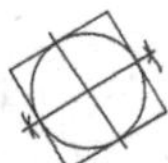

Fig. 49-11

Hexagon Inside a Circle

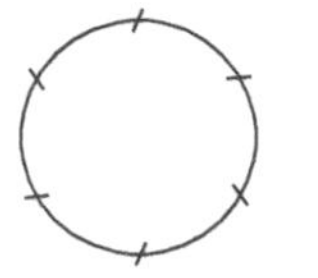
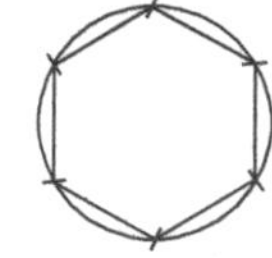

Fig. 49-12

1. Using the radius of the circle, step off arcs around the circle.
2. Connect the points where the arcs intersect the circle.

Hexagon Outside a Circle

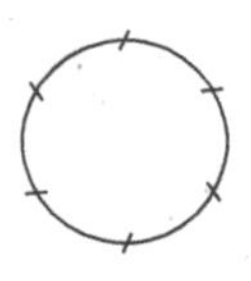
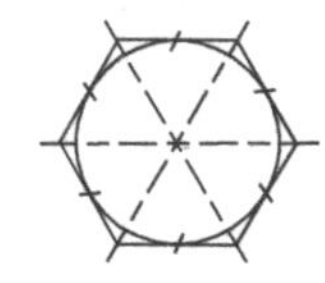

Fig. 49-13

1. Using the radius of the circle, step off arcs around the circle.
2. Construct a tangent line at each intersecting point.

Steps similar to these may be followed for constructing other polygons inside and outside of circles.

Dimensions of Polygons and Circles

Triangle

$E = \text{side} \times .57735$

$D = \text{side} \times 1.1547 = 2E$

$\text{Side} = D \times .866$

$C = E \times .5 = D \times .25$

Square

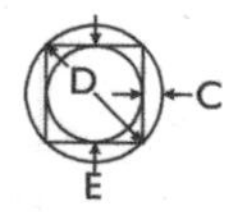

$E = \text{side} = D \times .7071$

$D = \text{side} \times 1.4142 = \text{Diagonal}$

$\text{Side} = D \times .7071$

$C = D \times .14645$

Pentagon

$E = \text{side} \times 1.3764 = D \times .809$

$D = \text{side} \times 1.7013 = E \times 1.2361$

$\text{Side} = D \times .5878$

$C = D \times .0955$

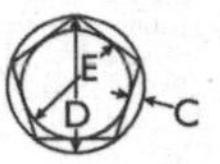

Hexagon

$E = \text{side} \times 1.71321 = D \times .866$

$D = \text{side} \times 2 = E \times 1.1547$

$\text{Side} = D \times .5$

$C = D \times .067$

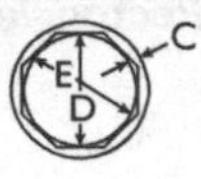

Octagon

$E = \text{side} \times 2.4142 = D \times .9239$

$D = \text{side} \times 2.6131 = E \times 1.0824$

$\text{Side} = D \times .3827$

$C = D \times .038$

Areas of Geometric Shapes

Triangle

$$\text{area (A)} = \frac{bh}{2}$$

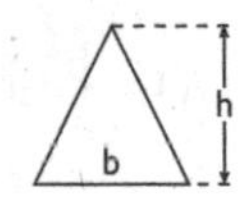

Square

$$\text{area (A)} = b^2$$

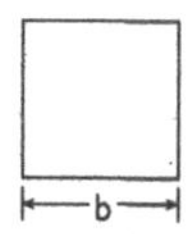

Rectangle

$$\text{area (A)} = ab$$

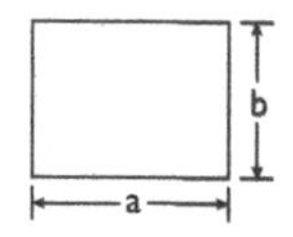

Parallelogram

$$\text{area (A)} = ah$$

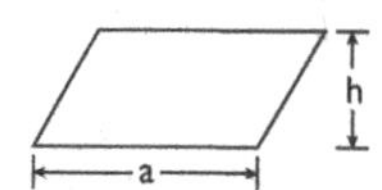

Trapezoid

$$\text{area (A)} = \frac{h}{2}(a + b)$$

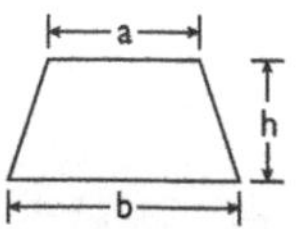

Trapezium

$$\text{area (A)} = \tfrac{1}{2}\begin{bmatrix} b(H+h) \\ +ah+cH \end{bmatrix}$$

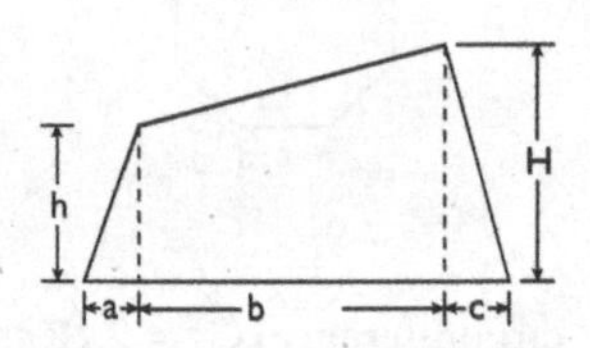

Regular Pentagon

$$\text{area (A)} = 1.720\ a^2$$

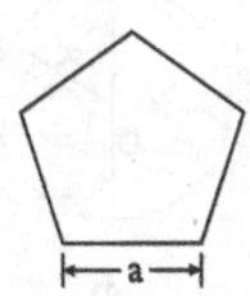

Regular Hexagon

$$\text{area (A)} = 2.598\ a^2$$

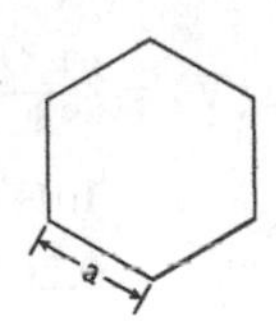

Regular Octagon

$$\text{area (A)} = 4.828\ a^2$$

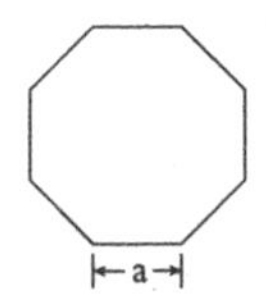

Circle

$$\text{circumference (C)} = 2\pi R$$
$$= \pi D$$
$$\text{area (A)} = \pi R^2$$

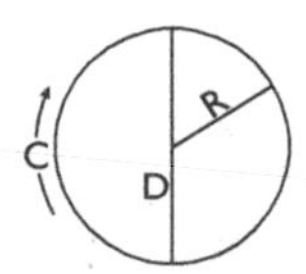

Volumes of Solid Geometry Shapes

Cone

$$\text{area (A)} = \pi RS$$
$$= R\sqrt{R^2 + h^2}$$
$$\text{volume (V)} = \frac{\pi R^2 h}{3}$$
$$= 1.047R^2 h$$
$$= 0.2618D^2 h$$

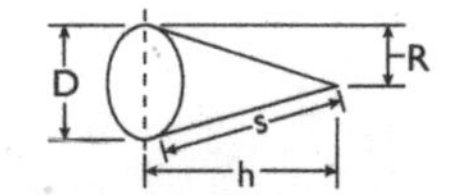

Cylinder

$$\text{cylindrical surface} = \pi Dh$$

$$\text{total surface} = 2\pi R(R + h)$$

$$\text{volume }(V) = \pi R^2 h$$

$$= \frac{c^2 h}{4\pi}$$

Sphere

$$\text{area }(A) = 4\pi R^2$$

$$= \pi D^2$$

$$\text{volume }(V) = \frac{4}{3}\pi R^3$$

$$= 1/6\pi D^3$$

Cube

$$\text{area }(A) = 6b^2$$

$$\text{volume }(V) = b^3$$

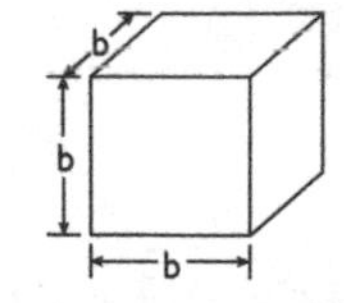

Rectangular Solid

$$\text{area (A)} = 2(ab + bc + ac)$$

$$\text{volume (V)} = abc$$

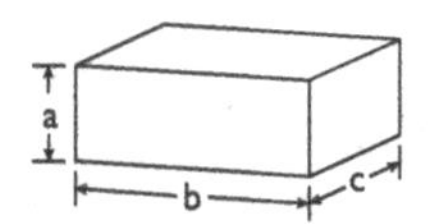

Ring of Rectangular Cross Section

$$\text{volume (V)} = \frac{\pi c}{4}\left(D^2 - d^2\right)$$

$$= \left(\frac{D + d}{2}\right)\pi bc$$

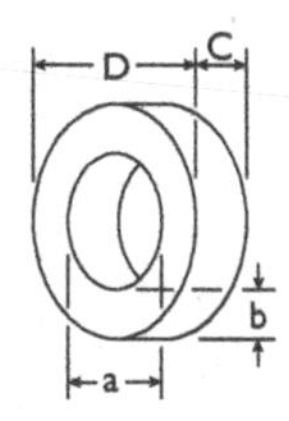

50. SHOP TRIGONOMETRY

Right-Angle Triangles

All triangles are made up of six parts—three sides and three angles. A right-angle triangle is one having one angle of 90°. A 90° angle is termed a *right angle.* The three sides of a right-angle triangle are called the *side opposite, side adjacent,* and *hypotenuse.* The hypotenuse is always the side directly across from, or opposite, the 90° angle. The other two sides are opposite one angle and adjacent to the other. Therefore, they may be called either opposite or adjacent sides in reference to the two angles (Fig. 50-1).

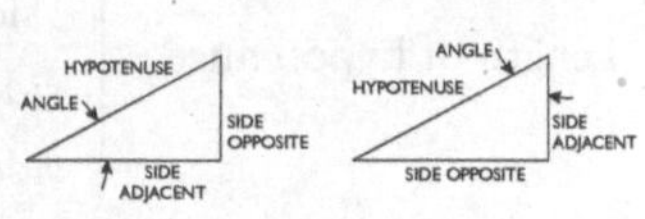

Fig. 50-1

The Trigonometric Functions

Trigonometric calculations employ numerical values called *trigonometric functions.* These values represent the ratios between the sides of triangles and are identified by the names *sine, cosine, tangent, cotangent, secant,* and *cosecant.* Each angle has a specific numerical value for each of its functions. These values are given in tables of trigonometric functions.

When one of the angles of a right-angle triangle (other than the 90° angle) and the length of one of the sides are known, the length of the other sides may be determined by use of the appropriate formula:

$$\text{Length of side opposite} = \begin{cases} \text{hypotenuse} \times \text{sine} \\ \text{hypotenuse} \div \text{cosecant} \\ \text{side adjacent} \times \text{tangent} \\ \text{side adjacent} \div \text{cotangent} \end{cases}$$

$$\text{Length of side adjacent} = \begin{cases} \text{hypotenuse} \times \text{cosine} \\ \text{hypotenuse} \div \text{secant} \\ \text{side opposite} \times \text{cotangent} \\ \text{side opposite} \div \text{tangent} \end{cases}$$

$$\text{Length of hypotenuse} = \begin{cases} \text{side opposite} \times \text{cosecant} \\ \text{side opposite} \div \text{sine} \\ \text{side adjacent} \times \text{secant} \\ \text{side adjacent} \div \text{cosine} \end{cases}$$

When the lengths of two sides of a right-angle triangle are known, the angles may be determined in two steps using the trigonometric functions of the angles.

1. Using the appropriate formula, calculate the numerical value of the function of the angle.
2. Using a table of trigonometric functions, find the angle that corresponds to the function calculated by formula.

$$\text{Sine} = \frac{\text{Side Opposite}}{\text{Hypotenuse}} \qquad \text{Cotangent} = \frac{\text{Side Adjacent}}{\text{Side Opposite}}$$

$$\text{Cosine} = \frac{\text{Side Adjacent}}{\text{Hypotenuse}} \qquad \text{Secant} = \frac{\text{Hypotenuse}}{\text{Side Adjacent}}$$

$$\text{Tangent} = \frac{\text{Side Opposite}}{\text{Side Adjacent}} \qquad \text{Cosecant} = \frac{\text{Hypotenuse}}{\text{Side Opposite}}$$

Example

Use this formula:

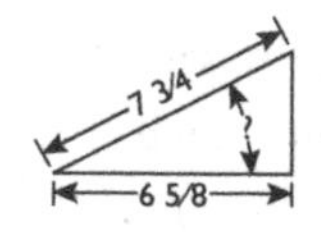

$$\text{Cosine} = \frac{\text{Side Adjacent}}{\text{Hypotenuse}}$$

$$\text{Cosine} = \frac{6\frac{5}{8}}{7\frac{3}{8}} \text{ or } \frac{6.625}{7.750} = .85483$$

From "the trigonometric functions table: *Angle = 31° 15½"*.

Chief SOH-CAH-TOA (A Memory Tool)

Because the sine, cosine, and tangent are used quite frequently, a simple memory tool can be employed to help remember them. The "Indian Chief" SOH-CAH-TOA gives the formulas of Sine = Opposite divided by Hypotenuse; Cosine = Adjacent divided by Hypotenuse; and Tangent = Opposite divided by Adjacent.

Pythagorean Rule—The Sum of the Squares

The Pythagorean rule can be used for all right-angle triangles. When the lengths of two sides of a right-angle triangle are known, the length of the third side may be determined by the use of the sum-of-the-squares formula. It states that the square of the hypotenuse is equal to the sum of the squares of the other two sides. The basic formula is commonly written this way:

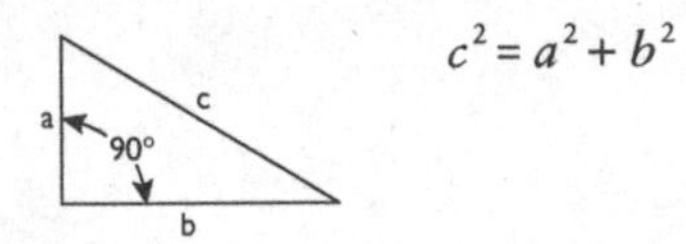

$$c^2 = a^2 + b^2$$

To find the length of the third side of a right-angle triangle when the lengths of two sides are known, the known values are substituted in the appropriate equation.

$c = \sqrt{a^2 + b^2}$ $\quad$ $a = \sqrt{c^2 - b^2}$ $\quad$ $b = \sqrt{c^2 - a^2}$

Example

Find the length of *c* when *a* is 9 and *b* is 11. Use this formula:

$$c = \sqrt{a^2 + b^2}$$

$$c = \sqrt{(9 \times 9) + (11 \times 11)} \text{ or } \sqrt{81 + 121} \text{ or } \sqrt{202}$$

$$c = \sqrt{202} \text{ or approximately } 14\tfrac{7}{12}$$

Any triangle having sides with a 3-4-5 length ratio is a right-angle triangle.

$$c^2 = a^2 + b^2$$

$$5^2 = 3^2 + 4^2$$

$$25 = 9 + 16$$

$$25 = 25$$

APPENDIX

Natural Trigonometric Functions

Degree	Sine	Cosine	Tangent	Secant	Degree	Sine	Cosine	Tangent	Secant
0	.00000	1.0000	.00000	1.0999	46	.7193	.6947	1.0355	1.4395
1	.01745	.9998	.01745	1.0001	47	.7314	.6820	1.0724	1.4663
2	.03490	.9994	.03492	1.0006	48	.7431	.6691	1.1106	1.4945
3	.05234	.9986	.05241	1.0014	49	.7547	.6561	1.1504	1.5242
4	.06976	.9976	.06993	1.0024	50	.7660	.6428	1.1918	1.5557
5	.08716	.9962	.08749	1.0038	51	.7771	.6293	1.2349	1.5890
6	.10453	.9945	.10510	1.0055	52	.7880	.6157	1.2799	1.6243
7	.12187	.9925	.12278	1.0075	53	.7986	.6018	1.3270	1.6616
8	.1392	.9903	.1405	1.0098	54	.8090	.5878	1.3764	1.7013
9	.1564	.9877	.1584	1.0125	55	.8192	.5736	1.4281	1.7434
10	.1736	.9848	.1763	1.0154	56	.8290	.5592	1.4826	1.7883
11	.1908	.9816	.1944	1.0187	57	.8387	.5446	1.5399	1.8361
12	.2079	.9781	.2126	1.0223	58	.8480	.5299	1.6003	1.8871
13	.2250	.9744	.2309	1.0263	59	.8572	.5150	1.6643	1.9416
14	.2419	.9703	.2493	1.0306	60	.8660	.5000	1.7321	2.0000
15	.2588	.9659	.2679	1.0353	61	.8746	.4848	1.8040	2.0627
16	.2756	.9613	.2867	1.0403	62	.8829	.4695	1.8807	2.1300
17	.2924	.9563	.3057	1.0457	63	.8910	.4540	1.9626	2.2027
18	.3090	.9511	.3249	1.0515	64	.8988	.4384	2.0503	2.2812
19	.3256	.9455	.3443	1.0576	65	.9063	.4226	2.1445	2.3662
20	.3420	.9397	.3640	1.0642	66	.9135	.4067	2.2460	2.4586
21	.3584	.9336	.3839	1.0711	67	.9205	.3907	2.3559	2.5598
22	.3746	.9272	.4040	1.0785	68	.9272	.3746	2.4751	2.6695
23	.3907	.9205	.4245	1.0864	69	.9336	.3584	2.6051	2.7904
24	.4067	.9135	.4452	1.0946	70	.9397	.3420	2.7475	2.9238
25	.4226	.9063	.4663	1.1034	71	.9455	.3256	2.6042	3.0715

(continued)

Natural Trigonometric Functions (continued)

Degree	Sine	Cosine	Tangent	Secant	Degree	Sine	Cosine	Tangent	Secant
26	.4384	.8988	.4877	1.1126	72	.9511	.3090	3.0777	3.2361
27	.4540	.8910	.5095	1.1223	73	.9563	.2924	3.2709	3.4203
28	.4695	.8829	.5317	1.1326	74	.9613	.2756	3.4874	3.6279
29	.4848	.8746	.5543	1.1433	75	.9659	.2588	3.7321	3.8637
30	.5000	.8660	.5774	1.1547	76	.9703	.2419	4.0108	4.1336
31	.5150	.8572	.6009	1.1663	77	.9744	.2250	4.3315	4.4454
32	.5299	.8480	.6249	1.1792	78	.9781	.2079	4.7046	4.8097
33	.5446	.8387	.6494	1.1924	79	.9816	.1908	5.1446	5.2408
34	.5592	.8290	.6745	1.2062	80	.9848	.1736	5.6713	5.7588
35	.5736	.8192	.7002	1.2208	81	.9877	.1564	6.6138	6.3924
36	.5878	.8090	.7265	1.2361	82	.9903	.1392	7.1154	7.1853
37	.6018	.7986	.7536	1.2521	83	.9925	.12187	8.1443	8.2055
38	.6157	.7880	.7813	1.2690	84	.9945	.10453	9.5144	9.5668
39	.6293	.7771	.8098	1.2867	85	.9962	.08716	11.4301	11.474
40	.6428	.7660	.8391	1.3054	86	.9976	.06976	14.3007	14.335
41	.6561	.7547	.8693	1.3250	87	.9986	.05234	19.0811	19.107
42	.6691	.7431	.9004	1.3456	88	.9994	.03490	28.6363	28.654
43	.6820	.7314	.9325	1.3673	89	.9998	.01745	57.2900	27.299
44	.6947	.7193	.9657	1.3902	90	1.0000	Inf.	Inf.	Inf.
45	.7071	.7071	1.0000	1.4142					

Metric and English Equivalent Measures

Measures of Length

Metric		English
1 meter	=	39.37 inches, or 3.28083 feet, or 1.09361 yards
0.3048 meter	=	1 foot
1 centimeter	=	0.3937 inch
2.54 centimeters	=	1 inch
1 millimeter	=	0.03937 inch
25.4 millimeters	=	1 inch
1 kilometer	=	1093.61 yards, or 0.62137 mile

Measures of Weight

Metric		English
1 gram	=	15.432 grains
0.0648 gram	=	1 grain
28.35 grams	=	1 ounce avoirdupois
1 kilogram	=	2.2046 pounds
0.4536 kilogram	=	1 pound
1 metric ton 1000 kilograms	=	0.9842 ton of 2240 pounds 9.68 cwt. 2204.6 pounds
1.016 metric tons 1016 kilograms	=	1 ton of 2240 pounds

Measures of Capacity

Metric		English
1 liter (= 1 cubic decimeter)	=	61.023 cubic inches 0.03531 cubic foot 0.2642 gallons (American) 2.202 pounds of water at 62°F
28.317 liters	=	1 cubic foot
3.785 liters	=	1 gallon (American)
4.543 liters	=	1 gallon (Imperial)

English Conversion Table

Length			
Inches	×	0.0833	= feet
Inches	×	0.02778	= yards
Inches	×	0.00001578	= miles
Feet	×	0.3333	= yards
Feet	×	0.0001894	= miles
Yards	×	36.00	= inches
Yards	×	3.00	= feet
Yards	×	0.0005681	= miles
Miles	×	63360.00	= inches
Miles	×	5280.00	= feet
Miles	×	1760.00	= yards
Circumference of circle	×	0.3188	= diameter
Diameter of circle	×	3.1416	= circumference
Area			
Square inches	×	0.00694	= square feet
Square inches	×	0.0007716	= square yards
Square feet	×	144.00	= square inches
Square feet	×	0.11111	= square yards
Square yards	×	1296.00	= square inches
Square yards	×	9.00	= square feet
Dia. of circle squared	×	0.7854	= area
Dia. of sphere squared	×	3.1416	= surface
Volume			
Cubic inches	×	0.0005787	= cubic feet
Cubic inches	×	0.00002143	= cubic yards
Cubic inches	×	0.004329	= U.S. gallons
Cubic feet	×	1728.00	= cubic inches
Cubic feet	×	0.03704	= cubic yards
Cubic feet	×	7.4805	= U.S. gallons
Cubic yards	×	46656.00	= cubic inches
Cubic yards	×	27.00	= cubic feet
Diameter of sphere cubed	×	0.5236	= volume
Weight			
Grains (avoirdupois)	×	0.002286	= ounces
Ounces (avoirdupois)	×	0.0625	= pounds
Ounces (avoirdupois)	×	0.00003125	= tons
Pounds (avoirdupois)	×	16.00	= ounces

Weight			
Pounds (avoirdupois)	×	0.01	= hundred-weight
Pounds (avoirdupois)	×	0.0005	= tons
Tons (avoirdupois)	×	32000.00	= ounces
Tons (avoirdupois)	×	2000.00	= pounds
Energy			
Horsepower	×	33000	= foot-pounds per minute
British thermal units	×	778.26	= foot-pounds
Ton of refrigeration	×	200	= British thermal units per minute
Pressure			
Pounds per square inch	×	2.31	= feet of water (60°F)
Feet of water (60°F)	×	0.433	= pounds per square inch
Inches of water (60°F)	×	0.0361	= pounds per square inch
Pounds per square inch	×	27.70	= inches of water (60°F)
Inches of mercury (60°F)	×	0.490	= pounds per square inch
Power			
Horsepower	×	746	= watts
Watts	×	0.001341	= horsepower
Horsepower	×	42.4	= British thermal units per minute
Water Factors (at point of greatest density—39.2°F)			
Miners inch (of water)	×	8.976	= U.S. gallons per minute
Cubic inches (of water)	×	0.57798	= ounces
Cubic inches (of water)	×	0.036124	= pounds
Cubic inches (of water)	×	0.004329	= U.S. gallons
Cubic inches (of water)	×	0.003607	= English gallons
Cubic feet (of water)	×	62.425	= pounds
Cubic feet (of water)	×	0.03121	= tons
Cubic feet (of water)	×	7.4805	= U.S. gallons
Cubic inches (of water)	×	6.232	= English gallons

(continued)

English Conversion Table *(continued)*

Water Factors (at point of greatest density—39.2°F)			
Cubic foot of ice	×	57.2	= pounds
Ounces (of water)	×	1.73	= cubic inches
Pounds (of water)	×	26.68	= cubic inches
Pounds (of water)	×	0.01602	= cubic feet
Pounds (of water)	×	0.1198	= U.S. gallons
Pounds (of water)	×	0.0998	= English gallons
Tons (of water)	×	32.04	= cubic feet
Tons (of water)	×	239.6	= U.S. gallons
Tons (of water)	×	199.6	= English gallons
U.S. gallons	×	231.00	= cubic inches
U.S. gallons	×	0.13368	= cubic feet
U.S. gallons	×	8.345	= pounds
U.S. gallons	×	0.8327	= English gallons
U.S. gallons	×	3.785	= liters
English gallons (Imperial)	×	227.41	= cubic inches
English gallons (Imperial)	×	0.1605	= cubic feet
English gallons (Imperial)	×	10.02	= pounds
English gallons (Imperial)	×	1.201	= U.S. gallons
English gallons (Imperial)	×	4.546	= liters

Metric Conversion Table

Length			
Millimeters	×	0.03937	= inches
Millimeters	÷	25.4	= inches
Centimeters	×	0.3937	= inches
Centimeters	÷	2.54	= inches
Meters	×	39.37	= inches (Act. Cong.)
Meters	×	3.281	= feet
Meters	×	1.0936	= yards
Kilometers	×	0.6214	= miles
Kilometers	÷	1.6093	= miles
Kilometers	×	3280.8	= feet

Area			
Square Millimeters	×	0.00155	= square inches
Square Millimeters	÷	645.2	= square inches
Square Centimeters	×	0.155	= square inches
Square Centimeters	÷	6.452	= square inches
Square Meters	×	10.764	= square inches
Square Kilometers	×	247.1	= acres
Hectares	×	2.471	= acres

Volume			
Cubic Centimeters	÷	16.387	= cubic inches
Cubic Centimeters	÷	3.69	= fluid drams (U.S.P.)
Cubic Centimeters	÷	29.57	= fluid ounces (U.S.P.)

Volume			
Cubic Meters	×	35.314	= cubic feet
Cubic Meters	×	1.308	= cubic yards
Cubic Meters	×	264.2	= gallons (231 cubic inches)
Liters	×	61.023	= cubic inches (Act. Cong.)
Liters	×	33.82	= fluid ounces (U.S.J.)
Liters	×	0.2642	= gallons (231 cubic inches)
Liters	÷	3.785	= gallons (231 cubic inches)
Liters	÷	28.317	= cubic feet
Hectoliters	×	3.531	= cubic feet
Hectoliters	×	2.838	= bushels (2150.42 cubic inches)
Hectoliters	×	0.1308	= cubic yards
Hectoliters	×	26.42	= gallons (231 cubic inches)

Weight			
Grams	×	15.432	= grains (Act. Cong.)
Grams	÷	981	= dynes
Grams (water)	÷	29.57	= fluid ounces
Grams	÷	28.35	= ounces avoirdupois
Kilograms	×	2.2046	= pounds
Kilograms	×	35.27	= ounces avoirdupois

(continued)

Metric Conversion Table *(continued)*

Weight			
Kilograms	×	0.0011023	= tons (2000 pounds)
Tonneau (Metric ton)	×	1.1023	= tons (2000 pounds)
Tonneau (Metric ton)	×	2204.6	= pounds
Unit Weight			
Grams per cubic centimeter	÷	27.68	= pounds per cubic inch
Kilogram per meter	×	0.672	= pounds per foot
Kilogram per cubic meter	×	0.06243	= pounds per cubic foot
Kilogram per Cheval	×	2.235	= pounds per horsepower
Grams per liter	×	0.06243	= pounds per cubic foot
Pressure			
Kilograms per square Centimeter	×	14.223	= pounds per square inch
Kilograms per square Centimeter	×	32.843	= feet of water (60°F)
Atmospheres (International)	×	14.696	= pounds per square inch
Energy			
Joule	×	0.7376	= foot-pounds
Kilogram-meters	×	7.233	= foot-pounds
Power			
Cheval vapeur	×	0.9863	= horsepower
Kilowatts	×	1.341	= horsepower
Watts	÷	746.	= horsepower
Watts	×	0.7373	= foot-pounds per second

Standard Tables of Metric Measure — Linear Measure

Unit	*Value, m*	*Symbol or Abbreviation*
Micron	0.000001	μ
Millimeter	0.001	mm
Centimeter	0.01	cm
Decimeter	0.1	dm
Meter (unit)	1.0	m
Dekameter	10.0	dkm
Hectometer	100.0	hm
Kilometer	1,000.0	km
Myriameter	10,000.0	Mm
Megameter	1,000,000.0	

Volume

Unit	*Value, l*	*Symbol or Abbreviation*
Milliliter	0.001	ml
Centiliter	0.01	cl
Deciliter	0.1	dl
Liter (unit)	1.0	l
Dekaliter	10.0	dkl
Hectoliter	100.0	hl
Kiloliter	1,000.0	kl

Surface Measure

Unit	*Value, m^2*	*Symbol or Abbreviation*
Square millimeter	0.000001	mm^2
Square centimeter	0.0001	cm^2
Square decimeter	0.01	dm^2
Square meter (centiare)	1.0	m^2
Square decameter (are)	100.0	a^2
Hectare	10,000.0	ha^2
Square kilometer	1,000,000.0	km^2

(continued)

Standard Tables of Metric Measure — Linear Measure *(continued)*

Mass

Unit	*Value, g*	*Symbol or Abbreviation*
Microgram	0.000001	µg
Milligram	0.001	mg
Centigram	0.01	cg
Decigram	0.1	dg
Gram (unit)	1.0	g
Dekagram	10.0	dkg
Hectogram	100.0	hg
Kilogram	1,000.0	kg
Myriagram	10,000.0	Mg
Quintal	100,000.0	q
Ton	1,000,000.0	

Cubic Measure

Unit	*Value, m^3*	*Symbol or Abbreviation*
Cubic micron	10^{-10}	μ^3
Cubic millimeter	10^{-9}	mm^3
Cubic centimeter	10^{-6}	cm^3
Cubic decimeter	10^{-3}	dm^3
Cubic meter	1	m^3
Cubic decameter	10^3	dkm^3
Cubic hectometer	10^6	hm^3
Cubic kilometer	10^9	km^3

Fractional Number and Letter Drill Sizes in Decimals

Drill Size	*Decimal*	*Drill Size*	*Decimal*	*Drill Size*	*Decimal*	*Drill Size*	*Decimal*
80	.0135	42	.0935	13/64	.2031	X	.3970
79	.0145	3/32	.0938	6	.2040	Y	.4040
1/64	.0156	41	.0960	5	.2055	13/32	.4062
78	.0160	40	.0980	4	.2090	Z	.4130
77	.0180	39	.0995	3	.2130	27/64	.4219

Drill Size	Decimal	Drill Size	Decimal	Drill Size	Decimal	Drill Size	Decimal
76	.0200	38	.1015	7/32	.2188	7/16	.4375
75	.0210	37	.1040	2	.2210	29/64	.4531
74	.0225	36	.1065	1	.2280	15/32	.4688
73	.0240	7/64	.1094	A	.2340	31/64	.4844
72	.0250	35	.1100	15/64	.2344	1/2	.5000
71	.0260	34	.1110	B	.2380	33/64	.5156
70	.0280	33	.1130	C	.2420	17/32	.5312
69	.0292	32	.1160	D	.2460	35/64	.5469
68	.0310	31	.1200	1/4	.2500	9/16	.5625
1/32	.0312	1/8	.1250	E	.2500	37/64	.5781
67	.0320	30	.1285	F	.2570	19/32	.5938
66	.0330	29	.1360	G	.2610	39/64	.6094
65	.0350	28	.1405	17/64	.2656	5/8	.6250
64	.0360	9/64	.1406				
63	.0370	27	.1440	H	.2660	41/64	.6406
62	.0380	26	.1470	I	.2720	21/32	.6562
61	.0390	25	.1495	J	.2770	43/64	.6719
60	.0400	24	.1520	K	.2810	11/16	.6875
59	.0410	23	.1540	9/32	.2812	45/64	.7031
58	.0420	5/32	.1562	L	.2900	23/32	.7188
57	.0430	22	.1570	M	.2950	47/64	.7344
56	.0465	21	.1590	19/64	.2969	3/4	.7500
3/64	.0469	20	.1610	N	.3020	49/64	.7656
55	.0520	19	.1660	5/16	.3125	25/32	.7812
54	.0550	18	.1695	O	.3160	51/64	.7969
53	.0595	11/64	.1719	P	.3230	13/64	.8125
1/16	.0625	17	.1730	21/64	.3281	53/64	.8281
52	.0635	16	.1770	Q	.3320	27/32	.8438
51	.0670	15	.1800	R	.3390	55/64	.8594
50	.0700	14	.1820	11/32	.3438	7/8	.8750
49	.0730	13	.1850	S	.3480	57/64	.8906
48	.0760	3/16	.1875	T	.3580	29/32	.9062
5/64	.0781	12	.1890	23/64	.3594	59/64	.9219
47	.0785	11	.1910	U	.3680	15/16	.9375
46	.0810	10	.1935	3/8	.3750	51/64	.9531
45	.0820	9	.1960	V	.3770	31/32	.9688
44	.0860	8	.1990	X	.3970	63/64	.9844
43	.0890	7	.2010	25/64	.3906	1	1.0000

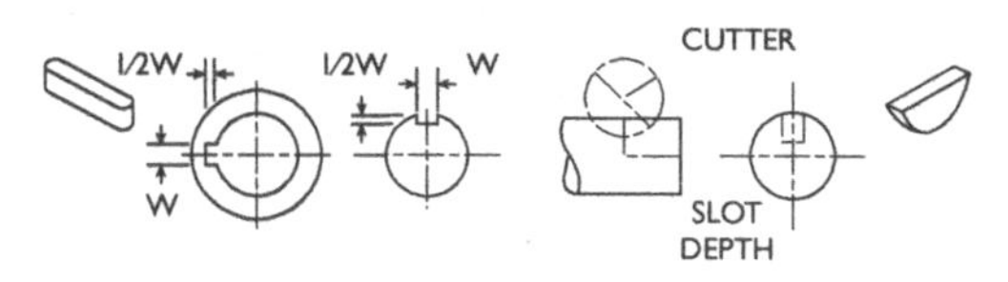

Keyway Data

		Woodruff Keyways*			
Shaft Diameter	*Square Keyways*	*Key Number*	*Thickness*	*Cutter Diameter*	*Slot Depth*
0.500	1/8 × 1/16	404	0.1250	0.500	0.1405
0.562	1/8 × 1/16	404	0.1250	0.500	0.1405
0.625	5/32 × 5/64	505	0.1562	0.625	0.1669
0.688	3/16 × 3/32	606	0.1875	0.750	0.2193
0.750	3/16 × 3/32	606	0.1875	0.750	0.2193
0.812	3/16 × 3/32	606	0.1875	0.750	0.2193
0.875	7/32 × 7/64	607	0.1875	0.875	0.2763
0.938	1/4 × 1/8	807	0.2500	0.875	0.2500
1.000	1/4 × 1/8	808	0.2500	1.000	0.3130
1.125	5/16 × 5/32	1009	0.3125	1.125	0.3228
1.250	5/16 × 5/32	1010	0.3125	1.250	0.3858
1.375	3/8 × 3/16	1210	0.3750	1.250	0.3595
1.500	3/8 × 3/16	1212	0.3750	1.500	0.4535
1.625	3/8 × 3/16	1212	0.3750	1.500	0.4535
1.750	7/16 × 7/32				
1.875	1/2 × 1/4				
2.000	1/2 × 1/4				
2.250	5/8 × 5/16				
2.500	5/8 × 5/16				
2.750	3/4 × 3/8				
3.000	3/4 × 3/8				
3.250	3/4 × 3/8				
3.500	7/8 × 7/16				
4.000	1 × 1/2				

*The depth of a Woodruff Keyway is measured from the edge of the slot.

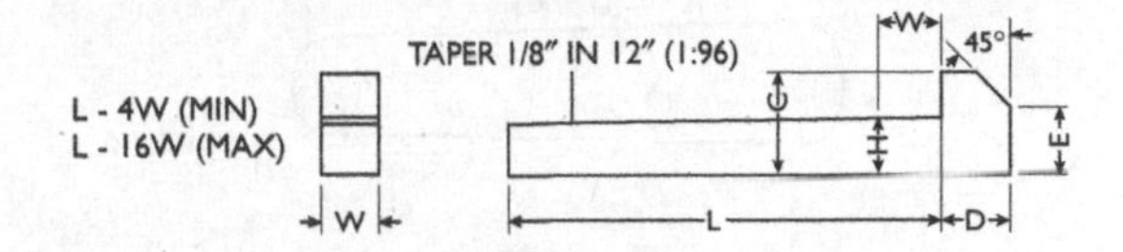

Dimensions of Standard Gib-head Keys, Square and Flat

	Square Type					*Flat Type*				
	Key		*Gib Head*			*Key*		*Gib Head*		
Diameters of Shafts	**W**	**H**	**C**	**D**	**E**	**W**	**H**	**C**	**D**	**E**
1/2–9/16	1/8	1/8	1/4	7/32	5/32	1/8	3/32	3/16	1/8	1/8
5/8–7/8	3/16	3/16	5/16	9/32	7/32	3/16	1/8	1/4	3/16	5/32
15/16–1 1/4	1/4	1/4	7/16	11/32	11/32	1/4	3/16	5/16	1/4	3/16
1 15/16–1 3/8	5/16	5/16	9/16	13/32	13/32	5/16	1/4	3/8	5/16	1/4
1 7/16–1 3/4	3/8	3/8	11/16	15/32	15/32	3/8	1/4	7/16	3/8	5/16
1 13/16–2 1/4	1/2	1/2	7/8	19/32	5/8	1/2	3/8	5/8	1/2	7/16
2 5/16–2 3/4	5/8	5/8	1 1/16	23/32	3/4	5/8	7/16	3/4	5/8	1/2
2 7/8–3 1/4	3/4	3/4	1 1/4	7/8	7/8	3/4	1/2	7/8	3/4	5/8
3 3/8–3 3/4	7/8	7/8	1 1/2	1	1	7/8	5/8	1 1/16	7/8	3/4
3 7/8–4 1/4	1	1	1 3/4	1 3/16	1 3/16	1	3/4	1 1/4	1	
13/16										
4 3/4–5 1/2	1 1/4	1 1/4	2	1 7/16	1 7/16	1 1/4	7/8	1 1/2	1 1/4	1
5 3/4–6	1 1/2	1 1/2	2 1/2	1 3/4	1 3/4	1 1/2	1	1 3/4	1 1/2	1 1/4

**ANSI B17.1 — 1934. Dimensions in inches.*

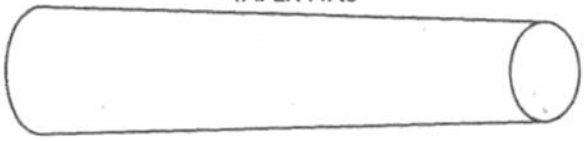

TAPER PINS

Taper Pins

Size Number of Pin	*Length of Pin*	*Large End of Pin*	*Small End of Reamer*	*Drill Size for Reamer*
0	1	0.156	0.135	28
1	1¼	0.172	0.146	25
2	1½	0.193	0.162	19
3	1¾	0.219	0.183	12
4	2	0.250	0.208	3
5	2¼	0.289	0.242	1/4
6	3¼	0.341	0.279	9/32
7	3¾	0.409	0.331	11/32
8	4½	0.492	0.398	13/32
9	5¼	0.591	0.482	31/64
10	6	0.706	0.581	19/32
11	7¼	0.857	0.706	23/32
12	8¾	1.013	0.842	55/64

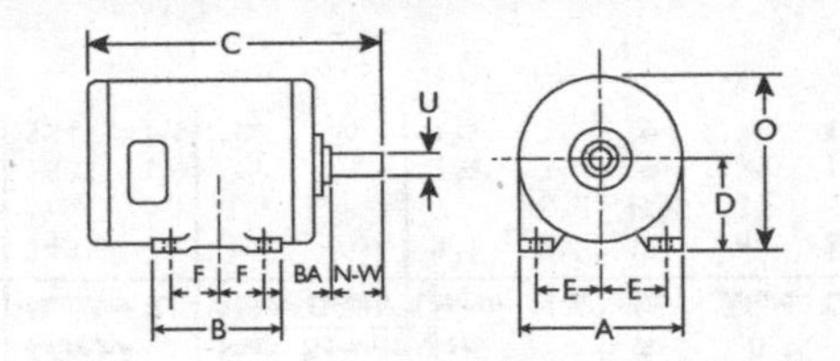

U Nema Motor Frame Dimensions

Horsepower Rating 3600	1800	1200	U Frame Number	U	Shaft Keyseat Width	Depth	Key Length	N-W	A Max.	B Max.	C	D	E	F	BA	O
1 1/2	1	3/4	182	7/8	3/16	3/32	1 3/8	2 1/4	8 7/8	6 1/2	12 1/8	4 1/2	3 3/4	2 1/4	2 3/4	8 15/16
2 & 3	1 1/2 & 2	1 & 1 1/2	184	7/8	3/16	3/32	1 3/8	2 1/4	8 7/8	7 1/2	13 1/8	4 1/2	3 3/4	2 3/4	2 3/4	8 15/16
5	3	2	213	1 1/8	1/4	1/8	2	3	10 3/8	7 1/2	15 5/16	5 1/4	4 1/4	2 3/4	3 1/2	10 7/16
7 1/2	5	3	215	1 1/8	1/4	1/8	2	3	10 3/8	9	16 13/16	5 1/4	4 1/4	3 1/2	3 1/2	10 7/16
10	7 1/2	5	254U	1 3/8	5/16	5/32	2 3/4	3 3/4	12 7/16	10 3/4	20 1/4	6 1/4	5	4 1/8	4 1/4	12 1/2
15	10	7 1/2	256U	1 3/8	5/16	5/32	2 3/4	3 3/4	12 7/16	12 1/2	22	6 1/4	5	5	4 1/4	12 1/2
20	15	10	284U	1 5/8	3/8	3/16	3 3/4	4 7/8	13 7/8	12 1/2	23 11/16	7	5 1/2	4 3/4	4 3/4	13 15/16
25	20		286U	1 5/8	3/8	3/16	3 3/4	4 7/8	13 7/8	12 1/2	25 3/16	7	5 1/2	5 1/2	4 3/4	13 15/16
	25	15	324U	1 7/8	1/2	1/4	4 1/4	5 5/8	15 7/8	14	26 3/8	8	6 1/4	5 1/4	5 1/4	15 15/16
30			324S	1 5/8	3/8	3/16	1 7/8	3 1/4	15 7/8	14	24	8	6 1/4	5 1/4	5 1/4	15 15/16

(continued)

U Nema Motor Frame Dimensions (continued)

Horsepower Rating			U Frame		Shaft Keyseat		Key		A	B						
3600	1800	1200	Number	U	Width	Depth	Length	N-W	Max.	Max.	C	D	E	F	BA	O
	30	20	326U	1 7/8	1/2	1/4	4 1/4	5 5/8	15 7/8	14	27 7/8	8	6 1/4	6	5 1/4	15 15/16
40			326S	1 5/8	3/8	3/16	1 7/8	3 1/4	15 7/8	14	25 1/2	8	6 1/4	6	5 1/4	15 15/16
	40	25	364U	2 1/8	1/2	1/4	5	6 3/8	17 5/8	15 1/4	29 3/16	9	7	5 5/8	5 7/8	17 13/16
50			364US	1 7/8	1/2	1/4	2	3 3/4	17 5/8	15 1/4	26 9/16	9	7	5 5/8	5 7/8	17 13/16
	50	30	365U	2 1/8	1/2	1/4	5	6 3/8	17 5/8	16 1/4	30 3/16	9	7	6 1/8	5 7/8	17 13/16
60			365US	1 7/8	1/2	1/4	2	3 3/4	19 3/4	16 1/4	27 9/16	9	7	6 1/8	5 7/8	17 13/16
	60	40	404U	2 3/8	5/8	5/16	5 1/2	7 1/8	19 3/4	16 1/4	32 7/16	10	8	6 1/8	6 5/8	19 7/8
75			404US	2 1/8	1/2	1/4	2 3/4	4 1/4	19 3/4	16 1/4	39 9/16	10	8	6 1/8	6 5/8	19 7/8
	75	50	405U	2 3/8	5/8	5/16	5 1/2	7 1/8	19 3/4	17 1/4	33 15/16	10	8	6 7/8	6 5/8	19 7/8
100			405US	2 1/8	1/2	1/4	2 3/4	4 1/4	19 3/4	17 1/4	31 1/16	10	8	6 7/8	6 5/8	19 7/8

T Nema Motor Frame Dimensions

Horsepower Rating			T Frame		Shaft Keyseat		Key		A	B						
3600	1800	1200	Number	U	Width	Depth	Length	N-W	Max.	Max.	C	D	E	F	BA	O
1 1/2	1	3/4	143T	7/8	3/16	3/32	1 3/8	2 1/4	7	6	12 5/8	3 1/2	2 3/4	2	2 1/4	7
2 & 1 1/2 & 1	145T	7/8	3/16	3/32	1 3/8	2 1/4	7	6	12 5/8	3 1/2	2 3/4	2 1/2	2 1/4	7	3	2
5	3	1 1/2	182T	1 1/8	1/4	1/8	1 3/4	2 3/4	9	6 1/2	12 3/4	4 1/2	3 3/4	2 3/4	2 3/4	9
7 1/2	5	2	184T	1 1/8	1/4	1/8	1 3/4	2 3/4	9	7 1/2	13 3/4	4 1/2	3 3/4	2 3/4	2 3/4	9

Horsepower Rating			T Frame		Shaft Keyseat		Key		A	B						
3600	1800	1200	Number	U	Width	Depth	Length	N-W	Max.	Max.	C	D	E	F	BA	O
10	7½	3	213T	1⅜	5/16	5/32	2⅜	3⅜	10½	7½	15 13/16	5¼	4¼	2¾	3½	10½
15	10	5	215T	1⅜	5/16	5/32	2⅜	3⅜	10½	9	17 5/16	5¼	4¼	3½	3½	10½
20	15	7½	254T	1⅝	⅜	3/16	2⅞	4	12½	10¾	20½	6¼	5	4⅛	4¼	12½
25	20	10	256T	1⅝	⅜	3/16	2⅞	4	12½	12½	22¼	6¼	5	5	4¼	12½
	25	15	284T	1⅞	½	¼	3¼	4⅝	14	12½	23 5/16	7	5½	4¾	4¾	14
30	25	15	284TS	1⅝	⅜	3/16	1⅞	3¼	14	12½	22	7	5½	4¾	4¾	14
	30	20	286T	1⅞	½	¼	3¼	4⅝	14	14	24⅞	7	5½	5½	4¾	14
40	30	20	286TS	1⅝	⅜	3/16	1⅞	3¼	14	14	23½	7	5½	5½	4¾	14
	40	25	324T	2⅛	½	¼	3⅞	5¼	16	14	26½	8	6¼	5¼	5¼	16
50	40	25	324TS	1⅞	½	¼	2	3¾	16	14	24⅝	8	6¼	5¼	5¼	16
	50	30	326T	2⅛	½	¼	3⅞	5¼	16	15½	27¾	8	6¼	6	5¼	16
60	50	30	326TS	1⅞	½	¼	2	3¾	16	15½	26⅛	8	6¼	6	5¼	16
	60	40	364T	2⅜	⅝	5/16	4¼	5⅞	18	15¼	28¾	9	7	5⅝	5⅞	18
75	60		364TS	1⅞	½	¼	2	3¾	18	15¼	26 9/16	9	7	5⅝	5⅞	18
	75	50	365T	2⅜	⅝	5/16	4¼	5⅞	18	16¼	29¾	9	7	6⅛	5⅞	18
100	75		365TS	1⅞	½	¼	2	3¾	18	16¼	27 9/16	9	7	6⅛	5⅞	18

Wire Gauge Standards

Decimal parts of an inch

Wire Gauge Number	American or Brown & Sharpe	Birmingham of Stubs Wire	Washburn & Moen on Steel Wire Gauge	American S. & W. Co.'s Music Wire	Imperial Wire Gauge	Stubs Steel Wire	U.S. Standard for Plate
00000	0.516549	0.500	0.4305	0.005	4.432		0.43775
0000	0.460	0.454	0.3938	0.006	0.400		0.40625
000	0.40964	0.425	0.3625	0.007	0.372		0.375
00	0.3648	0.380	0.3310	0.008	0.348		0.34375
0	0.32486	0.340	0.3065	0.009	0.324		0.3125
1	0.2893	0.300	0.2830	0.010	0.300	0.227	0.28125
2	0.25763	0.284	0.2625	0.011	0.276	0.219	0.265625
3	0.22942	0.259	0.2437	0.012	0.252	0.212	0.250
4	0.20431	0.238	0.2253	0.013	0.232	0.207	0.234375
5	0.18194	0.220	0.2070	0.014	0.212	0.204	0.21875
6	0.16202	0.203	0.1920	0.016	0.192	0.201	0.203125
7	0.14428	0.180	0.1770	0.018	0.176	0.199	0.1875
8	0.12849	0.165	0.1620	0.020	0.160	0.197	0.171875
9	0.11443	0.148	0.1483	0.022	0.144	0.194	0.15625
10	0.10189	0.134	0.1350	0.024	0.128	0.191	0.140625
11	0.090742	0.120	0.1205	0.026	0.116	0.188	0.125
12	0.080808	0.109	0.1055	0.029	0.104	0.185	0.109375
13	0.071961	0.095	0.0915	0.031	0.092	0.182	0.09375
14	0.064084	0.083	0.0800	0.033	0.080	0.180	0.078125

	Decimal parts of an inch						
Wire Gauge Number	American or Brown & Sharpe	Birmingham of Stubs Wire	Washburn & Moen on Steel Wire Gauge	American S. & W. Co.'s Music Wire	Imperial Wire Gauge	Stubs Steel Wire	U.S. Standard for Plate
15	0.057068	0.072	0.0720	0.035	0.072	0.178	0.0703125
16	0.05082	0.065	0.0625	0.037	0.064	0.175	0.0625
17	0.045257	0.058	0.0540	0.039	0.056	0.172	0.05625
18	0.040303	0.049	0.0475	0.041	0.048	0.168	0.050
19	0.03589	0.042	0.0410	0.043	0.040	0.164	0.04375
20	0.031961	0.035	0.0348	0.045	0.036	0.161	0.0375
21	0.028462	0.032	0.0317	0.047	0.032	0.157	0.034375
22	0.025347	0.028	0.0286	0.049	0.028	0.155	0.03125
23	0.022571	0.025	0.0258	0.051	0.024	0.153	0.028125
24	0.0201	0.022	0.0230	0.055	0.022	0.151	0.025
25	0.0179	0.020	0.0204	0.059	0.020	0.148	0.021875
26	0.01594	0.018	0.0181	0.063	0.018	0.146	0.01875
27	0.014195	0.016	0.0173	0.067	0.0164	0.143	0.0171875
28	0.012641	0.014	0.0162	0.071	0.0149	0.139	0.015625
29	0.011257	0.013	0.0150	0.075	0.0136	0.134	0.0140625
30	0.010025	0.012	0.0140	0.080	0.0124	0.127	0.0125
31	0.008928	0.010	0.0132	0.085	0.0116	0.120	0.0109375

(continued)

Wire Gauge Standards (continued)

	Decimal parts of an inch						
Wire Gauge Number	American or Brown & Sharpe	Birmingham of Stubs Wire	Washburn & Moen on Steel Wire Gauge	American S. & W. Co.'s Music Wire	Imperial Wire Gauge	Stubs Steel Wire	U.S. Standard for Plate
32	0.00795	0.009	0.0128	0.090	0.0108	0.115	0.01015625
33	0.00708	0.008	0.0118	0.095	0.0100	0.112	0.009375
34	0.006304	0.007	0.0104		0.0092	0.110	0.00859375
35	0.005614	0.005	0.0095		0.0084	0.108	0.0078125
36	0.005	0.004	0.0090		0.0076	0.106	0.00703125
37	0.004453		0.0085		0.0068	0.103	0.006640625
38	0.003965		0.0080		0.0060	0.101	0.00625
39	0.003531		0.0075		0.0052	0.099	
40	0.003144		0.0070		0.0048	0.097	

Metal Weights

Material	*Chemical Symbol*	*Weight, lb/in.³*	*Weight, lb/ft³*
Aluminum	Al	0.093	160
Antimony	Sb	0.2422	418
Brass	—	0.303	524
Bronze	—	0.320	552
Chromium	Cr	0.2348	406
Copper	Cu	0.323	450
Gold	Au	0.6975	1205
Iron (cast)	Fe	0.260	450
Iron (wrought)	Fe	0.2834	490
Lead	Pb	0.4105	710
Manganese	Mn	0.2679	463
Mercury	Hg	0.491	849
Molybdenum	Mo	0.309	534
Monel	—	0.318	550
Platinum	Pt	0.818	1413
Steel (mild)	Fe	0.2816	490
Steel (stainless)	—	0.277	484
Tin	Sn	0.265	459
Titanium	Ti	0.1278	221
Zinc	Zn	0.258	446

Uses of Carbon Steel

% Carbon	*Applications*
0.05 to 0.12	Chain, stampings, rivets, nails, pipe, welding stock, where very soft, plastic steel is needed
0.10 to 0.20	Very soft, tough steel; structural steels, machine parts; for case-hardened machine parts, screws
0.20 to 0.30	Better grade of machine and structural steel; gears, shafting, bars, bases, levers, etc.
0.30 to 0.40	Responds to heat treatment; connecting rods, shafting, crane hooks, machine parts, axles
0.40 to 0.50	Crankshafts, gears, axles, shafts, and heat-treated machine parts

(continued)

Uses of Carbon Steel *(continued)*

% Carbon	*Applications*
0.60 to 0.70	Low-carbon tool steel, used where a keen edge is not necessary, but where shock strength is wanted; drop hammer dies, set screws, locomotive wheels, screw drivers
0.70 to 0.80	Tough and hard steel; anvil faces, band saws, hammers, wrenches, cable wires, etc.
0.80 to 0.90	Punches for metal, rock drills, shear blades, cold chisels, rivet sets, and many hand tools
0.90 to 1.00	Used for hardness and high-tensile strength, springs, high-tensile wire, knives, axes, dies for all purposes
1.00 to 1.10	Drills, taps, milling cutters, knives, etc.
1.10 to 1.20	Used for all tools where hardness is a prime consideration; for example, ball bearings, cold-cutting dies, drills, wood-working tools, lathe tools, etc.
1.20 to 1.30	Files, reamers, knives, tools for cutting brass and wood
1.25 to 1.40	Used where a keen cutting edge is necessary; razors, saws, instruments, and machine parts where maximum resistance to wear is needed; boring and finishing tools

Commercial Pipe Sizes and Wall Thicknesses

This table lists the pipe sizes and wall thicknesses currently established as standard, or specifically:

1. The traditional standard weight, extra strong, and double extra strong pipe.

2. The pipe-wall thickness schedules listed in ANSI B36.10, which are applicable to carbon steel and alloys other than stainless steels.

3. The pipe-wall thickness schedules listed in ANSI B36.19, which are applicable only to stainless steels.

		Nominal Wall Thickness For														
		Schedule					Schedule			Schedule						
Nominal Pipe Size	Outside Diameter	5	10	20	30	Standard	40	60	Extra Strong	80	100	120	140	160	XX Strong
⅛	0.405	—	0.049	—	—	0.068	0.068	—	0.095	0.095	—	—	—	—	—
¼	0.540	—	0.065	—	—	0.088	0.086	—	0.119	0.119	—	—	—	—	—
⅜	0.675	—	0.065	—	—	0.091	0.091	—	0.126	0.126	—	—	—	—	—
½	0.840	—	0.083	—	—	0.109	0.109	—	0.147	0.147	—	—	—	0.187	0.294
¾	1.050	0.065	0.083	—	—	0.113	0.113	—	0.154	0.154	—	—	—	0.218	0.308
1	1.315	0.065	0.109	—	—	0.133	0.133	—	0.179	0.179	—	—	—	0.250	0.358
1¼	1.660	0.065	0.109	—	—	0.140	0.140	—	0.191	0.191	—	—	—	0.250	0.382
1½	1.900	0.065	0.109	—	—	0.145	0.145	—	0.200	0.200	—	—	—	0.281	0.400
2	2.375	0.065	0.109	—	—	0.154	0.154	—	0.218	0.218	—	—	—	0.343	0.436
2½	2.875	0.083	0.120	—	—	0.203	0.203	—	0.276	0.276	—	—	—	0.375	0.552
3	3.5	0.083	0.120	—	—	0.216	0.216	—	0.300	0.300	—	—	—	0.438	0.600
3½	4.0	0.083	0.120	—	—	0.226	0.226	—	0.318	0.318	—	—	—	—	—
4	4.5	0.083	0.120	—	—	0.237	0.237	—	0.337	0.337	—	0.438	—	0.531	0.674
5	5.563	0.109	0.134	—	—	0.258	0.258	—	0.375	0.375	—	0.500	—	0.625	0.750
6	6.625	0.109	0.134	—	—	0.280	0.280	—	0.432	0.432	—	0.562	—	0.718	0.864
8	8.625	0.109	0.148	0.250	0.277	0.322	0.322	0.406	0.500	0.500	0.593	0.718	0.812	0.906	0.875

(continued)

Commercial Pipe Sizes and Wall Thicknesses (continued)

		Nominal Wall Thickness For													
Nominal Pipe Size	Outside Diameter	Schedule					Schedule		Extra Strong	Schedule					XX Strong
		5	10	20	30	Standard	40	60		80	100	120	140	160	
10	10.75	0.134	0.165	0.250	0.307	0.365	0.365	0.500	0.500	0.593	0.713	0.843	1.000	1.125	—
12	12.75	0.156	0.180	0.250	0.330	0.375	0.406	0.562	0.500	0.687	0.843	1.000	1.125	1.312	—
14 OD	14.0	—	0.250	0.312	0.375	0.375	0.438	0.593	0.500	0.750	0.937	1.093	1.250	1.406	—
16 OD	16.0	—	0.250	0.312	0.375	0.375	0.500	0.656	0.500	0.843	1.031	1.218	1.438	1.593	—
18 OD	18.0	—	0.250	0.3											
12	0.438	0.375	0.562	0.750	0.500	0.937	1.156	1.375	1.562	1.781	—				
20 OD	20.0	—	0.250	0.375	0.500	0.375	0.593	0.812	0.500	1.031	1.281	1.500	1.750	1.968	—
22 OD	22.0	—	0.250	—	—	0.375	—	—	0.500	—	—	—	—	—	—
24 OD	24.0	—	0.250	0.375	0.562	0.375	0.687	0.968	0.500	1.218	1.531	1.812	2.062	2.343	—
26 OD	26.0	—	—	—	—	0.375	—	—	0.500	—	—	—	—	—	—
30 OD	30.0	—	0.312	0.500	0.625	0.375	—	—	0.500	—	—	—	—	—	—
34 OD	34.0	—	—	—	—	0.375	—	—	0.500	—	—	—	—	—	—
36 OD	36.0	—	—	—	—	0.375	—	—	0.500	—	—	—	—	—	—
42 OD	42.0	—	—	—	—	0.375	—	—	0.500	—	—	—	—	—	—

Contents of Lumber, in Board Feet, When Cross-Sectional Area and Length Are Known

[Number of board feet in various sizes for lengths given]

Size of Piece, in.	Length of Piece, ft								
	8	10	12	14	16	18	20	22	24
2 × 4	5⅓	6⅔	8	9⅓	10⅔	12	13⅓	14⅔	16
2 × 6	8	10	12	14	16	18	20	22	24
2 × 8	10⅔	13⅓	16	18⅔	21⅓	24	26⅔	29½	32
2 × 10	13⅓	16⅔	20	23⅓	26⅔	30	33⅓	36⅔	40
2 × 12	16	20	24	28	32	36	40	44	48
2 × 14	18⅔	23⅓	28	32⅔	37½	42	46⅔	51½	56
2 × 16	21½	26⅔	32	37½	42⅔	48	53⅓	58⅔	64
3 × 6	12	15	18	21	24	27	30	33	36
3 × 8	16	20	24	28	32	36	40	44	48
3 × 10	20	25	30	35	40	45	50	55	60
3 × 12	24	30	36	42	48	54	60	66	72
3 × 14	28	35	42	49	56	63	70	77	84
3 × 16	32	40	48	56	64	72	80	88	96
4 × 4	10⅔	13⅓	16	18⅔	21⅓	24	26⅔	29⅓	32
4 × 6	16	20	24	28	32	36	40	44	48
4 × 8	21½	26⅔	32	37⅓	42⅔	48	53⅓	58⅔	64
4 × 10	26⅔	33⅓	40	46⅔	53½	60	66⅔	73⅓	80
4 × 12	32	40	48	56	64	72	80	88	96
4 × 14	37⅓	46⅔	56	65⅓	74⅔	84	93⅓	102⅔	112
4 × 16	42⅔	53⅓	64	74⅔	85⅓	96	106⅔	117⅓	128
6 × 6	24	30	36	42	48	54	60	66	72
6 × 8	32	40	48	56	64	72	80	88	96
6 × 10	40	50	60	70	80	90	100	110	120
6 × 12	48	60	72	84	96	108	120	132	144
6 × 14	56	70	84	98	112	126	140	154	168
6 × 16	64	80	96	112	128	144	160	176	192
8 × 8	42⅔	53⅓	64	74⅔	85⅓	96	106⅔	117⅓	128
8 × 10	53⅓	66⅔	80	93½	106⅔	120	133⅓	146⅔	160
8 × 12	64	80	96	112	128	144	160	176	192
8 × 14	74⅔	93⅓	112	130⅔	149⅓	168	186⅔	205⅓	224
8 × 16	85⅓	106⅔	128	149⅓	170⅔	192	213⅓	234⅔	256
10 × 10	66⅔	83⅓	100	116⅔	133⅓	150	166⅔	183½	200
10 × 12	80	100	120	140	160	180	200	220	240
10 × 14	93½	116⅔	140	163⅓	186⅔	210	233⅓	256⅔	280
10 × 16	106⅔	133⅓	160	186⅔	213⅓	240	266⅔	296⅓	320
12 × 12	96	120	144	168	192	216	240	264	288
12 × 14	112	140	168	196	224	252	280	308	336
12 × 16	128	160	192	224	256	288	320	352	384

Bolting Dimensions for 150-lb Flanges

	150-lb Steel Flanges				
Nominal Pipe Size	Diameter of Bolt Circle	Diameter of Bolts	Number of Bolts	Length of Studs, 1/4-in. Raised Face	Bolt Length for 125-lb Cast-Iron Flanges
1/2	2 3/8	1/2	4	2 1/4	
3/4	2 3/4	1/2	4	2 1/4	
1	3 1/8	1/2	4	2 1/2	1 3/4
1 1/4	3 1/2	1/2	4	2 1/2	2
1 1/2	3 7/8	1/2	4	2 3/4	2
2	4 3/4	5/8	4	3	2 1/4
2 1/2	5 1/2	5/8	4	3 1/4	2 1/2
3	6	5/8	4	3 1/2	2 1/2
3 1/2	7	5/8	8	3 1/2	2 3/4
4	7 1/2	5/8	8	3 1/2	3
5	8 1/2	3/4	8	3 3/4	3
6	9 1/2	3/4	8	3 3/4	3 1/4
8	11 3/4	3/4	8	4	3 1/2
10	14 1/4	7/8	12	4 1/2	3 3/4
12	17	7/8	12	4 1/2	3 3/4

Blind Flanges

	150 lb		300 lb	
Nominal Pipe Size	Outside Diameter of Flange O	Thickness Q	Outside Diameter of Flange O	Thickness Q
1/2	3 1/2	7/16	3 3/4	9/16
3/4	3 7/8	1/2	4 5/8	5/8
1	4 1/4	9/16	4 7/8	11/16
1 1/4	4 5/8	5/8	5 1/4	3/4
1 1/2	5	11/16	6 1/8	13/16
2	6	3/4	6 1/2	7/8
2 1/2	7	7/8	7 1/2	1

	150 lb		300 lb	
Nominal Pipe Size	Outside Diameter of Flange O	Thickness Q	Outside Diameter of Flange O	Thickness Q
3	7½	15/16	8¼	1⅛
3½	8½	15/16	9	1 3/16
4	9	15/16	10	1¼
5	10	15/16	11	1⅜
6	11	1	12½	1 7/16
8	13½	1⅛	15	1⅝
10	16	1 3/16	17½	1⅞
12	19	1¼	20½	2

Welding Neck Flanges

	150 lb		300 lb	
Nominal Pipe Size	Outside Diameter of Flange O	Length Thru Hub Y	Outside Diameter of Flange O	Length Thru Hub Y
½	3½	1⅞	3¾	2 1/16
¾	3⅞	2 1/16	4⅝	2¼
1	4¼	2 3/16	4⅞	2 7/16
1¼	4⅝	2¼	5¼	2 9/16
1½	5	2 7/16	6⅛	2 11/16
2	6	2½	6½	2¾
2½	7	2¾	7½	3
3	7½	2¾	8¼	3⅛
3½	8½	2 13/16	9	3 3/16
4	9	3	10	3⅜
5	10	3½	11	3⅞
6	11	3½	12½	3⅞
8	13½	4	15	4⅜
10	16	4	17½	4⅝
12	19	4½	20½	5⅛
14	21	5	23	5⅝
16	23½	5	25½	5¾
18	25	5½	28	6¼

Standard Cast-Iron Companion Flanges and Bolts (For working pressures up to 125 lb/in.^{2} steam, 175 lb/in.^{2} WOG)

Size, in.	Diameter of Flange, in.	Bolt Circle, in.	Number of Bolts	Size of Bolts, in.	Length of Bolts, in.
3/4	3 1/2	2 1/2	4	3/8	1 3/8
1	4 1/4	3 1/8	4	1/2	1 1/2
1 1/4	4 5/8	3 1/2	4	1/2	1 1/2
1 1/2	5	3 7/8	4	1/2	1 3/4
2	6	4 3/4	4	5/8	2
2 1/2	7	5 1/2	4	5/8	2 1/2
3	7 1/2	6	4	5/8	2 1/2
3 1/2	8 1/2	7	8	5/8	2 1/2
4	9	7 1/2	8	5/8	2 3/4
5	10	8 1/2	8	3/4	3
6	11	9 1/2	8	3/4	3
8	13 1/2	11 3/4	8	3/4	3 1/4
10	16	14 1/4	12	7/8	3 1/2
12	19	17	12	7/8	3 3/4
14	21	18 3/4	12	1	4 1/4
16	23 1/2	21 1/4	16	1	4 1/4

Extra Heavy Cast-Iron Companion Flanges and Bolts (For working pressure up to 250 lb/in.^{2} steam, 400 lb/in.^{2} WOG)

Pipe Size, in.	Diameter of Flanges	Diameter of Bolt Circle	Number of Bolts	Diameter of Bolts	Length of Bolts
1	4 7/8	3 1/2	4	5/8	2 1/4
1 1/4	5 1/4	3 7/8	4	5/8	2 1/2
1 1/2	6 1/8	4 1/2	4	3/4	2 1/2
2	6 1/2	5	8	5/8	2 1/2
2 1/2	7 1/2	5 7/8	8	3/4	3
3	8 1/4	6 5/8	8	3/4	3 1/4
3 1/2	9	7 1/4	8	3/4	3 1/4
4	10	7 7/8	8	3/4	3 1/2

Pipe Size, in.	Diameter of Flanges	Diameter of Bolt Circle	Number of Bolts	Diameter of Bolts	Length of Bolts
5	11	9¼	8	¾	3¾
6	12½	10⅝	12	¾	3¾
8	15	13	12	⅞	4¼
10	17½	15¼	16	1	5
11	20½	17¾	16	1⅛	5½
14 OD	23	20¼	20	1⅛	5¾
16 OD	25½	22½	20	1¼	6

Feet Head of Water to Pounds per Square Inch

Feet Head	Pounds per Square Inch	Feet Head	Pounds per Square Inch
1	.43	100	43.31
2	.87	110	47.64
3	1.30	120	51.97
4	1.73	130	56.30
5	2.17	140	60.63
6	2.60	150	64.96
7	3.03	160	69.29
8	3.46	170	76.63
9	3.90	180	77.96
10	4.33	200	86.62
15	6.50	250	108.27
20	8.66	300	129.93
25	10.83	350	151.58
30	12.99	400	173.24
40	17.32	500	216.55
50	21.65	600	259.85
60	25.99	700	303.16
70	30.32	800	346.47
80	34.65	900	389.78
90	38.98	1000	433.00

Note: One foot of water at 62°F equals 0.433 pound pressure per square inch. To find the pressure per square inch for any feet head not given in the table above, multiply the feet head by 0.433.

Water Pressure to Feet Head

Feet Head	Pounds per Square Inch	Feet Head	Pounds per Square Inch
1	2.31	100	230.90
2	4.62	110	253.98
3	6.93	120	277.07
4	9.24	130	300.16
5	11.54	140	323.25
6	13.85	150	346.34
7	16.16	160	369.43
8	18.47	170	392.52
9	20.78	180	415.61
10	23.09	200	461.78
15	34.63	250	577.24
20	46.18	300	692.69
25	57.72	350	808.13
30	69.27	400	922.58
40	92.36	500	1154.48
50	115.45	600	1385.39
60	138.54	700	1616.30
70	161.63	800	1847.20
80	184.72	900	2078.10
90	207.81	1000	2309.00

Note: One pound of pressure per square inch of water equals 2.309 feet of water at 62°F. Therefore, to find the feet head of water for any pressure not given in the table above, multiply the pressure pounds per square inch by 2.309.

Boiling Points of Water at Various Pressures

Vacuum, in. Hg	Boiling Point	Vacuum, in. Hg	Boiling Point
29	76.62	7	198.87
28	99.93	6	200.96
27	114.22	5	202.25
26	124.77	4	204.85
25	133.22	3	206.70
24	140.31	2	208.50
23	146.45	1	210.25
22	151.87	Gauge Pounds	
21	156.75	0	212.

Vacuum, in. Hg	Boiling Point	Vacuum, in. Hg	Boiling Point
20	161.19	1	215.6
19	165.24	2	218.5
18	169.00	4	224.4
17	172.51	6	229.8
16	175.80	8	234.8
15	178.91	10	239.4
14	181.82	15	249.8
13	184.61	25	266.8
12	187.21	50	297.1
11	189.75	75	320.1
10	192.19	100	337.9
9	194.50	125	352.9
8	196.73	200	387.9

Total Thermal Expansion of Piping Material, in./100 ft above 32°F

Temperature, °F	Carbon and Carbon Moly Steel	Cast Iron	Copper	Brass and Bronze	Wrought Iron
32	0	0	0	0	0
100	0.5	0.5	0.8	0.8	0.5
150	0.8	0.8	1.4	1.4	0.9
200	1.2	1.2	2.0	2.0	1.3
250	1.7	1.5	2.7	2.6	1.7
300	2.0	1.9	3.3	3.2	2.2
350	2.5	2.3	4.0	3.9	2.6
400	2.9	2.7	4.7	4.6	3.1
450	3.4	3.1	5.3	5.2	3.6
500	3.8	3.5	6.0	5.9	4.1
550	4.3	3.9	6.7	6.5	4.6
600	4.8	4.4	7.4	7.2	5.2
650	5.3	4.8	8.2	7.9	5.6
700	5.9	5.3	9.0	8.5	6.1
750	6.4	5.8			6.7
800	7.0	6.3			7.2
850	7.4				
900	8.0				
950	8.5				
1000	9.0				

Specific Gravity of Gases (At 60°F and 29.92 in. Hg)

Dry air (1 ft^3 at 60°F and 29.92 in. Hg weigh 0.07638 lb)		1.000
Acetylene	C_2H_2	0.91
Ethane	C_2H_6	1.05
Methane	CH_4	0.554
Ammonia	NH_2	0.596
Carbon dioxide	CO_2	1.53
Carbon monoxide	CO	0.967
Butane	C_4H_{10}	2.067
Butene	C_4H_8	1.93
Chlorine	Cl_2	2.486
Helium	He	0.138
Hydrogen	H_2	0.0696
Nitrogen	N_2	0.9718
Oxygen	O_2	1.1053

Specific Gravity of Liquids

Liquid	*Temperature, °F*	*Specific Gravity*
Water (1 ft^3 weighs 62.41 lb)	50	1.00
Brine (Sodium Chloride 25%)	32	1.20
Pennsylvania Crude Oil	80	0.85
Fuel Oil No. 1 and 2	85	0.95
Gasoline	80	0.74
Kerosene	85	0.82
Lubricating Oil SAE 10-20-30	115	0.94

Tempering and Heat Colors

	Color	*Degrees*	
		Fahrenheit	*Centigrade*
Temper Colors	Faint straw 440	400	205
	Straw	440	225
	Deep straw	475	245
	Bronze	520	270

	Color	Degrees	
		Fahrenheit	*Centigrade*
	Peacock blue	540	280
	Full blue	590	310
	Light blue	640	340
Heat Colors	Faint red	930	500
	Blood red	1075	580
	Dark cherry	1175	635
	Medium cherry	1275	690
	Cherry	1375	745
	Bright cherry	1450	790
	Salmon	1550	840
	Dark orange	1680	890
	Orange	1725	940
	Lemon	1830	1000
	Light yellow	1975	1080
	White	2200	1200

Ampacity of Copper Wires

	Not More than Three Conductors in Conduit, Cable, or Buried Directly in the Earth		*Single Conductors in Free Air*		
Wire Size	*Type TW*	*Types RH, RHW, THW*	*Type TW*	*Types RH, RHW, THW*	*Weather-proof*
	A	B	C	D	E
14	15*	15*	15*	15*	30
12	20*	20*	20*	20*	35
10	30*	30*	30*	30*	50
8	40	50	60	70	70
6	55	65	80	95	95
4	70	85	105	125	125
2	95	115	140	170	175
1/0	125	150	195	230	235

(continued)

Ampacity of Copper Wires (continued)

	Not More than Three Conductors in Conduit, Cable, or Buried Directly in the Earth		Single Conductors in Free Air		
Wire Size	**Type TW**	**Types RH, RHW, THW**	**Type TW**	**Types RH, RHW, THW**	**Weather-proof**
	A	**B**	**C**	**D**	**E**
2/0	145	175	225	265	275
3/0	165	200	260	310	320

Note: To find the ampacity of other sizes, conductor metals, or insulations, consult Tables 310–16 to 310–19 in your copy of the Code.

** In the Code, the ampacities of these wires are shown higher than this, but a footnote limits their overcurrent protection to the figures given here.*

Index

G

M

S

T

W